REPLICATING AND REPAIRING
THE GENOME
From Basic Mechanisms to
Modern Genetic Technologies

REPLICATING AND REPAIRING
THE GENOME

From Basic Mechanisms to
Modern Genetic Technologies

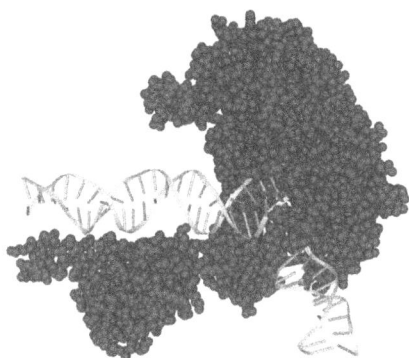

Kenneth N Kreuzer
Duke University, USA

World Scientific

NEW JERSEY · LONDON · SINGAPORE · BEIJING · SHANGHAI · HONG KONG · TAIPEI · CHENNAI · TOKYO

Published by

World Scientific Publishing Co. Pte. Ltd.
5 Toh Tuck Link, Singapore 596224
USA office: 27 Warren Street, Suite 401-402, Hackensack, NJ 07601
UK office: 57 Shelton Street, Covent Garden, London WC2H 9HE

British Library Cataloguing-in-Publication Data
A catalogue record for this book is available from the British Library.

Cover: The protein-DNA complex on the front cover is an image of bacteriophage T7 DNA polymerase (red) complexed with a DNA primer-template (gold), as resolved by a combination of X-ray crystallography and cryo-electron microscopy by Gao *et al.* (2019) (see Further Reading in Chapter 2 for complete citation; PDB ID 6N7W). The protein-DNA complex on the back cover and spine is an image of the human uracil-DNA glycosylase (red) complexed with a short duplex DNA (gold) with a uracil-like residue, as resolved by X-ray crystallography by Parikh *et al.* (1998) (see Further Reading in Chapter 10 for complete citation; PDB ID 1SSP). Both images were prepared using the web-based visualization system iCn3D (Wang *et al.*, 2019; full citation and web site address on page viii) with coordinates from the Protein Data Bank (www.rcsb.org; Berman *et al.*, 2000; full citation on page viii).

REPLICATING AND REPAIRING THE GENOME
From Basic Mechanisms to Modern Genetic Technologies

ISBN 978-981-121-569-8 (hardcover)
ISBN 978-981-121-570-4 (ebook for institutions)
ISBN 978-981-121-571-1 (ebook for individuals)

For any available supplementary material, please visit
https://www.worldscientific.com/worldscibooks/10.1142/11700#t=suppl

Typeset by Stallion Press
Email: enquiries@stallionpress.com

Dedication

This book is dedicated to two outstanding scientists who contributed greatly to the field during their lives and strongly impacted my career. Unfortunately, both passed away much too soon. As my graduate school mentor, Nick Cozzarelli taught me a great deal about conducting experimental science and sustaining a critical attitude about advances in the field. For many years, Tao Hsieh was a wonderful colleague with whom I enjoyed discussions, sharing teaching and other responsibilities, and collaborating. Both Nick and Tao were master biochemists with a phenomenal understanding of both DNA replication and DNA topology, and both are sorely missed.

Preface

Thousands of scientists studying DNA replication and repair over the last half century have uncovered an amazing wealth of information using numerous experimental systems, and thereby illuminated the very reactions that are necessary for life on our planet. However, this wealth of information is complex and dense, particularly since the fields are rife with terminology that often differs between experimental systems, and since many of the pathways are interconnected with other cellular pathways. Thus, the fields are difficult for a newcomer to appreciate and master. The intent of this book is to describe the logic and mechanisms involved in replication and repair within a reasonable and concise framework, avoiding much of the dense information and terminology that can make the field seem opaque.

While the main text will generally not attempt to provide the experimental designs and details behind the science, inserts called "How did they test that?" are included at the end of each chapter to illustrate some of the beautiful experimental approaches that uncovered key aspects of DNA replication and repair. These can be jumping-off points for readers to delve into primary literature in the field.

I should note that there are recent outstanding textbooks and specialized compendiums that present the fields in much finer and complete detail. These are listed in the Further Reading section of Chapter 1 (and other Chapters), and are highly recommended for those who will engage more deeply in specific areas of DNA replication and repair. The Further Reading sections also contain citations to a few key papers from the primary literature, including those highlighted in the "How did they test that?" inserts, as well as review articles that cover Chapter topics in more detail.

The internet provides many valuable resources relevant to the topics covered, along with sites for conducting literature searches (PubMed, Google Scholar, etc). Many informative animations of key processes can be found on YouTube (www.youtube.com). The Nobel Prize web site (www.nobelprize.org) has more information on the various Nobel Prizes highlighted in this book, and in some cases, the Nobel Prize lectures associated with the Prizes.

Regarding protein function, structure and nomenclature, the following resources are very useful. The Universal Protein Resource (www.UniProt.org) is a collection of databases and information organized collaboratively by the European Bioinformatics Institute, the Swiss Institute of Bioinformatics, and the Protein Information Resource. The web site is a front end for vast amounts of information on protein nomenclature, sequences, relationships, and functions, as well as relevant citations. The Protein Data Bank (www.rcsb.org)[1] provides access to structures of proteins and protein-DNA complexes, with a highly interactive interface that allows many different structural display options, 360° rotations, zooming, and the key primary reference(s) relevant for the particular protein or protein-DNA complex. A free web-based structure visualization system with many advantages (including the ability to save and share particular views as a URL) has recently been deployed and is also highly recommended: iCn3D (I-see-in-3D)[2]; see web site https://www.ncbi.nlm.nih.gov/Structure/icn3d/docs/icn3d_about.html.

---

[1]Berman, H.M., Westbrook, J., Feng, Z., Gilliland, G., Bhat, T.N., Weissig, H., Shindyalov, I.N., and Bourne, P.E. (2000). The Protein Data Bank. *Nucleic Acids Research 28*:235–242.

[2]Wang, J., Youkharibache, P., Zhang, D., Lanczycki, J., Geer, R.C., Madej, T., Phan, L., Ward, M., Lu, S., Marchler, G.H., Wang, Y., Bryant, S.H., Geer, L.Y., and Marchler-Bauer, A. (2019). iCn3D, a web-based 3D viewer for sharing 1D/2D/3D representations of biomolecular structures. *Bioinformatics*, doi:10.1093/bioinformatics/btz502.

Acknowledgments

I am deeply grateful to Bruce Alberts and Paul Modrich, who provided detailed and extremely helpful feedback on this book, and to Charles C. Richardson, who provided invaluable feedback on Chapter 2. The text was significantly improved thanks to their insightful comments. I also thank Zakiya Whatley and Beverley Fermor for valuable suggestions that helped make the book more accessible to newcomers to the field. Finally, I thank the Duke University Biochemistry Department for support and for providing an outstanding environment in which to pursue my research interests in DNA replication, repair and recombination, as well as teach and mentor students.

Contents

About the Author

Kenneth Kreuzer, PhD, is Emeritus Professor of Biochemistry at Duke University. Ken grew up in Western New York, received his Bachelor of Science degree in Biology from MIT (1974) and his PhD in Genetics from the University of Chicago (1978). His graduate research provided some of the earliest evidence that the antibacterial quinolones act by poisoning the enzyme DNA gyrase, and his postdoctoral research at the University of California, San Francisco, identified replication origins of bacteriophage T4. As an independent investigator for three decades at Duke University, Kreuzer and his lab made contributions concerning cytotoxic mechanisms of topoisomerase inhibitors, the role of R-loops in DNA replication, connections between homologous recombination and DNA replication, and mechanisms of double-strand break repair. This research was supported by multiple grants from the NIH and other sources and resulted in over 100 publications. Most of these papers were co-authored by students and postdocs in his lab, who all went on to varied and successful careers of their own.

Ken is a Fellow of the American Academy of Microbiology and of the American Association for the Advancement of Science, and served as Interim Chair of the Duke Biochemistry Department (2007–2010). He has taken a special interest in mentoring and teaching throughout his academic career. He served as Director of the Duke University Summer Research Opportunities Program

(1996–2003), the Duke Cell and Molecular Biology Graduate Program (2001–2006), the Duke Post-baccalaureate Research Education Program (PREP; 2003–2008), and as Co-Director of the Duke BioCoRE Program (Initiative to Maximize Student Development; 2013–2015). In honor of his graduate mentoring and diversity efforts, Ken has received the Samuel DuBois Cook Community Betterment Award, the Faculty Award for Graduate Teaching in the Duke Basic Biomedical Sciences, and the Duke University President's Diversity Award. Since retiring from running an active lab in 2015, Ken and his wife Bev live in the mountains of Southwest Virginia with their dogs and ponies. In addition to writing this book, he has been enjoying woodworking, outdoor activities (golfing, hiking and kayaking), and spending quality time with friends in a beautiful rural setting.

Chapter 1

The challenges of maintaining and duplicating the genome

1.1 Introduction

The human genome consists of about 3.3 billion base pairs of DNA spread between the 23 chromosomes. With the exception of the sex chromosomes in males, all the chromosomes are present in two copies, one from each parent, and thus the total sequence length in a human cell is about 6.6 billion base pairs. This DNA sequence complexity determines the precise amino acid sequences of many thousands of proteins, the nucleotide sequence of numerous structural RNAs, and myriad regulatory signals that dictate when genes are turned off and on in response to organism development, nutritional conditions, and stress or disease states. The size of the genome in base pairs can be appreciated by comparing it to the average length of one of the Harry Potter books, roughly 0.7-million alphabet letters. It would take something like 5000 Harry Potter-length books to print out one copy of the information stored in the human genome!

Another way to appreciate the length and complexity of our DNA is to think about all human DNA in your body at any one time. Estimates of the total number of human cell nuclei in an average adult are in the range of 7 trillion.[1] The 6.6 billion base pairs of DNA in the human cell nucleus has a length of about 2 m, and so the human DNA in your body would be about 14 trillion (1.4 × 10^{13}) m long if it were lined up end to end. Well, 14 trillion m equals 14 billion (1.4 × 10^{10}) km, and consider that the distance from the earth to the sun is about 0.15 billion (1.5 × 10^7) km. Your DNA could therefore stretch to the sun and back about 45 times!

In spite of its 6.6-billion base-pair length, the sequence of the genome is passed down to daughter cells with remarkable accuracy in every cell division during a human lifetime. Current estimates of the error rate for this process are in the range of one per billion (10^9) to one per ten billion base pairs replicated — it sounds even more impressive when this is expressed as an accuracy rate — 99.9999999% of the base pairs are copied correctly (at one error per billion). Comparing again to the Harry Potter books mentioned above, this would be like copying the 5000 imaginary Harry Potter books while making no more than a few mistakes — the vast majority of the 5000 books would be perfect. We will see that the impressive accuracy of genome replication results from a combination of a very accurate DNA replication machine combined with powerful DNA repair pathways, most of which restore the original nucleotide sequence and thereby avoid errors. Furthermore, as we will see, the DNA replication and repair machineries provide a remarkable resistance to DNA damage from both endogenous toxins like oxygen radicals and exogenous sources like the UV in sunlight and the many environmental chemicals that can damage DNA.

The extreme accuracy of the replication process is even more impressive when one considers the speed at which the cellular DNA can be replicated prior to cell division. Rapidly growing human cells

---

[1] The total number of human cells is difficult to estimate and somewhat controversial, but a recent compilation put the number at about 37 trillion. About 80% of human cells are red blood cells that lack a nucleus, leaving about 7 trillion nucleus-containing cells.

can double about once every 24 hours. Ignoring the fact that only part of this time is used for DNA replication, this still translates to a rate of replication (for the entire genome) of nearly 50,000 nucleotides per second! Some other eukaryotes display even faster replication rates. For example, very early Drosophila fruit fly embryos replicate their genome in about 8 minutes, for an overall rate of over 600,000 nucleotides per second. These very high rates of replication can only be achieved by using multiple replication machines simultaneously, similar to having 5000 people each copying one of the imaginary Harry Potter books to complete the job in a shorter period of time. If you consider a single replication machine in the human system, the rate of replication is on the order of 30 base pairs per second, while bacterial replication machineries can achieve a rate of about 1000 base pairs per second. If human cell replication utilized only a single replication machine, it would take roughly 7 years to replicate all the DNA in one cell at the rate of 30 base pairs per second.

1.2 The double-helical structure of DNA and the logic of replication and repair

In 1953, Watson and Crick published one of the most famous papers in biology, "A Structure for Deoxyribose Nucleic Acid" (see Further Reading at the end of this chapter). Critical research during the years leading up to this paper had provided evidence that DNA is the genetic material, and that the chemical composition of DNA consisted of purine and pyrimidine bases, a sugar called deoxyribose, and phosphates. A very important clue was that, for any particular DNA, the fractions of adenine and thymine bases were found to be identical, as were the fractions of guanine and cytosine bases (see "How did they test that" at the end of this chapter). Based on these prior studies and Rosalind Franklin's X-ray crystallographic data, Watson and Crick deduced that the structure of DNA consisted of an elegant double helix with two linear strands wound around each other in opposite directions.

1962 Nobel Prize in Physiology or Medicine

This prize was awarded to **Maurice Wilkins**, **James Watson**, and **Francis Crick** for discovering the double helix structure of DNA and inferring the significance of this structure for information transfer and replication of the genetic material.
https://www.nobelprize.org/prizes/medicine/1962/summary/

Before delving into the implications of this structure for the logistics of replication, it is important to appreciate the DNA molecule. Individual strands of the double helix consist of a string of deoxyribonucleoside (or simply nucleoside in common parlance) monophosphates linked to each other in a linear fashion. As illustrated in Figure 1.1A–1.1C, a deoxyribonucleoside consists of a purine or pyrimidine base linked, via one of its nitrogen atoms, to the 1′-C atom of the deoxyribose sugar. This kind of linkage is called an N-glycosidic bond. The four bases in DNA are the purines adenine and guanine, and the pyrimidines thymine and cytosine. The corresponding deoxyribonucleosides, with sugar linked to the base, are named deoxyadenosine, deoxythymidine, deoxyguanosine, and deoxycytidine (the RNA nucleosides are adenosine, uridine, guanosine, and cytidine). When a deoxyribonucleoside is also linked to one or more phosphate groups, it is called a deoxyribonucleotide (also referred to as deoxyribonucleoside monophosphate in the case of a single phosphate, deoxyribonucleoside diphosphate for two phosphates, etc.). The four common deoxyribonucleoside monophosphates that form the subunits of DNA are shown in Figure 1.1D (along with simple representations that will be used in figures throughout the book).

As mentioned above, the repeating unit in a DNA strand consists of deoxyribonucleoside monophosphate. Within each DNA strand, the monophosphate groups link together adjacent deoxyribose units by bonding to the 3′-C atom of one sugar and the 5′-C atom of its neighbor (corresponding to 3′-OH and 5′-OH groups in the unlinked deoxyribonucleoside; see Figure 1.1B). The linear nature of the strand results from the repeating sugar-phosphate units,

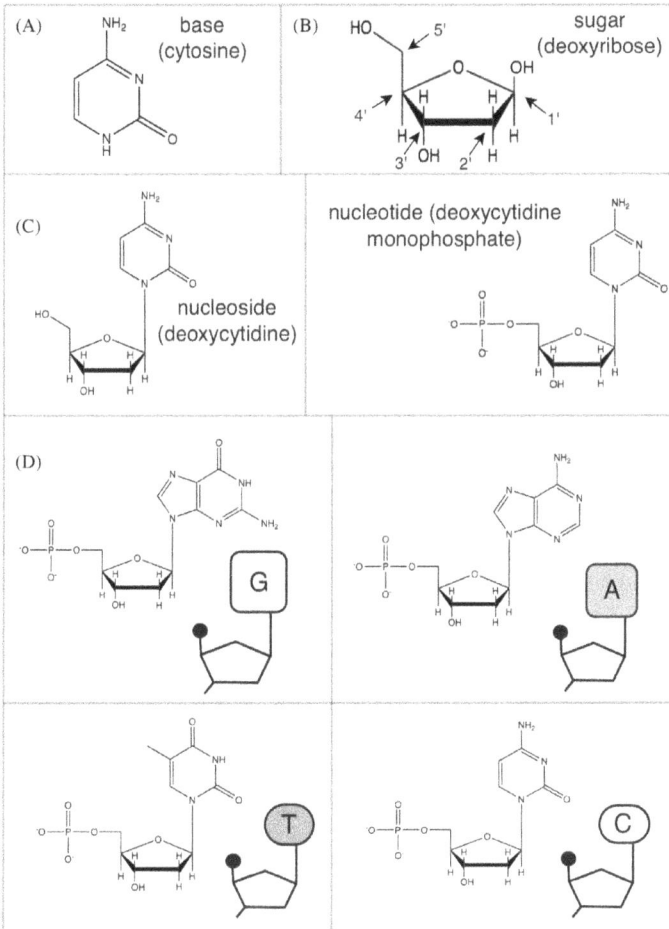

Figure 1.1. Bases, nucleosides, and nucleotides. DNA is composed of repeating chains of linked nucleotides. The basic components of each nucleotide are a base (A), the sugar deoxyribose (B) (ribose in RNA), and phosphates (one phosphate por residue in DNA, three in the triphosphate precursor). The free base cytosine is shown in panel A, and the sugar deoxyribose in panel B, with the numbered carbon positions indicated. The combination of a base and sugar is a nucleoside, such as deoxycytidine (panel C, left), with the base connected to the 1'-carbon of the sugar. The addition of one or more phosphates to the 5'-carbon of the nucleoside creates a nucleotide, such as deoxycytidine monophosphate (panel C, right). The four major nucleoside monophosphates in DNA are shown in panel D, along with a diagrammatic representation of each that will be used in various figures throughout this book. (Hydrogens on the carbon atoms in the bases and on the 5'-C of the sugars are not shown.)

called the sugar-phosphate backbone, while the bases are appended off of the sugar residues (Figure 1.2, right panel). The nature of the phosphodiester bond gives each DNA strand its directionality — the deoxyribonucleoside residue at each end of the DNA strand will have either a 3'-OH or a 5'-OH group that is not linked to a neighbor, and these residues define the 3' end or the 5' end.

In the double helix of DNA, the two involved strands are in opposite orientation, or "anti-parallel," with one sugar-phosphate backbone oriented in the 5' to 3' direction and the other 3' to 5' (Figure 1.2). The inherent beauty of the DNA double helix is that the four bases partner with each other from opposite strands — adenine (A) always pairs with thymine (T) and guanine (G) always pairs with cytosine (C) — the four canonical base pairs in duplex DNA (A:T, T:A, G:C, and C:G).[2] This pairing is made possible in the context of the two strands winding, in a right-handed helical fashion, around each other about once every 10.5 base pairs (Figure 1.2, left panel, middle, and bottom). The paired bases are on the interior of the duplex, while the sugar-phosphate backbone dominates the exterior.

As shown in Figure 1.3, the pairing of the bases involves a particular arrangement of hydrogen bonds across the helix (three bonds for G:C or C:G and two for A:T or T:A). The interior bases of the double helix are "stacked" on top of each other, providing stability. The overall shape of the helix provides two grooves, called the major and minor groove, which expose the edges of the bases and allow proteins to recognize particular sequences without needing to unwind the two strands of the double helix (Figure 1.2, bottom).

With this appreciation of the structure of DNA, we can now consider how this structure impacts the processes of DNA replication and repair. In their 1953 paper, Watson and Crick proposed how the double-helical structure resolves the puzzle of the duplication of the genetic material, with their famous line: "*It has not escaped our notice that the specific pairing we have postulated immediately suggests a possible copying*

---

[2] When discussing DNA or a particular DNA sequence, the nucleotide residues are often referred to simply as the base designations A, G, C, and T, rather than the more cumbersome dA, dG, dC, or dT, and without explicitly indicating the phosphate linkages. The deoxy designations will be used in situations where both RNA and DNA nucleotide residues are relevant and need to be distinguished.

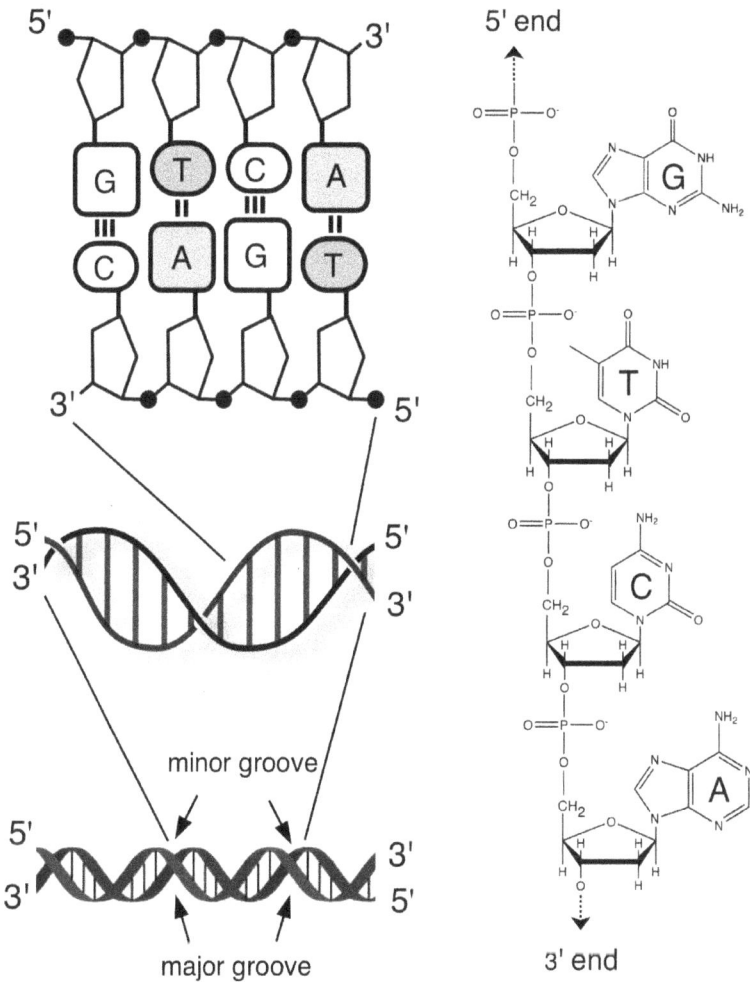

Figure 1.2. The structure of DNA. Three depictions of segments of duplex DNA are shown at different scales on the left. The top diagram shows the two antiparallel strands, with proper base pairing indicated. Note that A/T base pairs have two hydrogen bonds while G/C base pairs have three (short vertical lines). The middle diagram shows approximately one turn of the double helix, with longer vertical lines each indicating one base pair (approximately 10 base pairs per turn of the helix). The bottom diagram shows a longer stretch of duplex DNA, with major and minor grooves indicated. The chemical structure of the strand segment at the top left is shown on the right side of the figure. (Hydrogens on the carbon atoms in the bases are not shown.)

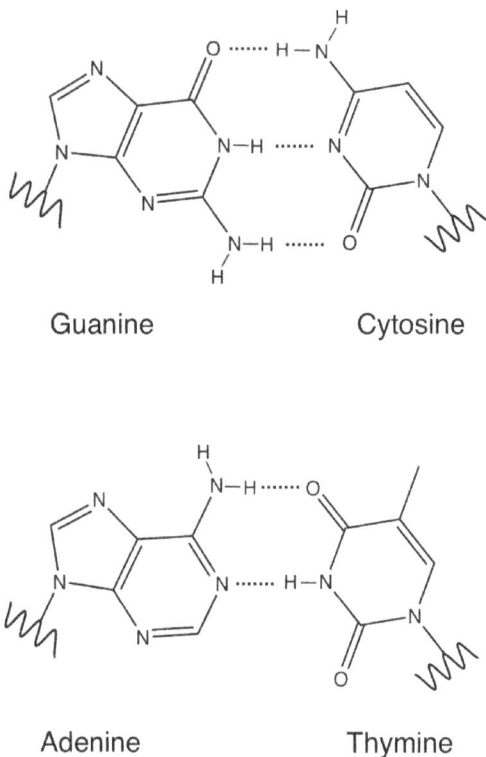

Figure 1.3.   The two major base pairs in DNA. The hydrogen bonds within the guanine-cytosine and adenine-thymine base pairs are shown as dotted lines. The N-glycosidic bonds that connect each base to its respective deoxyribose sugar are indicated by the solid line ending in a squiggle. (Hydrogens on the carbon atoms in the bases are not shown.)

mechanism for the genetic material." They realized that the two strands of DNA contain, in a sense, redundant information — either strand could direct an entirely new double helix, identical to the parental duplex, by using the rules of base pairing to recreate the opposite strand. If both single strands of the parental are used in this way as templates for their new partner strand, the result is two daughter duplexes that have the same sequence as each other and as the parental duplex. Note that each daughter duplex consists of one strand of parental DNA and one strand of new DNA, hence the process is characterized as being "semi-conservative" (one strand conserved, one new).

Many DNA repair reactions rely on the information redundancy in the DNA duplex. Damaged or incorrect bases on one of the two strands of DNA can be corrected by using the template information in the other strand to replace the derelict bases. This allows remarkable stability of the DNA sequence in the face of myriad forms of DNA damage, induced by physical agents such as UV and X-rays and the countless chemicals that can damage DNA in diverse ways.

The information redundancy in DNA is a bit like having a complete backup copy of all the information on your computer, except that the backup is not a silent library that is only accessed in times of need. The backup is built right into the structure of the DNA molecule — neither strand has a primary or secondary information role, rather both strands have equivalent importance. For example, along the length of any particular chromosome, the "template" strand that is used to direct the synthesis of messenger RNA for protein synthesis differs for different proteins.

The double-helical nature of the DNA molecule has several dramatic implications for the logic of DNA replication. First, the fact that the two strands are aligned in opposite directions means that the mechanisms to replicate the two strands need to be somewhat distinct, assuming replication proceeds in an orderly direction. The only alternative would be to completely unwind the entire molecule and start replication with two single strands — this strategy is not employed in cells, although we will see that it forms the basis of the powerful method of *polymerase chain reaction* (PCR; see Section 15.2). Second, the winding of the two strands around each other every 10.5 base pairs introduces an incredible topological problem that must be solved for successful replication and cell division. For example, the two strands of the entire content of human DNA in one cell are wound around each other over 600 million times, and yet the two daughter molecules must be completely disentangled from each other for cell division to be successful (i.e., the two daughter cells each receive a full complement of the genome). Third, in a related issue, the winding of the strands implies that something has to spin rather quickly during the process of unwinding the DNA for replication. Considering the bacterial replication process, with its rate of 1000 base pairs replicated per second, either the DNA or the proteins engaged in unwinding/replication need to crank up a spin rate of nearly 6000 rpm ((1000 bp/second × (1 revolution/10.5 bp)) × 60 seconds/minute), faster than the turbine of some jet engines. Adding to the complexity is the fact that multiple replication machineries act on different regions of the same DNA molecule simultaneously.

1.3 The key functions needed for the process of DNA replication

The next several chapters will consider detailed aspects of the proteins and reactions involved in DNA replication. To set the stage for these chapters, let us first consider the very basic functions needed for replication — the basic reactions that must occur for successful genome duplication.

The heart of any DNA replication reaction is the ability to polymerize the deoxynucleoside monophosphates in the new daughter strand, a reaction catalyzed by enzymes called "DNA polymerases." DNA polymerases use the rules of base pairing to accurately select which deoxynucleotide to insert into a growing chain (Figure 1.4A). As shown in the figure, the two parental strands

---

Figure 1.4. (Figure on Facing Page) The basic synthetic step in DNA synthesis. The reaction catalyzed by DNA polymerases involves the addition, sequentially, of single nucleotide residues to a DNA chain. The new residue is always added to the free 3′ end of a preexisting primer, and hence the direction of chain growth is 5′ to 3′. DNA polymerase nearly always inserts the nucleotide which correctly base pairs with the opposing base on the template strand, in this case, a C residue opposite a template G residue (panel A). The precursors for DNA synthesis are nucleoside triphosphates, with monophosphate incorporated into the DNA and pyrophosphate released. Panel B shows the reaction in more detail, highlighting the chemistry of nucleotide addition within the dotted boxes. The 3′-OH group on the primer chain performs a nucleophilic attack on the first (α) phosphate of the incoming nucleoside triphosphate, releasing pyrophosphate, and cementing the bridge (the phosphodiester bond) between the newly added residue and the growing primer chain. Notice that the polymer product of the reaction on the right has a free 3′-OH group on the newly added residue, allowing this 3′-OH group to serve as the attacking group for the next nucleotide addition.

(A)

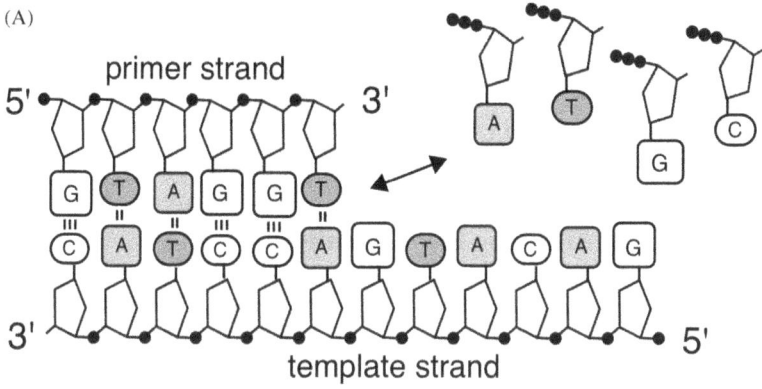

primer strand

5'   3'

template strand

3'   5'

(B)

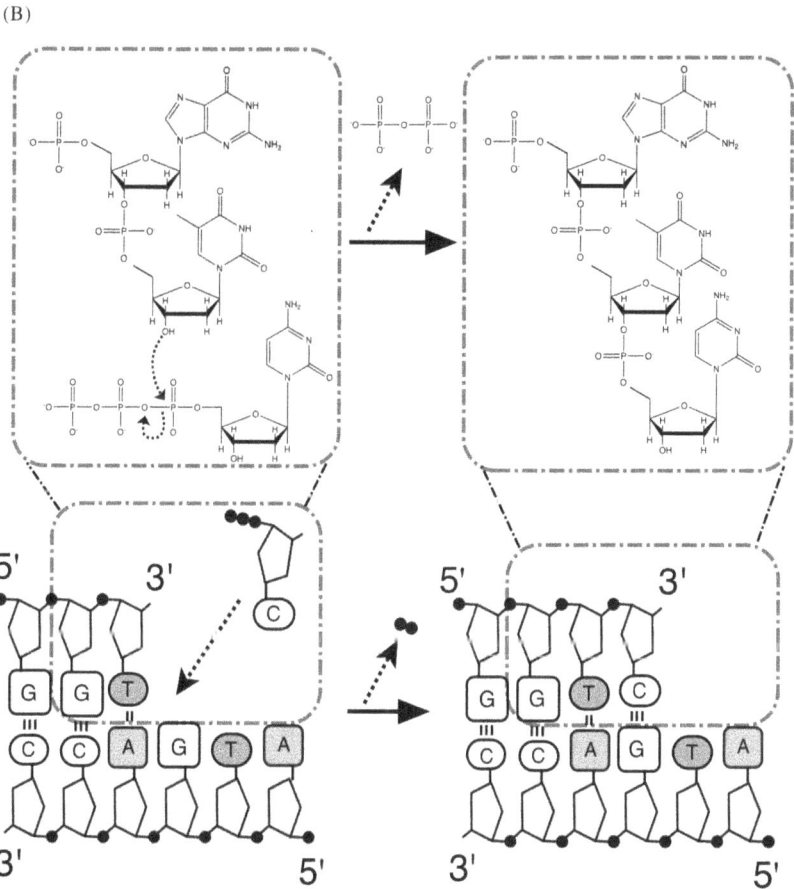

5'   3'

3'   5'

5'   3'

3'   5'

have distinct roles — the chain to which residues are added is called the "primer" and the chain used for testing the base pair is called the "template," since it is the template for the information in the newly synthesized region.

The substrates for this extension reaction are nucleoside triphosphates, with the triphosphate attached to the 5′-C of the deoxyribose. The reaction involves a phosphoryl transfer reaction in which the 3′-OH group at the primer terminus engages in a nucleophilic attack on the first phosphate group (α) of the high-energy nucleoside triphosphate (Figure 1.4B). The reaction results in cleavage of the linkage between the first and second (β) phosphate groups, with the 3′-OH of the primer becoming linked to the α phosphate of the incoming residue. The β and γ phosphates are released in the reaction, still linked to each other as pyrophosphate. Note that the primer has now been extended by one residue, and the 3′-OH group of the newly added residue is now the new 3′ end of the growing primer strand. The next cycle of nucleotide addition will utilize this 3′-OH group to direct the nucleophilic attack of the new nucleoside triphosphate in the polymerase reaction. Successive rounds of the addition result in extensive chain growth in the 5′ to 3′ direction.

Nearly all DNA polymerases require a preexisting primer and extend new chains in the 5′ to 3′ direction (see Section 4.9 for the sole exception). The consequences of these characteristics will be evident at many points in this book. The most immediate consequence relates to the difference in replicating the two strands. As the DNA replication machinery replicates a parental duplex from one location to another, one of the strands can be synthesized continuously in the 5′ to 3′ direction, potentially by a single DNA polymerase molecule that continuously adds residues, and this strand is called the "leading strand." The other strand, however, requires a DNA polymerase that travels in the opposite direction with respect to its template strand (Figure 1.5). Since the synthesis of this strand is somewhat delayed compared to the leading strand, it is referred to as the "lagging strand." Note also that the lagging strand cannot be made in one continuous process (unless

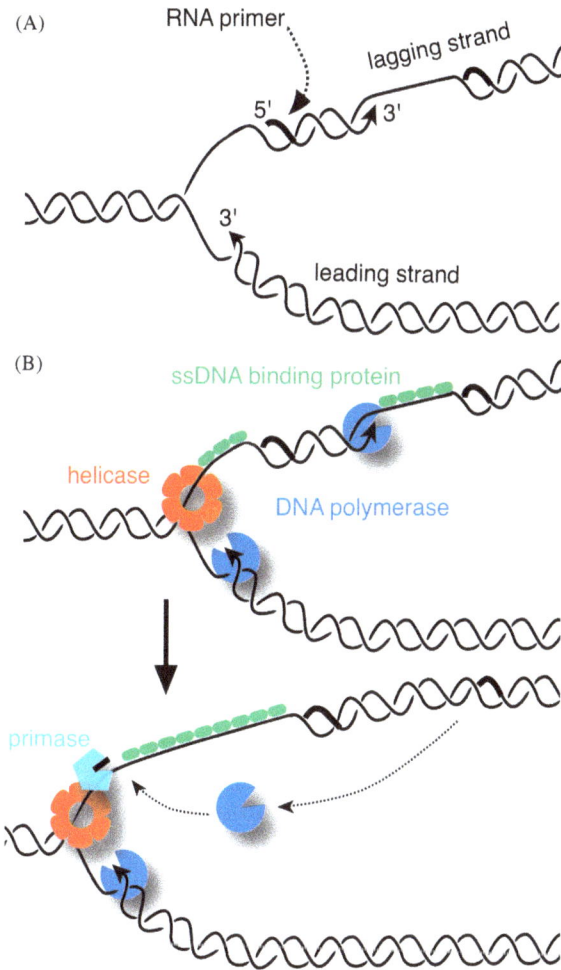

Figure 1.5. Simplified schematic of a replication fork. The new leading strand can be synthesized continuously due to the 5′ to 3′ directionality of DNA polymerases, but the new lagging strand must be synthesized discontinuously in short Okazaki fragments, which are later joined together. The major protein players at the replication fork include DNA polymerase (dark blue), helicase (orange), primase (light blue), and ssDNA-binding protein (green). See text for further discussion.

the process is started only after the entire leading strand is completed). Instead, the lagging strand is synthesized in short pieces, which are joined together after synthesis to form the complete daughter strand. These short pieces are called "Okazaki fragments," after the scientist who discovered them in the 1960s.[3]

While one particular protein species carries the active site for DNA synthesis and can rightly be called the DNA polymerase itself, every replicative DNA polymerase acts within a complex of multiple protein species. The additional protein subunits greatly modulate the activity of the polymerase, for example, allowing the different behavior of DNA polymerase on the leading and lagging strand.

As mentioned above, DNA polymerases generally require a preexisting primer to add deoxynucleotide residues. How then can DNA synthesis be started at the locations where leading-strand synthesis begins, and the multiple locations where Okazaki fragments begin? It turns out that RNA polymerases do not have this requirement for a preexisting primer, even though they share many of the other characteristics of DNA polymerases (e.g., synthesizing product in the 5′ to 3′ direction, using the rules of base pairing, using triphosphate nucleotide substrates, releasing pyrophosphate, etc.). RNA polymerases involved in transcription begin their RNA synthesis at sequence elements called promoters, linking two nucleoside triphosphates in the first step (Figure 1.6). From there on, additional nucleoside monophosphates are added in a reaction essentially identical to that in DNA synthesis mentioned above.

Getting back to DNA replication, the second major function needed for the functional replication machinery is a special form of RNA polymerase that provides short RNA primers, which are extended by the DNA polymerases (Figure 1.5). In most systems, these primases provide the primer for the leading-strand polymerase reaction at the sites where replication begins, called replication origins. They also function repeatedly during lagging-strand synthesis to prime the synthesis of every Okazaki fragment. Note also that these short stretches of RNA residues are removed prior to the completion

---

[3] See Okazaki *et al.* (1968) in Further Reading at the end of this chapter.

(A)

(B)

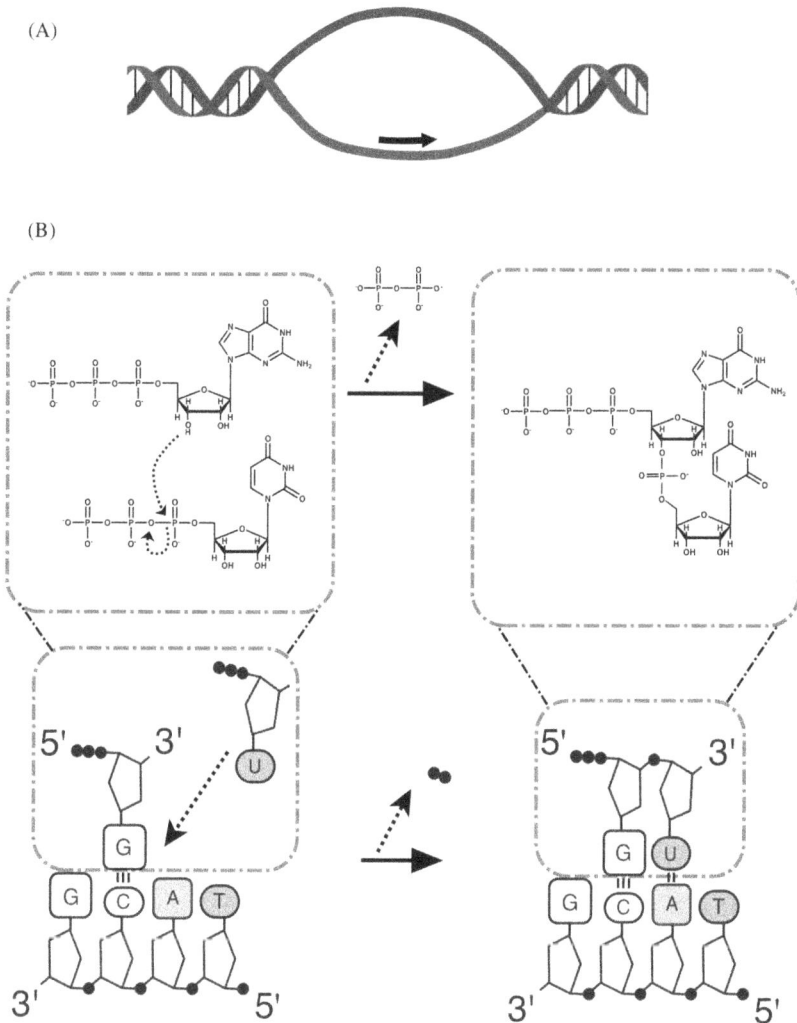

Figure 1.6. Initiation of transcription. RNA polymerase initiates RNA synthesis within an unwound bubble region of DNA that is formed by the enzyme at promoters (panel A). The synthetic process begins with the formation of a phosphodiester bond between two ribonucleoside triphosphates, with the initial product consisting of a dinucleotide with triphosphate at the 5′ end (panel B).

of DNA replication; the enzymes that perform this function will be discussed in later chapters.

As discussed above, the base pairs in a DNA duplex are interior to the molecule, with the two or three hydrogen bonds of the base pairs buried deep in the molecule. Furthermore, the two daughter duplexes after replication have one strand each from the parent molecule and one new strand. It is therefore obvious that the DNA duplex must be separated into its two constituent strands as replication occurs. This process is called DNA unwinding to reflect the loss of the helical turns of the two strands around each other. While some DNA polymerases are capable of driving this DNA unwinding reaction, the replication machinery has a more efficient type of enzyme that carries out this important function, called a "DNA helicase." DNA helicases unwind the duplex DNA into its two constituent strands while hydrolyzing a ribonucleotide triphosphate (usually ATP) to fuel the reaction (Figure 1.5). While we discuss the enzymes of DNA synthesis separately, this is a good time to point out that the three enzymes introduced so far, DNA polymerases, primases, and helicases, function together and interact physically, influencing the activities of each other in profound ways. Indeed, these and additional proteins involved in DNA replication form an intricate and well-coordinated machine for the duplication of DNA (Chapters 2–4).

During the synthesis of the short Okazaki fragments on the lagging strand as replication proceeds, segments of the lagging-strand template are exposed as single strands following helicase action (Figure 1.5). Single-stranded DNA (ssDNA) is more vulnerable to various forms of damage, and also in a sense more disorganized than duplex DNA, which has a regular helical structure. This leads us to the next important protein in the replication machinery, one that does not have enzymatic function but rather binds to ssDNA and helps protect and organize it. This protein, called ssDNA-binding protein, binds to transient regions of ssDNA as replication proceeds and is critical for the proper functioning of the replication machinery. There is evidence from prokaryotic replication systems that ssDNA-binding protein also plays a key organizational role in "spooling" the ssDNA into a more regular structure. As will be discussed later in the book, ssDNA-binding protein also plays many key roles in DNA repair and other cellular pathways.

As already mentioned, RNA primers become incorporated into the 5′ ends of new DNA strands during replication. These RNA residues must be removed before completion of DNA synthesis, and this job is completed by nucleases of different kinds. To avoid leaving a single-stranded gap at the site of every primer, a DNA polymerase is employed to extend the 3′ end of the preceding Okazaki fragment until it reaches the 5′ end created by nuclease action. Finally, the two adjacent fragments need to be joined together, and this step employs a special enzyme called "DNA ligase" (see Chapters 2–4 for more detailed discussion of Okazaki fragment processing).

As mentioned above, the two parental strands in a DNA duplex are wound around each other once every 10.5 base pairs. Unwinding of these two parental strands during DNA replication has the potential to leave the two daughter duplexes in a hopelessly tangled state, for example with the two daughter duplexes of each human chromosome wrapped around each other millions of times. This potential problem relates to the higher order structure, or topology, of the DNA molecule. To solve this problem, all cells have enzymes, called "DNA topoisomerases" that alter the topology of DNA in various ways. As described in Chapter 7, DNA topoisomerases are critical for completing DNA replication properly, and inhibition of the appropriate topoisomerase does indeed lead to a hopeless tangling of the two daughter duplexes as cells try to divide. DNA topoisomerases are also important during other processes, including transcription. Intriguingly, these enzymes are the targets for a number of chemotherapeutic drugs, including commonly used anticancer and antibacterial agents.

Two additional pathways, described in Chapters 5 and 6, are also needed to successfully complete DNA replication. Because of the inherent directionality of DNA polymerases, the very ends of chromosomes in eukaryotic cells require a special mechanism of replication and a different set of proteins from the normal replication machinery. These chromosome ends also have a special structure, called a "telomere," and one of the key enzymes in the replication of chromosome ends is a specialized DNA polymerase called "telomerase." Finally, a special repair pathway called "mismatch repair" increases the fidelity of DNA replication well beyond that which can be achieved with the replication machinery itself; mismatch repair occurs just after DNA replication.

1.4 Repairing and tolerating damage to the DNA molecule

In spite of its central role in heredity, DNA is prone to diverse forms of damage. The most frequent form of damage involves loss of a purine or pyrimidine base, which occurs thousands of times a day in human cells. Bases are also altered within the DNA molecule into a different form, for example, the relatively frequent deamination of a cytosine residue into uracil. Numerous chemical and physical agents promote these and many other forms of damage to the DNA molecule. Interestingly, even the endogenous chemicals within a cell can damage DNA, for example, the essential cofactor S-adenosylmethionine can methylate adenine residues in DNA, and metabolic oxygen radicals cause various forms of oxidative DNA damage. The forms and causes of DNA damage will be elaborated in more detail in Chapter 8.

Conceptually, the simplest form of DNA repair involves chemical reversal of the damage to restore a completely intact DNA that is indistinguishable from the DNA prior to damage. This direct damage reversal pathway can occur with only a few specific forms of DNA damage and relies on special proteins that evolved specifically for the reversal reaction (see Chapter 9). For example, *ultra*violet light (UV) leads to covalent linkages between two adjacent pyrimidines, and enzymes called "DNA photolyases" can reverse these linkages dependent on the adsorption of light. Another group of direct damage-reversal pathways acts on DNA bases that have been damaged by the addition of an alkyl group. A number of proteins can recognize specific forms of alkylated bases and directly reverse the alkylation, releasing the normal undamaged base as a product.

As mentioned above, the information redundancy on the two strands of the DNA duplex provides a fundamental opportunity to repair DNA damage without losing any genetic information. As long as the damage is on only one strand of a DNA duplex, the offending segment of DNA can be removed and replaced by a DNA polymerase reaction that uses the intact, undamaged opposite strand as a template. This forms the basis of several critical repair pathways, including base and nucleotide excision repair (Chapter 10) and mismatch repair (Chapter 6).

When the damage to the DNA molecule involves both strands, such as a *double-strand break* (DSB), the information redundancy in the two strands is irrelevant. Nonetheless, cells have evolved surprisingly robust pathways to repair DSBs and other forms of damage that impact both strands. As will be described in Chapter 11, one set of DSB repair pathways uses homologous DNA as a template to generate an error-free product. In diploid eukaryotic cells, for example, a break in the maternally derived chromosome can be repaired with the assistance of the paternally derived chromosome. Similarly, after the process of DNA replication, a break in one of the two duplex replication products (called sister chromosomes) can be repaired with the assistance of the other. A second set of DSB repair pathways, called nonhomologous end joining, simply reconnects two broken ends without using any template to try to ensure correct alignment. This pathway often involves the addition or removal of some bases from the DNA ends, leading to frequent mutations at the site of the break.

When considering the various forms of DNA damage, it is critical to distinguish between DNA damage and mutation. As we have seen above, DNA damage can often be reversed or repaired in such a way that the repaired DNA molecule is identical to the starting DNA, with no changes in the DNA sequence. On the other hand, a mutation in DNA is a heritable change in the DNA sequence. A particular form of DNA damage might, at some frequency, lead to a mutation at the site of damage, but the damage itself is not a mutation. It is also important to note that the accumulation of a certain amount of DNA damage might lead directly to cell death, for example, because the cell can no longer replicate its DNA successfully. On the other hand, a small subset of mutations might also cause cell death or other untoward effects by a completely different mechanism, for example, by changing or destroying the activity of an important cellular enzyme. While detailed mechanisms of mutagenesis (creation of mutations) are beyond the scope of this book, we discuss in later chapters certain mutation pathways in the context of DNA replication errors and specific DNA repair pathways.

The distinction between damage and mutation is particularly striking and relevant when discussing the special class of translesion DNA polymerases. These polymerases have the surprising ability to synthesize opposite a damaged base without using that base as a template. Obviously, this can lead to a mutation, and indeed these DNA polymerases were discovered as causing a subset of mutations in cells. Nonetheless, translesion polymerases assist the cell in surviving DNA damage, as will be discussed in Chapter 12.

In the closing two chapters of this book, the basic knowledge about replication and repair pathways/proteins will be used to appreciate certain human disease states and also the burgeoning field of biotechnology. Numerous human diseases, particularly cancer, involve heritable (germ line) or newly formed (somatic) mutations in the genes that encode proteins involved in DNA replication and repair (Chapter 14). Also, many of the tools used in modern genetic technologies are completely dependent on proteins involved in replication and repair, often derived from simple bacterial systems (Chapter 15).

1.5 Summary of key points

- Human genome of 6.6 billion base pairs is replicated with very high accuracy.
- Double helical structure of DNA allows semiconservative replication and provides an embedded "backup copy" of the genome.
- DNA polymerases synthesize new chains in the 5′ to 3′ direction, necessitating different replication mechanisms on the leading and lagging strands.

- Key proteins in DNA replication include DNA polymerase, primase, helicase, and ssDNA binding protein.
- Additional processes important for DNA replication include Okazaki fragment processing, mismatch repair, resolving topological problems, and telomere replication (when the chromosome is linear).
- Multiple DNA repair pathways remove DNA damage, often avoiding the generation of mutations, which alter the DNA sequence.
- Proteins involved in DNA replication and repair are prominent in human diseases and as tools in modern genetic technologies.

Further Reading

Alberts, B., Johnson, A., Lewis, J., Morgan, D., Raff, M., Roberts, K., & Walter, P. (2015). *Molecular Biology of the Cell* (6th ed.). New York, NY: Garland Science, Taylor and Francis Group.

Bansal, M. (2003). DNA structure: Revisiting the Watson-Crick double helix. *Current Science, 85*(11), 1556–1563.

Bell, S. D., Mechali, M., & DePamphilis, M. L. (2013). *DNA Replication.* Cold Spring Harbor, NY: Cold Spring Harbor Press.

Chargaff, E., Lipshitz, R., & Green, C. (1952). Composition of the desoxypentose nucleic acids of four genera of sea-urchin. *J Biol Chem, 195*(1), 155–160.

DePamphilis, M. L., & Bell, S. D. (2011). *Genome Duplication.* New York, NY: Garland Science.

Friedberg, E. C., Elledge, S. J., Lehmann, A. R., Lindahl, T., & Muzi-Falconi, M. (2014). *DNA Repair, Mutagenesis, and other Responses to DNA Damage. A Subject Collection from Cold Spring Harbor Perspectives in Biology.* Cold Spring Harbor, NY: Cold Spring Harbor Press.

Friedberg, E. C., Walker, G. C., Siede, W., Wood, R. D., Schultz, R. A., & Ellenberger, T. (2006). *DNA Repair and Mutagenesis* (2nd ed.). Washington, DC: ASM Press.

Frixione, E., & Ruiz-Zamarripa, L. (2019). The "scientific catastrophe" in nucleic acids research that boosted molecular biology. *J Biol Chem, 294*(7), 2249–2255.

Haber, J. (2013). *Genome Stability: DNA Repair and Recombination.* New York, NY: Garland.

Kornberg, A., & Baker, T. A. (1992). *DNA Replication* (2nd ed.). New York, NY: W.H. Freeman.

Kowalczykowski, S., Hunter, N., & Heyer, W.-D. (2016). *DNA Recombination.* Cold Spring Harbor, NY: Cold Spring Harbor Press.

Okazaki, R., Okazaki, T., Sakabe, K., Sugimoto, K., & Sugino, A. (1968). Mechanism of DNA chain growth. I. Possible discontinuity and unusual secondary structure of newly synthesized chains. *Proc Natl Acad Sci U S A, 59*(2), 598–605.

Watson, J. D., & Crick, F. H. (1953). Molecular structure of nucleic acids: A structure for deoxyribose nucleic acid. *Nature, 171*(4356), 737–738.

How did they test that?
The base composition of DNA and Chargaff's rule

Studies of DNA in the first half of the 1900s uncovered the basic units of the molecule, namely the sugar, phosphate, and four distinct bases (for a fascinating review of this period of research, see Frixione and Ruiz-Zamarripa, 2019). At the time, DNA was not suspected to be the genetic material — it seemed too simple. The dominant model was the "tetranucleotide hypothesis," in which the four bases were linked to each other in a simple ring structure. This structural proposal was based on crude data that suggested roughly equal amounts of the four bases in preparations of DNA. Erwin Chargaff and his colleagues conducted very careful and detailed studies that disproved this hypothesis and provided key evidence that helped Watson and Crick arrive at the correct structure of duplex DNA. Chargaff and his associates meticulously purified DNA from various sources, carefully eliminating RNA contamination. Next, they hydrolyzed the DNA samples and determined the fractions of each of the four bases in the resulting mixtures. Without going into the chemical details, the determinations were done as follows. First, the base mixtures were subjected to paper chromatography, in which the various bases migrated to different positions along paper strips soaked in various chemical solutions. Four small areas on the paper strips, containing each of the purified bases, were cut out, and the four bases were eluted into separate test tubes. These were then analyzed by UV spectroscopy, which provided the adsorption maxima (revealing base identity) and intensities (revealing amounts). In one study, Chargaff *et al.* (1952) measured the base composition from sperm DNA isolated from four different sea urchin species (Table 1.1). In each of the four samples, the proportions of adenine and thymine were equal (within experimental error), as were the proportions of guanine and cytosine (Table 1.1). However, the proportions of other pairs of bases were clearly disparate. This data clearly disprove the tetranucleotide hypothesis mentioned above. The equality of adenine/thymine and guanine/cytosine, which was seen in disparate species (Table 1.2), is the so-called "Chargaff rule."

Table 1.1.   Base content of sea urchin sperm DNAs.

Species	% Adenine	% Thymine	% Guanine	% Cytosine
Psammechinus miliaris	29.0 (±0.7)	30.0 (±0.1)	16.3 (±0.3)	16.7 (±0.2)
Paracentrotus lividus	30.3 (±0.6)	29.9 (±0.5)	16.2 (±0.3)	15.9 (±0.5)
Echinocardium cordatum	30.0 (±0.7)	31.0 (±0.6)	15.9 (±0.2)	17.3 (±0.2)
Arbacia lixula	28.2 (±0.5)	28.7 (±0.4)	17.5 (±0.2)	18.1 (±0.3)

Base values are in mole percent; data from Chargaff *et al.* (1952).

Table 1.2.   Base content of diverse species.

Species	% Adenine	% Thymine	% Guanine	% Cytosine
Maize	26.8	27.2	22.8	23.2
Octopus	33.2	31.6	17.6	17.6
Chicken	28.0	28.4	22.0	21.6
Rat	28.6	28.4	21.4	20.5
Human	29.3	30.0	20.7	20.0

Base values are in mole percent; data from Chargaff as cited in Bansal (2003).

Chapter 2

The simple DNA replication system of a bacterial virus

2.1 Why the interest in a bacterial virus?

The DNA replication systems of prokaryotic and eukaryotic cells are quite complex, with over 20 proteins needed to replicate the genome of the model prokaryote *Escherichia coli*. This complexity undoubtedly relates to the relatively large genome sizes of these cells, the need for a very low error rate particularly given the large genome, the need to replicate the genome once and only once per cell cycle, and the need to carefully couple DNA replication to other cellular events such as cell division.

Because of the complexity of cellular replication systems, many scientists were initially drawn to study DNA replication in simpler viral systems. The rationale was that the key functions in DNA replication would be more evident, the required proteins easier to identify and isolate, and the biochemical mechanisms easier to decipher in a system with fewer interacting parts (i.e., replication proteins). Viruses of bacterial cells, called bacteriophages or phages, provide such simple model replication systems. Some viruses co-opt the host replication machinery, while others encode their own replication proteins. For those viruses that encode their own replication proteins, the number of involved proteins is indeed smaller than for host cell DNA replication.

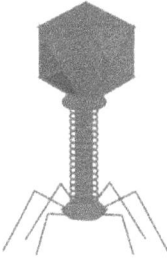

1969 Nobel Prize in Physiology or Medicine

This prize was awarded to **Max Delbruck, Alfred D. Hershey** and **Salvador E. Luria** for their studies on bacterial viruses, elucidating important aspects of their genetic structure and replication mechanisms.

https://www.nobelprize.org/prizes/medicine/1969/summary/

The bacteriophage called T7 turned out to provide a particularly good system for detailed study in that only four proteins are needed to form a fully functional replisome (a few other proteins are involved in initiating and finalizing the process; see below). Studies in this simple T7 system have been very productive, leading to high-resolution structures of each of the replication proteins and the replisome, a complete reconstitution of the reaction with the purified proteins in vitro, and remarkably insightful studies on the dynamics of the replication process based on the in vitro system. These studies of the T7 replication system were spearheaded in the laboratory of Charles C. Richardson (Harvard Medical School). For the same reasons that the T7 system was experimentally tractable, it is also a great system to introduce many of the concepts and mechanisms in DNA replication, and that will be the topic of this chapter.

There is another reason for interest in the T7 replication system. Surprisingly, replication of phage T7 DNA has been found to be very similar to the replication of mitochondrial DNA in human cells. The T7 replication proteins (as well as T7 RNA polymerase) are structurally homologous to the corresponding proteins in human mitochondria, and the basic mechanisms involved in the replication of the two genomes have clear parallels. This has led to the proposal that the genes for mitochondrial replication proteins have their evolutionary origin in a bacterial virus. It is interesting to speculate about how the genes from a bacterial virus could have been captured during the process by which a bacterial cell formed a symbiotic relationship with a primitive eukaryotic cell to form the precursors of mitochondria. Anyone interested in understanding how mitochondria replicate their DNA should certainly start by carefully learning all there is to know about phage T7 DNA replication.

2.2 The four proteins involved in T7 DNA replication

The T7 genome encodes only three proteins needed to form the phage replisome: a DNA polymerase, a combined helicase/primase protein, and an ssDNA-binding protein. The fourth protein involved in T7 DNA replication is a host-encoded protein, thioredoxin, which interacts with the T7 DNA polymerase in a 1:1 complex. For readers who delve into the primary literature on this topic, the T7 DNA polymerase is called gene *product* 5 (gp5), the helicase/primase is gp4, and the ssDNA-binding protein is gp2.5. For simplicity, in this book, we will use only the generic names that reflect the functions of these proteins.

The T7 DNA polymerase contains two major domains, an N-terminal exonuclease domain of 201 amino acids and a C-terminal polymerase domain of 503 amino acids. The structure of the polymerase domain resembles a partially closed right hand, with analogs of a palm, finger region, and thumb (Figure 2.1A). The active site for the polymerase reaction is within the palm subdomain, and the DNA (blue in Figure 2.1A) passes through the enzyme under the grip of the thumb. This right-hand architecture is common to nearly all DNA polymerases, even those belonging to different families.

The host protein thioredoxin binds to T7 DNA polymerase on an extension of the polymerase thumb (Figure 2.1A; thioredoxin in red). As will be discussed below, this binding enhances the activity of polymerase, allowing it to replicate a much longer stretch of DNA. As you might know, thioredoxin is a coenzyme involved in redox reactions in the cell, but this redox function of thioredoxin is not needed for T7 DNA replication. Instead, thioredoxin acts as a structural protein in its complex with T7 DNA polymerase, increasing polymerase processivity and configuring the polymerase to bind the DNA helicase (see below).[1]

---

[1] Bacterial thioredoxin is also involved in assembly of certain bacteriophages, and again, the redox function of the protein is not required. Thioredoxin has been found to bind some 80 different bacterial proteins, suggesting that it may be involved as a structural component in additional processes.

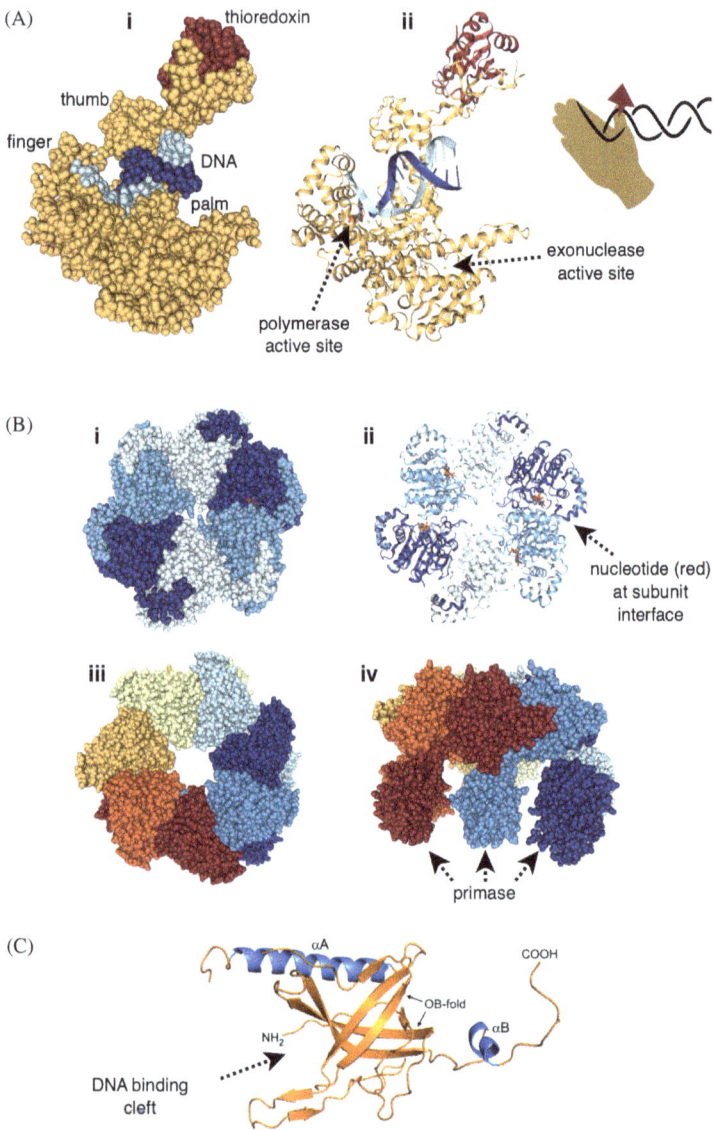

Figure 2.1. X-ray crystallographic structures of the three major replication proteins of bacteriophage T7. Like other DNA polymerases, the structure of T7 DNA polymerase (yellow) resembles a right hand, with palm, finger, and thumb domains (panel A). Appended to the thumb is a binding domain for the host thioredoxin protein (red); the primer-template DNA is in blue. These and many other structures throughout the book are from the RCSB Protein Database (www.rcsb.org; Berman *et al.*, 2000). Structure (i) is shown in the spacefill format and structure

When we counted only four different proteins in T7 DNA replication, we stretched the truth just a bit. It turns out that the gene encoding the T7 helicase/primase protein actually makes two closely related proteins in roughly equal amounts. One of these is missing the N-terminal 63 amino acids, but otherwise consists of the same exact amino acid sequence as the C-terminal remainder of the larger protein. These two forms result from the use of two different translation initiation codons during translation (in the same reading frame). As will be discussed below, the functional form of the helicase/primase during replication is a hexamer (6-mer). Some evidence suggests that the functional hexamer has three subunits of each of the two protein species in vivo; however, this stoichiometry is not required for assembly of helicase complexes in vitro.

The helicase function of the helicase/primase protein is carried out by the C-terminal segment, while the primase function resides in the N-terminal region. The smaller form of the protein discussed above is missing a portion of the primase region and, correspondingly, lacks primase activity.

←

Figure 2.1. (**Figure on Facing Page**) (ii) in the cartoon format, which represents α-helices as helices made from flat ribbons and β-sheets as flat ribbons that terminate with an arrow. Both images are PDB structure 1SKR (Li *et al.*, 2004). The T7 helicase (panel B) crystallized in several different forms. The image shown in structure (i) (spacefill) and (ii) (cartoon) is a hexamer of a truncated (helicase-only) form, with alternating subunits shown in different shades of blue. The bound nucleotide is shown in red in structure (ii), highlighting the active sites for nucleotide hydrolysis between adjacent subunits. These two images are from the RCSB PDB (www.rcsb.org) of PDB ID 1EOJ (Singleton *et al.*, 2000). Full-length helicase/primase without DNA crystallizes as a heptamer. The heptameric forms in structure (iii) and (iv) (both spacefill, 90° rotation between the two) are from the RCSB PDB (www.rcsb.org) of PDB ID 1Q57 (Toth *et al.*, 2003). Other closely related forms will be discussed later in this chapter. The T7 ssDNA-binding protein (yellow with α-helices in blue) is composed of an OB fold and two short α helices (panel C). This protein structure (PDB 1JE5) image is reproduced from Kulczyk and Richardson (2016), with permission from Elsevier; permission conveyed by Copyright Clearance Center, Inc. The location of the DNA-binding cleft is indicated based on the structure of a complex between an evolutionarily related ssDNA binding protein and DNA (Cernooka *et al.*, 2017).

X-ray crystallography has revealed several related structures of full-length or truncated versions of the T7 helicase/primase. A truncated form with only the helicase domain was found to form a beautiful hexameric ring or donut structure, with a hole in the middle large enough to accommodate ssDNA (Figure 2.1B, images i and ii). The full-length helicase/primase in the absence of DNA forms heptamers (7-mers), and a crystal structure of the heptamer again reveals a ring with a hole in the middle (Figure 2.1B, images iii and iv). Image iii shows the helicase domains of the heptamer, looking down on the protein from the C-terminal side. Rotating the complex into the plane of the paper by 90°, the primase domains are seen to extend downwards off the helicase heptamer (image iv). As mentioned above, the native complex during bacteriophage T7 infections may well contain three each of the full-length and N-terminally truncated subunits, but no structure yet exists for this hybrid form of the complex. We will return to structural variations in the T7 helicase/primase complex when we discuss the mechanism of helicase unwinding and the function of the replisome complex.

The fourth protein is the T7 ssDNA-binding protein, a 232-amino acid protein with a fairly simple structure (Figure 2.1C). The central part of the protein consists of a β-sheet with five strands oriented in a characteristic pattern called an OB-fold (*o*ligosaccharide/oligonucleotide-*b*inding fold). As the fold name implies, this is a common structural motif that a variety of proteins use for binding ssDNA, which occupies a cleft with the OB-fold at the base (Figure 2.1C). The C-terminal region of T7 ssDNA-binding protein, which is quite acidic, extends away from the body of the protein and is known to interact with the other T7 replication proteins (see below).

2.3 Activities of T7 DNA polymerase in the replisome

Many features of T7 DNA polymerase are conserved in other replicative DNA polymerases, making the enzyme a good model. The enzyme tightly grips the primer-template using all three subdomains (palm, finger, and thumb; Figure 2.1A). As introduced in Chapter 1, nearly all DNA polymerases require a primer for extension, and the single-stranded template is used to determine which of the four DNA nucleotides will be added to the end of the primer by base pairing (Figure 1.4). The active site of the enzyme, where the new nucleotide is added, is within the palm subdomain near the base of the finger subdomain (Figure 2.1A). The active site region is tightly organized to strongly favor the correct base pairing of the incoming nucleotide and disfavor the three possible mispairs, contributing to the high fidelity of the enzyme.

The structure and geometry of the active site of the polymerase are very adept at aligning the incoming nucleotide for proper base pairing and excluding the three possible incorrect nucleotides (which don't fit the active site region very well). Detailed biochemical and structural studies have provided beautiful insights into the details of this base selection process, which is beyond the scope of this chapter. The fidelity of the polymerase step is very high but not perfect, and an incorrect base is inserted roughly once every 20,000 or so incorporation cycles. Incorrect base insertion is probably impossible to avoid. For example, tautomer forms of the DNA bases exist transiently at a very low level and these can pair with the "wrong" template base and thereby fit reasonably well within the polymerase active site. Without some additional mechanism to correct these misinsertions, mutation rates would be very high and complex cells may never have evolved.

The T7 DNA polymerase illustrates one of the major correction mechanisms/pathways that explain the extremely low mutation rate seen in genome replication (see Chapter 1). The exonuclease domain of the polymerase, introduced above, plays the major function of correcting most of the misinsertions from the polymerization reaction, increasing the fidelity by one to two orders of magnitude (Figure 2.2). This activity is therefore referred to as "proofreading exonuclease." As the "exo" in the name implies, exonucleases attack DNA from one of the two ends (from outside). The proofreading exonuclease in T7 and other replicative polymerases is a 3′ to 5′ exonuclease, because it attacks DNA from the 3′ end, which of course is the end of the growing strand found at the polymerase active site.

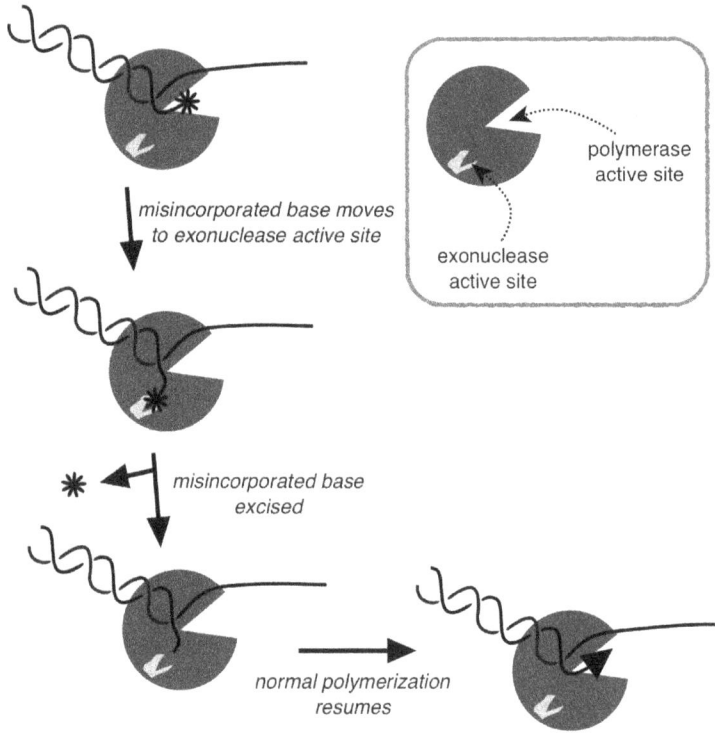

Figure 2.2. Excision of misincorporated residue by DNA polymerase. Replicative DNA polymerases generally have a 3′ to 5′ exonuclease activity that removes the vast majority of misincorporated residues. In the case of T7 DNA polymerase, the exonuclease activity resides in the same protein chain as the DNA polymerase activity, but the two active sites are some distance from each other. Upon misincorporation, the growing 3′ end is displaced from the polymerase active site to the exonuclease active site, where the 3′ terminal residue is excised (as a monophosphate). Synthesis resumes when the corrected 3′ end returns to the polymerase active site.

Notice in the polymerase structure (Figure 2.1A) that the exonuclease active site is far removed from the polymerase active site, and indeed the DNA primer-template within the complex must dramatically rearrange in order that a 3′ end relocates to the exonuclease active site and gets clipped. How does the enzyme favor exonuclease action when an incorrect base is incorporated? Incorporation of an incorrect base will invariably cause mispairing and thereby destabilize the double helix in its vicinity. For example, the end of the primer would become destabilized and fray when the rare tautomeric form of the incorporated base described above reverts to its normal form. Two forces then favor removal of the incorrect base. The exonuclease active site prefers a single-stranded end rather than one in a duplex primer-template, and the polymerase active site is inhibited from adding an additional base by a mispaired 3′ terminus in the primer-template (Figure 2.2). In this way, a misincorporated base has a high likelihood of ending up in the exonuclease active site, and once the terminal mismatched base is removed, the (correctly paired) duplex primer-template reforms and is favored to migrate back to the polymerase active site.

An important property of DNA polymerases, called processivity, is the ability to extend a primer (add a nucleotide residue) repeatedly without dissociating from the primer-template. A processive polymerase extends repeatedly, while a distributive polymerase incorporates a small number of nucleotide residues before dissociating and switching to a different primer-template. Processivity increases the rate of DNA replication because it eliminates the time-consuming process of recruiting a new polymerase whenever the replicating polymerase dissociates. It turns out that the processivity of all replicative polymerases is modulated extensively in order to carry out an efficient and concerted replication reaction. The most obvious need for modulating processivity is to accommodate the very different length requirements for replication on the leading versus the lagging strand.

The T7 DNA polymerase devoid of thioredoxin has quite a low processivity, incorporating only about 10–15 nucleotides before dissociating. However, the addition of thioredoxin dramatically increases processivity to an average of about 800 nucleotides per binding event. Concomitant with the increased processivity, the structure of the region of polymerase that binds thioredoxin is reorganized, and the polymerase complex interacts more extensively with the primer-template. Thioredoxin itself is situated over the duplex portion of the primer-template in the structure of the complex (Figure 2.1A). One model for the increased processivity conferred by thioredoxin is that the thumb domain with thioredoxin folds down over the duplex portion of the primer-template after nucleotide addition, essentially encircling the duplex region transiently. This mechanism would be somewhat analogous to that used in cellular DNA replication, where specialized proteins called sliding clamps tether the DNA polymerase to its template (see Chapters 3 and 4).

Although the increase in processivity conferred by thioredoxin is impressive, it is not sufficient to account for the rapid copying of the 40,000-base pair viral genome. Indeed, the processivity of T7 DNA polymerase is further enhanced by the helicase/primase hexamer when the proteins are appropriately arranged for leading-strand synthesis (Figure 2.3, also see below). In this case, DNA polymerase extending the leading strand continues for approximately 5000 nucleotides, on average, before dissociating from the nascent 3′ end. Remarkably, however, this dissociation does not significantly impede replication. There is another DNA polymerase binding site on the helicase/primase,[2] and the dissociating polymerase can transiently bind to this site and then quickly return to the leading-strand 3′ end. In all, the helicase/primase increases DNA polymerase

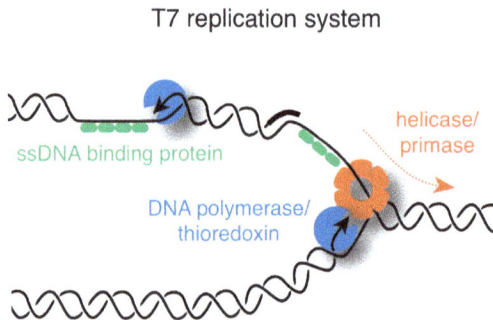

T7 replication system

Figure 2.3. T7 replication fork, the "traditional" view. The T7 replication fork is depicted with two DNA polymerase molecules (blue), the combined helicase/primase (orange) and the ssDNA-binding protein (green). The direction of helicase movement is indicated by the dotted arrow.

---

[2] Thioredoxin also plays a key role here, configuring the DNA polymerase so it can bind the second site on the helicase/primase.

processivity to greater than 17,000 nucleotides, likely sufficient to copy the entire T7 genome.

2.4 Mechanisms of unwinding by the T7 helicase

The two major domains of the T7 helicase/primase protein have distinct functions — one catalyzes the unwinding of the parental DNA duplex (helicase) and the other provides RNA primers for lagging-strand synthesis of Okazaki fragments (primase). As we see in Chapters 3 and 4, cellular replication systems utilize two distinct proteins for these two functions, although even in those cases the two proteins interact functionally and communicate with each other.

Focusing first on DNA unwinding activity, T7 helicase translocates in the 5′ to 3′ direction along a single strand of DNA, unwinding any duplex region that is encountered during this translocation. Recent structural studies support an attractive "hand-over-hand" model for this translocation. Helicase complexes with ssDNA were resolved into multiple hexameric forms that resembled lock-washers rather than symmetrical rings. The image at the top left of Figure 2.4A shows one such hexamer, with the discontinuity of the lock-washer in between the subunits labeled HelA and HelF.[3] The key structural transition in the model is that the subunit on the 5′ side of the discontinuity moves up to the 3′ end of the lock-washer, as the discontinuity shifts to the next interface (between HelE and HelF in the first step; Figures 2.4A and B). Given the 5′ to 3′ direction of movement along the DNA, the HelF subunit is essentially moving from the back to the front of the lock-washer. This movement comprises approximately 24 angstroms (10^{-10} meters) in the 5′ to 3′ direction. Long-range movement of the helicase along the DNA is accomplished by the sequential movement of individual subunits from the back to the front of the lock-washer, as the discontinuity essentially rotates around the ring (Figure 2.4B). Note that

---

[3]These names are arbitrary, in that the six subunits are identical, but help in explaining the subunit behavior during helicase movement along the DNA.

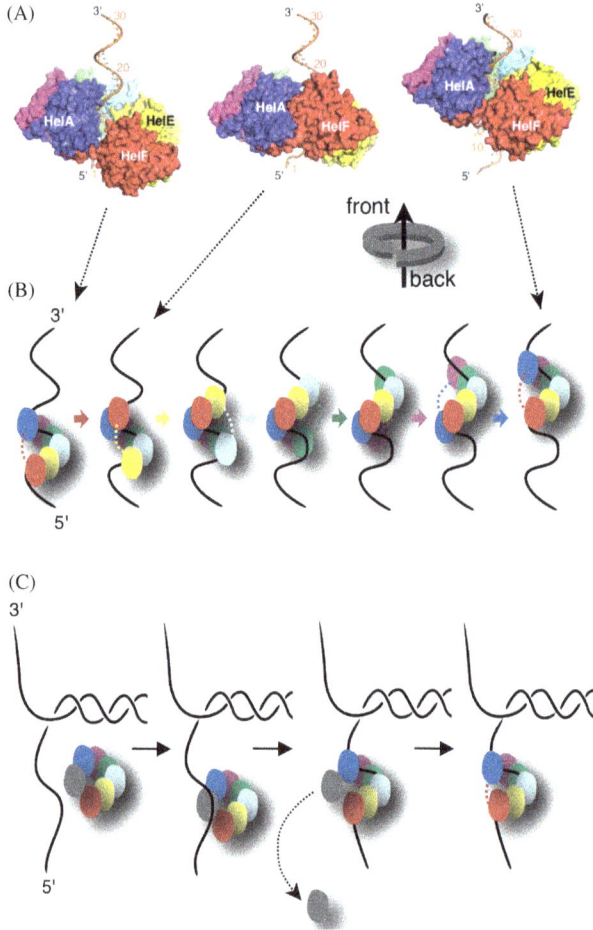

Figure 2.4. Movement and loading of the T7 replicative helicase. Recent structural studies strongly support a "hand-over-hand" mechanism for translocation of the T7 helicase along ssDNA (Gao *et al.*, 2019). In panel A, closely related helicase-DNA structures illustrate key features of the mechanism (see text for detailed discussion). Gao *et al.* (2019) refer to the six subunits as HelA, HelB, HelC, HelD, HelE, and HelF. The lock-washer image between panels A and B illustrates the overall structure, with DNA indicated by the arrow (pointing 5′ to 3′). The sequential movement along ssDNA is illustrated in the cartoon in panel B. The dotted lines represent the inter-subunit contacts and are only shown in this cartoon for the junction across the discontinuity of the lock-washers. The arrows to the right of each cartoon figure are in the same color as the subunit that is moving up the hexamer structure. Notice that the overall hexamer is moving in the 5′ to 3′ direction along the DNA.

the structure at the end of six cycles of subunit movement is identical to the structure at the start, except that the hexamer has advanced along the ssDNA.

One critical aspect of the model is the role of nucleotide hydrolysis. T7 helicase hydrolyzes dTTP to translocate along DNA. The sites of nucleotide binding are at the interface between adjacent subunits in the hexamer, and given the hexameric state of the protein, there are six potential active sites around the ring (Figure 2.1B, image ii). Each subunit movement across the discontinuity is proposed to be driven by hydrolysis of the dTTP bound to that subunit, which then binds a new dTTP after its repositioning at the front of the lock-washer.

Another critical aspect of the hand-over-hand model is that each subunit has an extension that contacts the adjacent subunit, and this extension reaches across the discontinuity of the lock-washer. The extension is most easily seen in Figure 2.4A images as the red segment (HelF) at the bottom of the blue subunit (HelA), but each subunit contacts its neighbor in this manner. In the cartoon in Figure 2.4B, extension across the discontinuity of the lock-washer is depicted as the dotted line (the other five extensions are not shown in this cartoon for simplicity). The extensions to the adjacent subunits are actually the regions in between the helicase and primase portions of the protein (see Figure 2.1B, image iv; also see Figure 2.1B, image i, where the inter-subunit contacts are very evident).

The ssDNA in these structures was found to be threaded helically through the middle of the hexamer, with each helicase subunit contacting the phosphodiester backbone of two adjacent nucleotides.

Figure 2.4. (**Figure on Facing Page**) Panel C depicts a ring-opening model for helicase loading. In solution without DNA, the T7 helicase forms heptamers, but hexamers are observed after loading onto ssDNA. The model proposes that one subunit of the heptamer is ejected as the ring rotates into the hexameric lock-washer form. Panel A is reproduced from Gao *et al.* (2019), with permission from the American Association for the Advancement of Science; permission conveyed by Copyright Clearance Center, Inc. Panel B was modified from a figure by Gao *et al.* (2019), and panel C from a figure by Kulczyk and Richardson (2016).

Six subunits are thus engaged with 12 adjacent nucleotides along the ssDNA, and the subunit that jumps from the back to the front of the lock-washer skips over the 10 intervening nucleotides.

The central hole of the hexameric helicase is not large enough to accommodate duplex DNA, and accordingly any complementary strand that is base-paired in front of the hexamer gets displaced as the enzyme translocates through that region. Translocation along a single strand of DNA is quite fast, greater than 100 nucleotides per second, but the enzyme slows down considerably (more than 10-fold) when it is also stripping a complementary strand from the DNA. This makes sense because DNA base pairs must be disrupted to displace the strand.

The stimulation of DNA polymerase activity by the helicase/primase complex was discussed above. The interaction of these two proteins is mutually beneficial, in that polymerase also stimulates the rate of helicase unwinding by about 10-fold. In a sense, the activity of each enzyme in the absence of the other only hints at its true capabilities. Evidently, these two enzymes form components of a well-coordinated molecular machine, the structure of which will be considered below.

The directionality of a replicative helicase is an important characteristic, because it dictates whether the enzyme is traveling along the leading-strand or the lagging-strand template. In the case of T7 helicase/primase, the 5′ to 3′ directionality places the protein on the lagging-strand template (Figure 2.3). This positions the attached primase on the strand where it needs to synthesize RNA primers for Okazaki fragment synthesis.

DNA helicases must be carefully controlled to prevent accidental unwinding of duplex genomic DNA. We will see in subsequent chapters that cellular replication systems have dedicated helicase-loading proteins, which are regulated to allow unwinding only at the time and place dedicated for genomic replication. Surprisingly, there is no such protein known in the case of phage T7. During in vitro reactions, the T7 helicase can load without any accessory proteins onto ssDNA, even if the DNA is circular. This result implies that either the hexamer is newly assembled around the

DNA or that hexamers are somehow loaded through a breach in the ring structure. A hint about the loading mechanism is that T7 helicase/primase readily forms a heptamer in solution without DNA (but a hexamer with ssDNA). One model for loading is that the heptamer loses one subunit as the DNA passes through the resulting void, with the hexamer then closing down around the DNA (Figure 2.4C).

The primase activity of the helicase/primase complex provides tetraribonucleotide primers for the initiation of Okazaki fragments on the lagging strand. Primers are made only at certain template sequences, with primers having the sequences 5'-ACCA-3', 5'-ACCC-3', and 5'-ACAC-3'. The catalytic reaction of primases is much like that of standard RNA polymerases. In both cases, the first step is the condensation of two nucleoside triphosphates into a dinucleotide with a 5' triphosphate end on the first residue, a phosphodiester linkage between the two, and a free hydroxyl on the 3' position of the second nucleotide residue. Subsequent nucleotide additions follow the paradigm of DNA/RNA polymerase nucleotide addition, with phosphodiester bond formation at the 3'-OH of the most recently incorporated nucleotide residue and release of pyrophosphate.

The interactions between T7 DNA polymerase and the helicase/primase complex facilitate the hand-off of the tetraribonucleotide primer to the DNA polymerase for Okazaki fragment extension. As we will see shortly, the events in lagging-strand synthesis, including primer formation, hand-off, and Okazaki fragment extension, are carefully coordinated in the overall replication process.

2.5 The replisome machine functions with a looped lagging-strand template

The inherent 5' to 3' directionality of DNA polymerase dictates that the leading- and lagging-strand products must be synthesized in opposite directions with respect to the parental double helix. Thus, for many years, models of the replication fork placed the lagging-strand polymerase quite far in space and uncoupled from the leading-strand polymerase (as in Figure 2.3). However, this is not an accurate view of the replication process.

A variety of experimental results instead show that the replisome functions as a unitary and well-coordinated multi-protein complex. Rather than being splayed apart from the leading strand, the lagging strand is looped around such that the two polymerases are bound together within the same protein complex (Figure 2.5). It is helpful to step through the cycle of lagging-strand synthesis to appreciate how this process works. As the synthesis of one Okazaki fragment is completed, the lagging-strand polymerase bumps into the 5′ end of the previously synthesized Okazaki fragment. This collision provides a signal that induces the DNA polymerase to dissociate from that segment of the lagging-strand. Note that the polymerase is still associated with the remainder of the replisome proteins upon dissociation of the completed Okazaki fragment. Evidence suggests that a second signal, perhaps acting as a backup mechanism, can also promote release of the lagging-strand segment by polymerase. This second signal is the synthesis of a new RNA primer for the next Okazaki fragment (Figure 2.5). Figure 2.5 does not differentiate whether the RNA primer is synthesized before or after DNA polymerase release, and likely either order is possible. In any case, the new RNA primer (primer 3 in the diagram), initially in contact with the helicase/primase, is now handed off to the DNA polymerase that just dissociated from the completed Okazaki fragment. Okazaki fragment extension by the DNA polymerase brings us back to where we started and completes the cycle (Figure 2.5). During replication of the T7 genome, which is roughly 40,000-base pairs long, this lagging-strand cycle occurs about 50 times on average. Because of the repeated cycle whereby the lagging-strand loop grows and shrinks, this mode of replication has been described as the "trombone model."

Several features of the trombone model are worthy of note (Figure 2.5). First, during much of the cycle, two different regions of the lagging-strand template are single stranded — part of the loop (a segment that was just unwound by the helicase) and the segment between the active lagging-strand polymerase and the

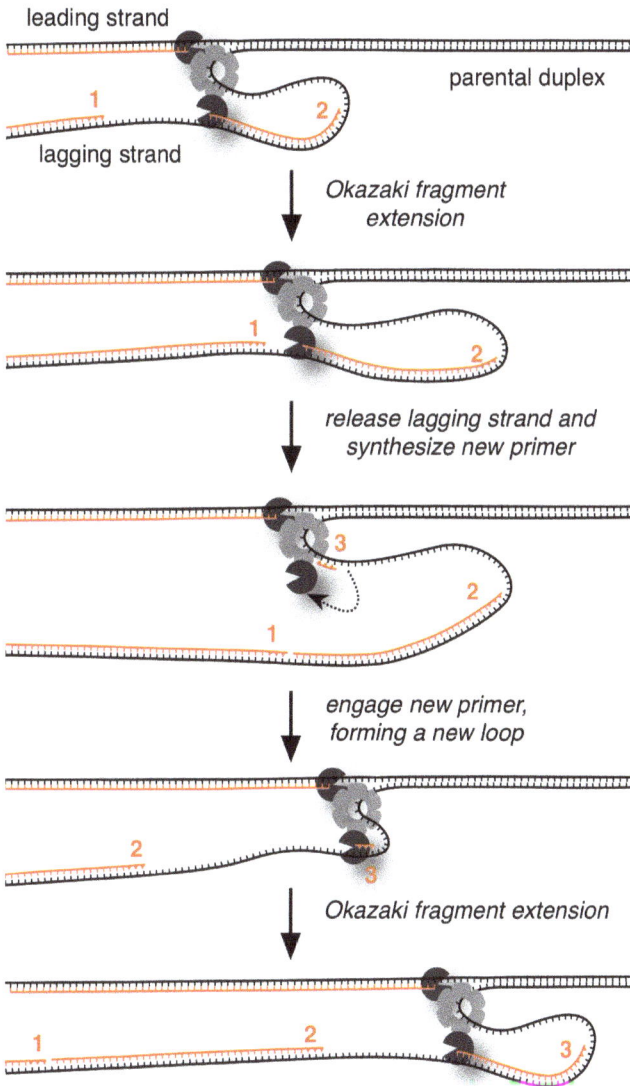

Figure 2.5. The trombone model of DNA replication. The figure depicts one round of Okazaki fragment synthesis, with only the two DNA polymerases and replicative helicase shown for simplicity. Newly synthesized DNA is in red, and RNA primers (also red) are numbered. Note that the overall structures at the top and bottom are identical, except that one additional Okazaki fragment has been synthesized and the leading strand is extended accordingly. See text for detailed discussion of the steps in this model.

previous Okazaki fragment. Second, the loop starts off as a very small single-stranded segment and grows to include a large stretch each of single- and double-stranded DNA. The duplex DNA within the loop consists of the portion of the current Okazaki fragment that has been completed. Third, each of the single-stranded regions is coated with the ssDNA-binding protein (not shown in Figure 2.5), and so this protein must rapidly exchange and re-equilibrate on the strands as replication proceeds. Fourth, the lagging-strand polymerase remains associated with the replisome for multiple rounds of Okazaki fragment synthesis, rather than exchanging after each cycle as would be expected with the splayed-out replication fork in Figure 2.3. Experimental evidence for this conclusion includes the finding that an ongoing replication reaction continues many rounds of Okazaki fragment synthesis after being diluted extensively, to the point where any remaining free DNA polymerase in solution could not bind to the complex. Fifth, replication of the leading and lagging strands occurs at essentially the same rates, reflecting coordinated synthesis of the two strands.

The looping model raises an important question — what happens if DNA polymerization on one of the strands is stalled or blocked? Does the polymerase on the other strand continue synthesizing DNA? This would create a dangerous situation, with a long stretch of ssDNA on the strand with the blocked polymerase and a complete uncoupling of the two DNA polymerases of the replisome. This question has been approached with a very special replication substrate using the T7 replication system (see "How did they test that?" at the end of this chapter). Polymerase extension on the lagging-strand template was blocked by incorporation of a special "chain-terminating" dideoxynucleotide only on that strand. Dramatically, synthesis of both the leading and lagging strands were strongly inhibited, demonstrating that the leading-strand polymerase halts when the lagging-strand polymerase is inhibited. Thus, the replisome is highly coordinated and the polymerases on the two strands somehow communicate with each other to maintain this coordination.

While a looped lagging strand is likely to be generally applicable to the process of DNA replication throughout biology, some of the Figures in the remainder of this book will show the "old-fashioned" splayed-out version for simplicity. Keep this in mind when you consider those simplified figures.

2.6 Structural model for the T7 replisome

A recent culmination of structural approaches provides a dramatic and informative three-dimensional model for the T7 replisome. One key study determined the structure of a complex of the helicase/primase with T7 DNA polymerase in the absence of DNA, using X-ray crystallography and other methods (Figure 2.6A). The helicase/primase was in a heptameric form, consistent with the DNA-free form discussed above. Somewhat surprisingly, the complex contained not two but three copies of the DNA polymerase! The authors speculated that the third DNA polymerase might serve as a spare, available to exchange with one of the actively synthesizing polymerases if the latter became blocked or disabled.

More recently, a method called cryo-electron microscopy (cryo-EM) was used to determine the structure of the T7 replisome complexed with DNA that mimics a replication fork (Figure 2.6B and cartoon in Figure 2.6C). In this case, the helicase/primase was in a hexameric form, as expected when bound to ssDNA (see above). As in the previous study using X-ray crystallography, three DNA polymerases were bound to the helicase/primase complex. In the cryo-EM study, the two active polymerases (leading and lagging strand) could be identified, and the third polymerase did appear as a spare, showing no engagement with a DNA primer/template.[4]

This stunning model of an active replication complex has several notable features that illuminate the replication process. Starting

---

[4] The third DNA polymerase visualized in these studies is bound to a different site in the helicase/primase complex than the secondary site discussed in Section 2.3; thus, a total of at least four potential DNA polymerase binding sites are involved in replication of T7 DNA.

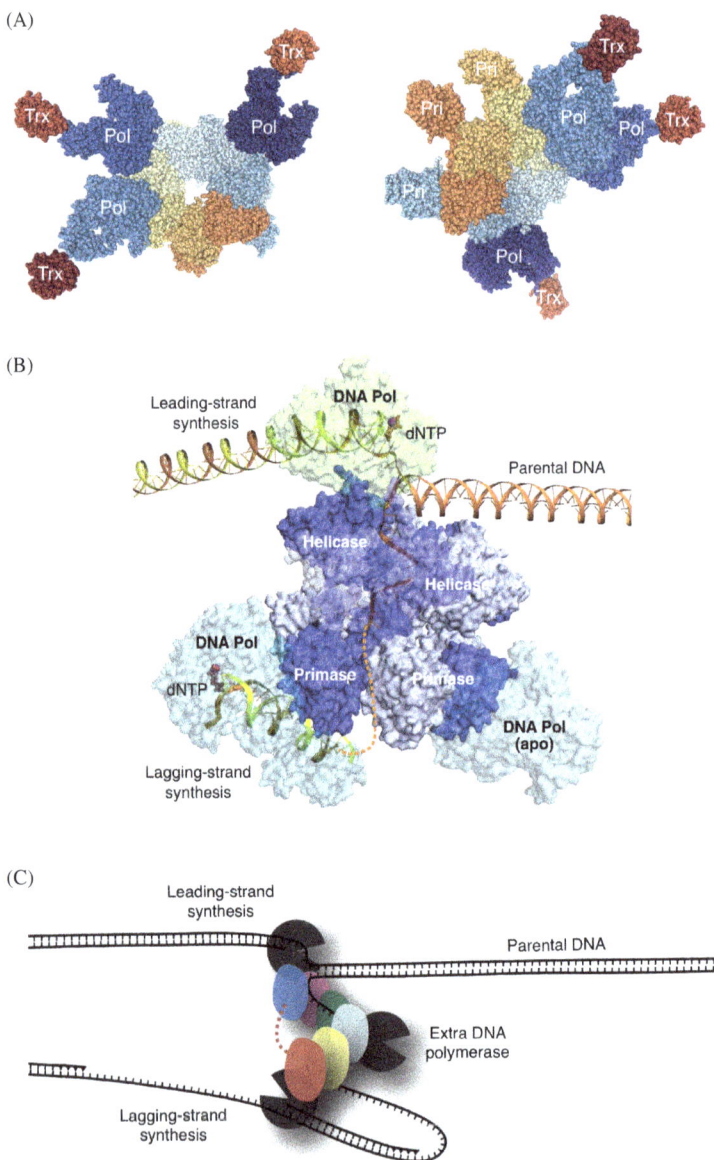

Figure 2.6. Current models for the structure of the T7 replisome. Two orientations of an X-ray crystal structure of the T7 replisome in the absence of DNA are shown in panel A. The central core is a heptameric helicase (varied colors) with three associated DNA polymerase/thioredoxin complexes (labeled Pol and Trx, respectively). Recall that the T7 helicase forms a heptamer rather than a hexamer in the absence of DNA. The image on the right provides a rotated view showing the underside with three of the primase domains (labeled Pri) extruding from the core.

at the parental DNA, the separation point of the parental duplex is actually within the leading-strand polymerase, not the helicase! This came as a surprise to those who expected that the unwinding enzyme would be at the frontline of an unwinding reaction. However, recall that unwinding by the helicase is greatly accelerated when it is in complex with polymerase, and the two proteins clearly work in concert with one another (see above).

The leading-strand template DNA, just past the unwinding point, enters the active site of the leading-strand polymerase (top DNA polymerase in Figure 2.6B). This leaves only a few nucleotides of ssDNA between the unwinding point and the point of leading-strand extension on that template strand.

In contrast, the lagging-strand template DNA bends away from the leading-strand polymerase, threads through the helicase/primase complex (12 nucleotides of DNA; see above), and emerges in the vicinity of the lagging-strand polymerase (bottom left polymerase in Figure 2.6B). As it exits the helicase/primase, the DNA traverses a protein-free region before entering the lagging-strand DNA polymerase, presumably allowing the formation and loss of the lagging-strand loops during active DNA replication.

Each of the three bound DNA polymerases contacts two primase domains in this replisome structure. Recall that an alternative form of helicase/primase lacking a portion of the primase domain is also produced in vivo. Future studies will undoubtedly investigate

---

Figure 2.6.   (**Figure on Facing Page**) The structures are from the RCSB PDB (www. rcsb.org) of PDB ID 5IKN (Wallen *et al.*, 2017). The replisome structure in panel B, containing replication fork DNA, was derived from a combination of structural methods. The hexameric form of the helicase is complexed with three DNA polymerases, one engaged with leading strand, one with lagging strand, and one "apo" form that is not engaged with DNA. The leading-strand polymerase and helicase collaborate in unwinding the parental DNA, and the template strand for lagging-strand synthesis threads through the helicase before engaging DNA polymerase. This image is reproduced from Gao *et al.* (2019), with permission from the American Association for the Advancement of Science; permission conveyed by Copyright Clearance Center, Inc. Panel C provides a cartoon tracing how the replisome structure in panel B relates to the trombone model of Figure 2.5 and the helicase movement pathway in Figure 2.4B.

whether a very similar replisome structure forms with three copies of full-length helicase/primase, three copies of truncated helicase/primase, and three copies of DNA polymerase.

Considering the positions of the leading- and lagging-strand polymerases, the third (unengaged) DNA polymerase (bottom right in Figure 2.6B) is in prime position to take over synthesis on the lagging strand should the lagging-strand polymerase become disabled or dissociate. It will be interesting to deduce the structural transitions that occur during this proposed polymerase-switching process.

With these remarkable advances in the T7 replication system, we are now in a position where we can contemplate how the trombone model plays out in the context of an actual three-dimensional structure of a protein complex of some 600,000 Daltons. Using the structural model (Figure 2.6B) and cartoon (Figure 2.6C) as guides, try to visualize how the parental, leading, and lagging strands slide through the complex during the various steps of DNA replication.[5]

2.7 T7 ssDNA-binding protein helps to organize the replisome

The transient but fairly extensive regions of ssDNA on the lagging-strand template are blanketed by the ssDNA-binding protein, which both protects and organizes the DNA. The dynamics of protein binding to these single-stranded regions is fascinating. As mentioned above, ssDNA-binding protein has an acidic C-terminal extension. Indeed, 15 out of 26 residues in this extension are aspartic or glutamic acid. When free in solution, the ssDNA-binding protein forms a dimer in which the acidic C-terminal extension of each monomer binds to the (basic) DNA-binding site within the body of the opposite subunit (Figure 2.7). Binding of the C-terminal extension to the DNA-binding site presumably shields the protein from binding other negatively charged molecules in the cell and modulates the affinity of the protein for ssDNA.

---

[5] A number of animated videos of the trombone model can be found online, including those at www.youtube.com.

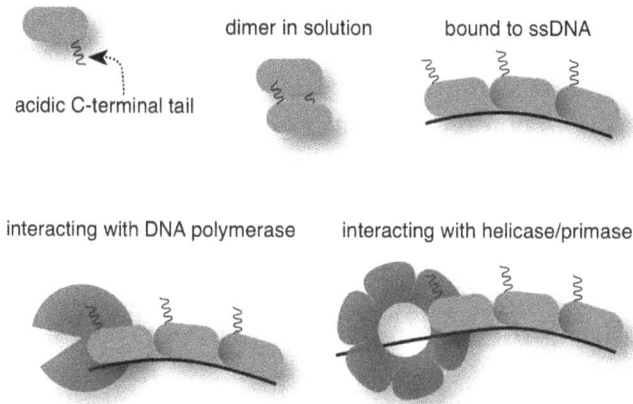

Figure 2.7.   Protein–protein interactions centered on the T7 ssDNA-binding protein. The acidic C-terminal tail of the protein is critical for interactions with itself and other proteins. In solution, the protein dimerizes via mutual interactions between the tail of one protein and the body of its partner. Once bound to ssDNA, the acidic C-terminal tail becomes available for interactions with the DNA polymerase and with the helicase/primase, as shown at the bottom of the figure.

The C-terminal extension of the ssDNA-binding protein is the region that binds the T7 DNA polymerase and helicase/primase. This implies that the dimer in solution has weakened interactions with these replication proteins. However, upon binding to ssDNA, the C-terminal extensions are displaced from the DNA binding sites and thereby become available for interactions with the other replication proteins (Figure 2.7). Overall, this simple but elegant architecture of the ssDNA-binding protein allows the protein to be readily available in solution to bind any ssDNA that is generated, and yet not sequester the much more limited amounts of replication proteins. Also, the architecture presumably ensures that free protein does not bind extensively to the polymerase and helicase/primase at the replication fork, allowing preferential binding of ssDNA-binding proteins that are bound to the ssDNA within the fork.

Through its interactions with the DNA polymerase and helicase/primase, the ssDNA-binding protein plays a central role in coordinating and regulating the functioning of the replication machine. The protein modulates the activities of both of these proteins and is critical for efficient lagging-strand synthesis. Indeed, when the ssDNA-binding protein is withheld from in vitro replication reactions, lagging-strand synthesis is strongly inhibited, with abnormally short Okazaki fragments, and coordination between leading- and lagging-strand synthesis is lost.

2.8 Finalizing the lagging-strand product

The reactions described above can provide a fully intact, duplex DNA molecule as the leading-strand product. However, the lagging-strand product will consist of one intact parental strand along with a product strand that has an interruption and a small segment of RNA where each Okazaki fragment was initiated. These imperfections are repaired following the main replication reaction to generate the fully intact DNA products (Figure 2.8).

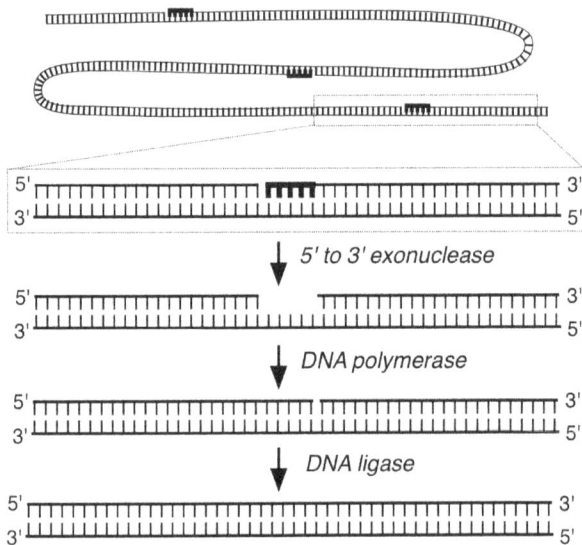

Figure 2.8. Okazaki fragment processing. The sequential action of a 5′ to 3′ exonuclease, DNA polymerase, and DNA ligase result in removal of the RNA primer and sealing of adjacent Okazaki fragments into an intact strand.

An additional T7 protein[6] is a specialized nuclease with two key activities. First, this protein is a 5′ to 3′ exonuclease and plays an important role in removing the RNA primers at the 5′ ends of each Okazaki fragment. Second, the protein has an activity called "flap endonuclease," which can cleave off a short segment of extra DNA or RNA at a nick. We will return to a discussion of flap endonucleases when we discuss eukaryotic DNA replication in Chapter 4.

In addition to processing Okazaki fragment ends, the T7 exonuclease also plays a key role in degrading the bacterial host DNA during T7 infection. T7-encoded nucleases degrade the host DNA in order to re-utilize the resulting nucleotide residues for synthesizing its own DNA. Since the T7 genome is less than 1% the length of the bacterial (*E. coli*) genome, this is an important source of precursors that can fuel the synthesis of many phage genomes.

The action of the T7 exonuclease on the 5′ ends of each Okazaki fragment temporarily leaves a short gap of ssDNA, which is filled in by DNA polymerase (Figure 2.8). Once the gap is bridged so that the product strand has two adjacent nucleotide residues, DNA ligase seals the nick and the lagging strand is complete. T7 encodes its own DNA ligase,[7] but T7 mutants lacking that enzyme can instead use the host DNA ligase for Okazaki fragment maturation.

2.9 Back to the beginning — How does T7 DNA replication initiate?

The elegant dynamics of the replication process discussed above begs the question of how the process is initiated. T7 DNA replication initiates at a defined location in the bacteriophage genome, where the replication machinery must be loaded. The so-called replication origin consists of two tandem promoters for T7 RNA polymerase-induced transcription and an AT-rich region of DNA just downstream (Figure 2.9A). Importantly, the RNA polymerase transcripts are used as the primer for leading-strand synthesis during in vitro reactions and presumably also in vivo.

---

[6] Called gp6 for the aficionado.

[7] Called gp1.3.

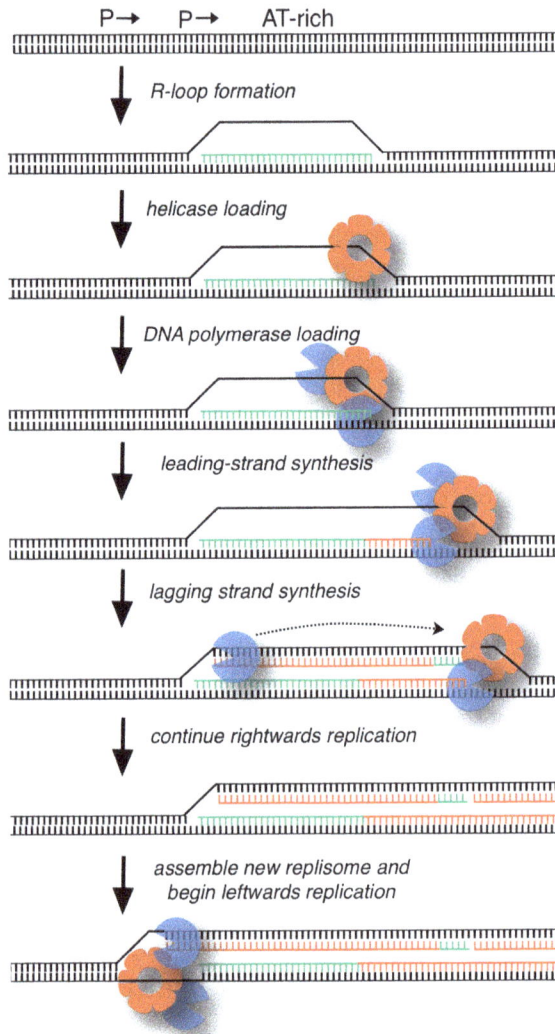

Figure 2.9. Model for initiation of bacteriophage T7 DNA synthesis. RNA polymerase generates a transcript from either of two transcriptional promoters in the origin region. In this model, the RNA transcript forms a stable RNA–DNA hybrid, called an R-loop. The displaced strand of DNA is used as a loading site for the helicase/primase, and the RNA transcript is used as primer for DNA synthesis in the rightward direction. The DNA polymerase that completes the first Okazaki fragment is associated with the rightward helicase complex (indicated by dotted arrow) and travels with that complex in the rightward direction (this diagram does not present the folded structure of the trombone model for simplicity sake).

Why is the AT-rich region just downstream of the promoters necessary for origin function? One model is that the relatively weak DNA base pairing in this AT-rich region allows the RNA transcript to form a transient but extended RNA–DNA hybrid (called R-loop) (Figure 2.9). This would displace a single strand of DNA that provides a prime target for loading of the helicase/primase complex, with subsequent loading and activation of both leading- and lagging-strand DNA polymerases (Figure 2.9). This would constitute a complete replisome like the ones discussed above, traveling away from the origin (in the rightward direction in Figure 2.9). Note that the lagging-strand DNA polymerase travels with the rightward replisome complex, as depicted by the dotted arrow in Figure 2.9 (and would actually be associated with that complex as discussed above). This initiation process has been recreated in vitro and requires the four T7 replication proteins along with the T7 RNA polymerase.

With the folding of the lagging strand around into a loop structure, one might expect that the replication-origin region would contain special features that create this loop. However, this is not the case. Very simple DNA substrates such as the one diagrammed in Figure 2.9 are converted into properly folded lagging-strand loops by the simple T7 replication machinery discussed above. Thus, the replisome complex discussed above has the intrinsic ability to create properly folded lagging-strand loops.

DNA replication from the T7 origin occurs in a "bidirectional" fashion, with one replisome generated in each of the two directions from the origin. Thus, after the first replisome is assembled and travels away from the origin as described above (the rightward direction in Figure 2.9), a second replisome complex is assembled for the leftward direction. In this case, the 3′ end of the first Okazaki fragment from the rightward direction is used as the

---

Figure 2.9. (**Figure on Facing Page**) After rightwards replication has exited the origin region, a new leftward replisome is assembled to complete the establishment of bidirectional DNA synthesis. New molecules of helicase and DNA polymerase assemble on the branched DNA at that site, with the 3′ end of the first (rightward) Okazaki fragment serving as primer for the leftward leading strand. In this figure, RNA residues are in green and newly synthesized DNA is in red.

primer for the new leading strand in the leftward direction (Figure 2.9). The precise pathway of protein loading for leftward direction is not yet clear, but the combination of a branch with a free 3′ end (from the first rightward Okazaki fragment) appears to be sufficient to trigger loading.

2.10 Summary of key points

- The bacteriophage T7 replisome consists of only four proteins: the helicase/primase, ssDNA-binding protein, and DNA polymerase with its host thioredoxin partner.
- T7 DNA polymerase has separate active sites for polymerization and proofreading exonuclease activities.
- Proofreading exonuclease greatly reduces the mutation rate by removing a large majority of misinserted nucleotide residues.
- Processivity of T7 DNA polymerase is increased by the thioredoxin protein partner, interaction with the T7 helicase/primase, and rapid rebinding of polymerase when it dissociates from the 3′ end.
- T7 helicase/primase unwinds DNA by translocating along a single-strand in a "hand-over-hand" manner, whereby each subunit of the hexamer shifts sequentially from the back to the front of the lock-washer-like structure.
- T7 DNA polymerase stimulates the unwinding activity of the helicase/primase complex, and thus the two proteins are mutually reinforcing.
- The helicase/primase complex travels along the lagging-strand template in the T7 system.
- The replisome functions with a looped lagging strand, and both leading- and lagging-strand polymerases are associated with the helicase/primase complex at the fork.
- The T7 replisome functions as a well-coordinated protein machine, with communication between the leading- and lagging-strand polymerases.
- A structural model of the functioning replisome shows the unwinding point of the parental DNA within the leading-strand

polymerase, a flexible DNA region that allows looping between the helicase/primase and the lagging-strand polymerase, and a third DNA polymerase that appears to function as a spare to allow polymerase switching on the lagging strand.

- All ssDNA at the replication fork is coated with ssDNA-binding protein, which helps to organize the replisome.
- Discontinuities in the lagging strand are repaired by the action of a specialized T7 nuclease, DNA polymerase, and DNA ligase.
- A transcript synthesized by T7 RNA polymerase is needed to initiate DNA replication, likely via R-loop formation.

Further Reading

Berman, H. M., Westbrook, J., Feng, Z., Gilliland, G., Bhat, T. N., Weissig, H., Shindyalov, I. N., & Bourne, P. E. (2000). The Protein Data Bank. *Nucleic Acids Res, 28*, 235–242.

Cernooka, E., Rumnieks, J., Tars, K., & Kazaks, A. (2017). Structural basis for DNA recognition of a single-stranded DNA-binding protein from Enterobacter phage Enc34. *Sci Rep, 7*, 15529, doi:10.1038/s41598-017-15774-y

Gao, Y., Cui, Y., Fox, T., Lin, S., Wang, H., de Val, N., Zhou, Z. H., & Yang, W. (2019). Structures and operating principles of the replisome. *Science, 363*(6429), eaav7003.

Hamdan, S. M., & van Oijen, A. M. (2010). Timing, coordination, and rhythm: Acrobatics at the DNA replication fork. *J Biol Chem, 285*(25), 18979–18983.

Holt, I. J., & Reyes, A. (2012). Human mitochondrial DNA replication. *Cold Spring Harb Perspect Biol, 4*(12), a012971.

Kulczyk, A. W., & Richardson, C. C. (2016). The replication system of bacteriophage T7. *Enzymes, 39*, 89–136.

Lee, J., Chastain, P. D., 2nd, Kusakabe, T., Griffith, J. D., & Richardson, C. C. (1998). Coordinated leading and lagging strand DNA synthesis on a minicircular template. *Mol Cell, 1*(7), 1001–1010.

Lee, S. J., & Richardson, C. C. (2011). Choreography of bacteriophage T7 DNA replication. *Curr Opin Chem Biol, 15*(5), 580–586.

Li, Y., Dutta, S., Doublie, S., Bdour, H. M., Taylor, J. S., & Ellenberger, T. (2004). Nucleotide insertion opposite a cis-syn thymine dimer by a replicative DNA polymerase from bacteriophage T7. *Nat Struct Mol Biol, 11*, 784–790.

Singleton, M. R., Sawaya, M. R., Ellenberger, T., & Wigley, D. B. (2000). Crystal structure of T7 gene 4 ring helicase indicates a mechanism for sequential hydrolysis of nucleotides. *Cell, 101*, 589–600.

Sinha, N. K., Morris, C. F., & Alberts, B. M. (1980). Efficient in vitro replication of double-stranded DNA templates by a purified T4 bacteriophage replication system. *J Biol Chem, 255*(9), 4290–4293.

Toth, E. A., Li, Y., Sawaya, M. R., Cheng, Y., & Ellenberger, T. (2003). The crystal structure of the bifunctional primase-helicase of bacteriophage T7. *Mol Cell, 12*, 1113–1123.

Wallen, J. R., Zhang, H., Weis, C., Cui, W., Foster, B. M., Ho, C. M. W., Hammel, M., Tainer, J. A., Gross, M. L., & Ellenberger, T. (2017). Hybrid methods reveal multiple flexibly linked DNA polymerases within the bacteriophage T7 replisome. *Structure, 25*, 157–166.

How did they test that?
Are leading- and lagging-strand synthesis coupled?

Lee, Chastain, Kusakabe, Griffith, and Richardson (1998) exploited a powerful "minicircle" assay and the DNA replication system of bacteriophage T7. The minicircle, constructed with chemically synthesized oligonucleotides, consists of a 70-bp circle with a 5′-single-stranded extension (panel A). The 3′ end at the branch point serves as the site for initiation of leading-strand synthesis, while lagging-strand synthesis initiates when primase sites are exposed during the first round of leading-strand synthesis. The substrate is replicated efficiently (panel B), producing duplex products that can exceed 20,000 bp (hundreds of times around the circle). A key feature is that the two strands differ greatly in nucleotide content — the leading-strand product contains a great excess of dG over dC residues, while the lagging-strand product is just the opposite (the only exception is from the primase sites that needed to be present). This sequence arrangement is useful for two reasons. First, synthesis of the two strands can be measured separately: about 95% of label from radioactive dGTP goes into the leading strand, while 95% of label from radioactive dCTP goes into the lagging strand. Second, synthesis on the two strands can be specifically inhibited with different chain-terminating nucleotides. For example, dideoxy CTP (ddCTP; structure shown in panel C) would be incorporated 20-fold more frequently into the lagging-strand product than the leading-strand product, and thereby preferentially inhibit lagging-strand synthesis.

In this experiment, DNA replication is monitored with either radioactive dCTP (panel D, left graph; lagging-strand synthesis) or radioactive dGTP (right graph; leading-strand synthesis). After 2.5 minutes of the reaction, ddCTP (non-radioactive) is added to inhibit lagging-strand synthesis. Note that the chain-terminating nucleotide is added at an 80-fold lower concentration than normal dCTP, and so the chain terminator is only incorporated in a small fraction of extensions opposite dG residues. The striking result is that ddCTP addition rapidly inhibits both leading- and lagging-strand synthesis at roughly equal efficiencies. Evidently, the leading-strand polymerase is somehow induced to stop synthesis when its lagging-strand partner in the replisome is obstructed.

Panels A and D in this Figure were reproduced from Lee *et al.* (1998), with permission from Elsevier; permission conveyed by Copyright Clearance Center, Inc.

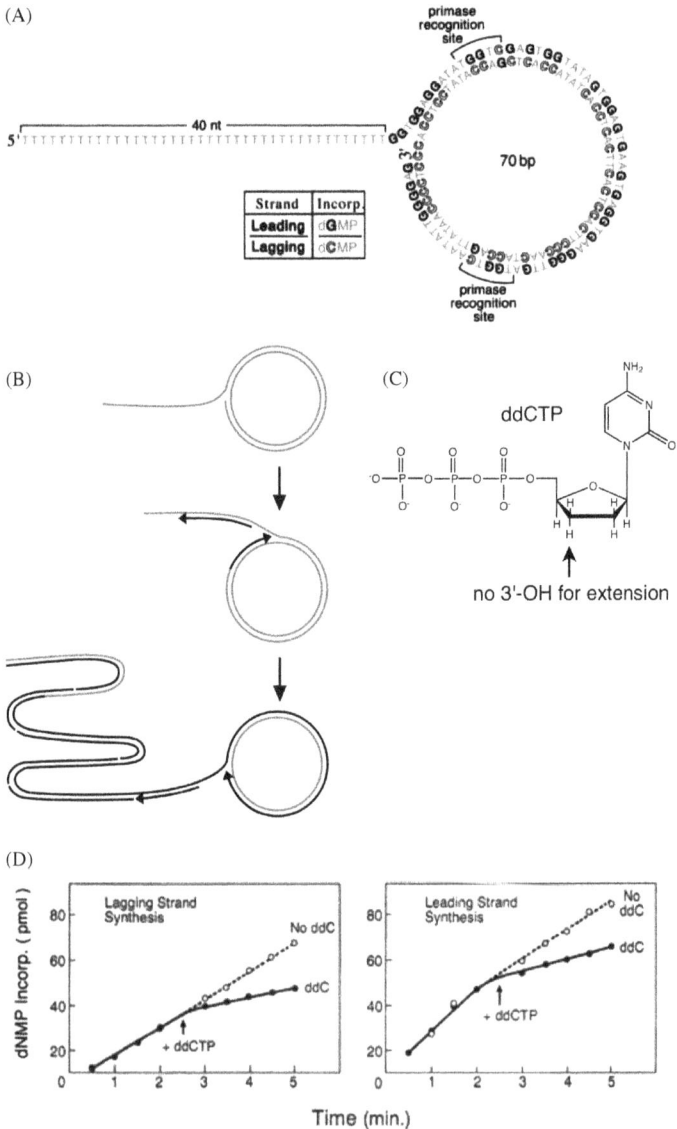

Chapter 3

The highly efficient replication system of bacteria

DNA replication in both prokaryotic and eukaryotic cells is significantly more complex than in bacteriophage T7, and yet the basic functions and enzymatic mechanisms are very similar. In addition, the structural architecture of many of the key replication proteins, such as DNA polymerases and helicases, are analogous and sometimes evolutionarily related. The distinctions evident in cellular systems include the use of many more protein subunits in the overall replication reaction. One such difference of great importance is the use of specialized proteins, called sliding clamps, which tether cellular replicative polymerases to their DNA template and orchestrate polymerase behavior.

The most extensively studied cellular replication system is that of the bacterium, *Escherichia coli*, and this remarkably efficient system will be the focus of this chapter. Under optimal conditions, *E. coli* can duplicate its 4.6-million base-pair chromosome with high accuracy at a speed that allows the cells to divide every 20 minutes. Many bacterial species have replication proteins that are homologous to those of *E. coli* and presumably replicate by similar mechanisms, with only minor variations. However, it is worth noting that more distantly related bacteria sometimes show significant deviations from the *E. coli* model. For example, certain gram-positive bacterial species such as *Bacillus subtilis* have an extra replicative polymerase with distinct properties. Without going into details, one of the *Bacillus subtilis* replicative polymerases is thought to extend the RNA primer on the lagging strand for a short distance, and then hand off the reaction to the other replicative polymerase (which also replicates the leading strand). As we will see in the next chapter, this kind of handoff is also a key feature of eukaryotic DNA replication.

3.1 The *E. coli* replisome from 30,000 feet

To a first approximation, the *E. coli* chromosome is replicated in its entirety by just two replisome complexes loaded at a single origin site on the chromosome (also see Chapter 5). These two replisomes traverse the circular chromosome in opposite directions at a speed of roughly 1000 base pairs per second, and meet up to complete replication in a special terminus region on the opposite side of the circle. This speed, while quite impressive, is not quite high enough to explain how the bacterial cell can divide every 20 minutes (1000 base pairs/second × 60 seconds/minute × 20 minutes/division cycle × 2 replication forks = 2.4 million base pairs, only about half the number in the genome). This riddle is solved by the fact that *E. coli* and other bacteria actually carry out multiple rounds of replication at the same time under conditions of rapid cell division. Thus, two or three replisomes may be following each other around the chromosome at any one time, and partially replicated chromosomes must be segregated to daughter cells at the time of cell division (see Chapter 5).

A key question about DNA replication in the context of a living cell is whether the replisome is motoring along a relatively fixed DNA molecule or whether the DNA is slithering through a relatively fixed replisome. In the latter case, the DNA could actually be transported or "pumped" from one region of the cell to another during the act of replication. Scientists have been able to approach this question by placing visible markers (fluorophores) on the *E. coli* replisome proteins or, in a different experiment, particular locations of the chromosomal DNA. Strikingly, the replication proteins tended to remain near the middle of the cell throughout the period of DNA replication, while specific locations of the chromosome migrated in the cell. As replication progressed, the two duplicated copies of any particular chromosomal region moved in opposite directions from near the center of the cell toward the two poles. In this way, the movement of the duplicated chromosomes in preparation for bacterial cell division is coordinated with the act of replication. The separation of the two duplicated copies of the chromosome is made very challenging by the fact that each duplicated copy contains one of the two strands of parental DNA, and these two parental strands were wrapped around each other once every 10 or so base pairs before replication occurred. We will return to these topological issues in Chapter 7.

3.2 The replicative DNA polymerase holoenzyme

Two different DNA polymerases are involved in *E. coli* DNA replication. For historical reasons, the main replicative polymerase for both leading- and lagging-strand synthesis is called DNA polymerase III, while the polymerase involved in processing Okazaki fragments is DNA polymerase I. DNA polymerase III can be isolated as a multi-protein complex called the "holoenzyme," and the composition of the holoenzyme reveals much about the overall replisome (Figure 3.1A, right). The complex contains two copies of the actual DNA polymerase, which is a three-subunit enzyme that is referred to as the DNA polymerase III "core" (Figure 3.1A, left). In the context of chromosomal replication, one of these core enzymes replicates the leading strand and the other replicates the lagging strand.[1] The holoenzyme also contains one copy of a multi-subunit complex that can load sliding clamps on the DNA, not surprisingly called the "clamp loader." As we will see below, repetitive clamp loading is a central feature of lagging-strand replication. The holoenzyme also

---

[1] Some evidence suggests that the functioning *Escherichia coli* holoenzyme contains three copies of DNA polymerase, including a spare like the one discussed in the T7 replisome in the previous chapter, but this point is still controversial.

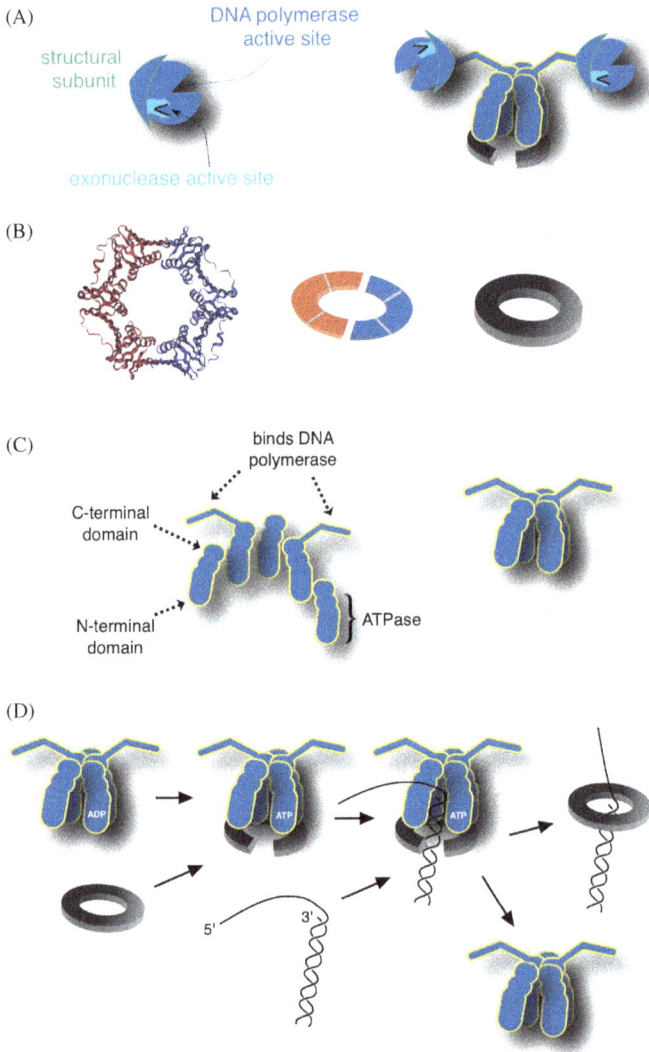

Figure 3.1. The *Escherichia coli* replicative DNA polymerase. The three-subunit core of the bacterial replicative polymerase (panel A, left) has active sites for polymerase and exonuclease activity (in separate subunits; the open mouths), and a structural subunit (the hat). The complete DNA polymerase complex, called holoenzyme, is diagrammed at the right of panel A and consists of two copies of the core polymerase along with clamp (panel B) and clamp loader (panel C). The sliding clamp is a ring-shaped protein consisting of two identical subunits (red and blue in panel B), each of which has three similar subdomains. The image of the crystal

contains the sliding clamp, along with two other subunits that play roles in the overall architecture of the complex.

The subunits of the holoenzyme are named after letters of the Greek alphabet. Detailed discussions of the roles of every subunit are beyond the scope of this chapter, but for those who engage the literature further, the names of the subunits are as follows: polymerase III core is composed of α, ε, and θ; the clamp loader is composed of τ, γ, δ, and δ′, and is also called the γ complex; the clamp is called β, and the architectural proteins are χ and ψ.

The three-subunit polymerase III core complex that replicates both the leading and lagging strand has separate subunits with DNA polymerase and proofreading exonuclease activity (unlike the combined activities within the major subunit of the bacteriophage T7 DNA polymerase). When misincorporation occurs during DNA synthesis by polymerase III core, the growing DNA chain with the errant nucleotide residue is therefore passed between the two protein subunits to excise the incorrect base and then back to the polymerase subunit to resume DNA synthesis. The third subunit of the polymerase core plays a structural role and assists the proofreading subunit, and in some bacterial species, this subunit is missing.

3.3 Dynamics of clamp loading in bacterial replication

The sliding clamp was uncovered as a protein factor that greatly increases the processivity of replicative DNA polymerase. Early studies of sliding clamps led to a remarkable discovery. When clamps were loaded onto circular DNA, they remained bound to the DNA

Figure 3.1. (**Figure on Facing Page**) structure of the clamp shown at the left of panel B is from the RCSB PDB (www.rcsb.org) of PDB ID 2POL (Kong *et al.*, 1992). The C-terminal face of the clamp, which binds both clamp loader and DNA polymerase core, is facing up in the cartoon diagram on the right of panel B. The clamp loader (panel C) consists of five subunits derived from three different genes, as described in the text. Two of the subunits, the ones with the extended arms, bind to the core of the DNA polymerase in the holoenzyme. The clamp-loading pathway is diagrammed in panel D. Parts of this figure were modified from Kelch *et al.* (2012).

indefinitely, but if the DNA was cut with a restriction enzyme to form linear DNA, the clamps quickly fell off (see "How did they test that?" at the end of this chapter). This led to the model that clamps encircle the DNA like donuts with a string through the donut hole and can readily slide along the DNA. The eventual solution of sliding clamp structures confirmed this view, showing a beautiful donut-like structure with a hole large enough to accommodate duplex DNA (Figure 3.1B).

A quick glance at the crystal structure of the *E. coli* sliding clamp reveals what appears to be a hexameric complex — six very similar structures make up the donut structure. However, there are actually only two subunits, each of which has three domains of similar structure. Several features of the clamp structure are notable. First, α helices line the central hole, providing strong electrostatic potential that allows water-mediated interactions with DNA. This arrangement allows free movement of the DNA through the hole in the clamp. Second, the C-termini of both subunits are exposed on one face of the clamp, and this is the site of interaction of the clamp with other proteins including the DNA polymerase and the clamp loader. Third, in spite of the similar structures of the three domains in each subunit, they have very different amino acid sequences. Finally, clamps from a bacteriophage called T4 and from eukaryotic cells have a quite similar structure with six domains in a donut shape, but in contrast, are formed from three subunits with two domains each (see Chapter 4).

As already introduced, the major function of the sliding clamp is to increase the processivity of the replicative DNA polymerase. The processivity of replicative polymerase is quite a challenging problem in DNA replication, since the leading strand is synthesized continuously for as many as millions of base pairs, while lagging-strand synthesis must be stopped, within each Okazaki fragment, after only a thousand or so base pairs (in bacteria) or a few hundred base pairs (in eukaryotes). The dynamics of the sliding clamp are critical for the careful activation and termination of DNA polymerase action at appropriate places in the replication fork to achieve this distinct behavior on the leading and lagging strands.

The clamp loader complex within the holoenzyme plays the key role of loading a new clamp for each Okazaki fragment (Figure 3.1C and 3.1D). During each round of Okazaki fragment synthesis, a clamp is loaded onto the short RNA primer made by primase (see below) and the clamp is then handed off to the replicative DNA polymerase. Being associated with clamp, the polymerase now has a high level of processivity so that it can synthesize the Okazaki fragment to completion.

To accomplish its loading reaction, the clamp loader must bind to the clamp when it is free in solution, deliver the clamp to the DNA target, and then lose its affinity for the clamp so that the clamp can associate with the DNA polymerase. During this reaction sequence, the clamp loader must direct the donut-shaped clamp through a key transformation — opening up the donut to allow DNA to enter the central hole and then allowing the donut circle to reform as the loader dissociates from the clamp (Figure 3.1D).

The structure of the clamp loader helps to explain the intricate gymnastics of this loading reaction. In both bacteria and eukaryotes, the clamp loader is a five-subunit complex, and each subunit has a related three-domain structure. The two N-terminal domains together have the structure characteristic of proteins of the AAA+ family of ATPases (named after the rather generic description, ATPases *a*ssociated with various cellular *a*ctivities), and the function of ATP binding and hydrolysis is to fuel the conformational changes needed to carry out the abovementioned steps in the loading reaction. Members of the AAA+ family typically adopt a hexameric ring structure, but the five clamp loader ATPase modules adopt a configuration like a hexamer missing one subunit. This reveals a C shape with a gap between two of the subunits and a hole near the bottom (Figure 3.1C and 3.1D). The remaining C-terminal domains of the clamp loader subunits form a tight pentameric cap at the top end of the clamp loader.

Upon binding to ATP, the clamp loader binds the C-terminal face of a free clamp on the bottom of the loader — the binding reaction pries open one of the subunit–subunit interfaces of the clamp, cracking the donut (Figure 3.1D). This gap in the clamp is sufficiently large to allow the passage of DNA. The clamp–clamp loader complex has a high affinity for the site where RNA primer has been synthesized, specifically binding to the 3′ end where DNA synthesis must begin (this is called the primer-template site). This binding occurs in an interesting fashion — the interior of the clamp loader is organized to bind duplex DNA lengthwise, but the cap forces a tight bend as the DNA approaches. SsDNA is much more flexible than duplex DNA, and so a primer-template junction easily forms the bent structure required, positioning the clamp loader at the correct site (Figure 3.1D). Notice that as the clamp loader binds the primer-template junction, the duplex DNA has been threaded into both the interior of the clamp loader and simultaneously through the crack in the clamp. Also notice that the clamp is positioned in a specific orientation, with the side that binds clamp loader (the C-terminal face) pointing in the same direction as the 3′ end of the primer. Once bound to the primer-template junction, the ATPase activity of the clamp loader is activated, and the ATP-free clamp loader loses its affinity for the primer-template site. As the clamp loader dissociates from the primer-template and clamp, the clamp springs back to its donut shape, which now encircles the DNA. DNA polymerase also binds to the C-terminal face of the clamp, and so the clamp is oriented correctly on the primer-template junction to bind the replicative polymerase, which can now proceed with Okazaki fragment synthesis.

As mentioned earlier, clamp loaders from both prokaryotes and eukaryotes consist of five subunits, each being an ATPase family member. The subunit composition of the bacterial clamp loader is interesting. Due to redundancy, only three genes encode the five subunits of the *E. coli* clamp loader. Two of these genes encode a single subunit each, while the remaining three subunits are all derived from a single gene called *dnaX*. Recall from Chapter 2 that one particular T7 gene encodes two different forms of the helicase protein in bacteriophage T7. In a related way, the *E. coli dnaX* gene encodes two distinct proteins, but by means of a different mechanism. In this case, ribosomes translating the *dnaX* mRNA sometimes engage in a process called programmed translational frameshifting, in which the reading frame is suddenly shifted to change the register of the triplet code downstream of the frameshift event. Shortly after the frameshift occurs, the ribosome encounters a stop codon, resulting in a short (47 kDa) form of the protein. Most of the translation events do not result in a frameshifting event, and these lead to the long (71 kDa) form of the protein. Both forms of the protein are competent for binding the clamp and the other subunits of the clamp-loader complex, but only the long form has the domain required for binding DNA polymerase and the replicative helicase. Recent biochemical experiments argue that the normal replisome contains two of the long forms and one short form, accounting for the three subunits encoded by *dnaX*. This stoichiometry is important, because the two long forms of the protein are able to contact two polymerase core complexes, one for the leading strand and one for the lagging strand. Note that the clamp loader thereby plays the key role of tethering and coordinating the two DNA polymerase core complexes within the replisome.

3.4 Coordinated action of helicase, primase, and DNA polymerase holoenzyme

As in the case of bacteriophage T7, the proteins in the bacterial replisome are carefully coordinated and function together as a protein machine (Figure 3.2A). The DNA polymerase holoenzyme described above interacts with the replicative helicase, directly linking the unwinding of the parental DNA strand with the synthesis of the leading strand. Thus, both clamp and the replicative helicase assist the leading-strand DNA polymerase in its rapid and processive synthesis reaction around the bacterial chromosome. The structure of an intact functioning *E. coli* replisome has yet to be deduced, and the number of interacting subunits is greater than that in the phage T7 system. For these reasons, our understanding of the *E. coli* replisome is not quite to the level of that of the T7 replisome.

The *E. coli* replicative helicase has many similarities to that of bacteriophage T7. It is a hexameric protein that unwinds DNA using the energy of nucleotide hydrolysis, and it tracks along ssDNA in the 5′ to

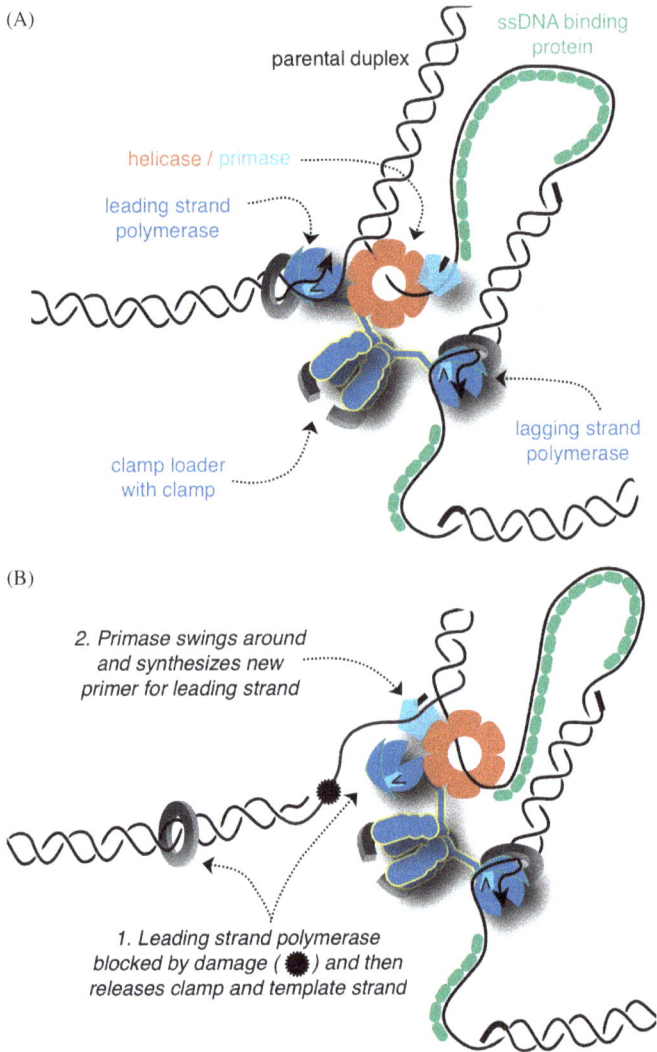

(A)

parental duplex

ssDNA binding protein

helicase / primase

leading strand polymerase

clamp loader with clamp

lagging strand polymerase

(B)

2. Primase swings around and synthesizes new primer for leading strand

1. Leading strand polymerase blocked by damage (●) and then releases clamp and template strand

Figure 3.2. The *Escherichia coli* replication machinery. The DNA polymerase holo-enzyme complex links the leading and lagging strands, with one copy of the core polymerase on each (panel A). The clamp-loading complex within the holoenzyme repeatedly loads sliding clamps (gray rings) as replication progresses. The helicase (orange) and primase (light blue) form a complex at the front of the replication machinery, unwinding the parental duplex. ssDNA-binding protein (green) covers all ssDNA on the lagging strand. On some occasions, blockage of the leading-strand polymerase can lead to re-priming of synthesis on the leading strand (panel B), rescuing an otherwise stalled replication fork.

3′ direction, placing it on the lagging-strand template (Figures 3.2A and 3.3A). It has a ring or donut-like shape, and the lagging-strand template passes through the hole as the leading-strand template is unwound and passes around the outside of the helicase. As in the T7 system, unwinding activity of *E. coli* replicative helicase is stimulated when it is associated with the DNA polymerase complex. This reveals a general feature in which replicative polymerases and helicases reciprocally stimulate activities of each other within the replication complex.

While the *E. coli* replicative helicase does not itself possess primase activity, it does associate with the *E. coli* primase enzyme, again placing the primase in the proper location to synthesize RNA primers for Okazaki fragment synthesis on the lagging strand (Figure 3.2A). The functional unit for binding primase consists of two helicase subunits, and so the hexameric helicase can bind no more than three primase monomers at a time. The bacterial primase, like that of T7, has a weak sequence selectivity, with a preference for initiating primers using the sequence 5′-AG. The bacterial enzyme differs from that of T7 in synthesizing longer RNA primers, roughly 10–12 nucleotides long.

The bacterial helicase has some additional distinctive features compared to the T7 helicase. The *E. coli* enzyme, like most other DNA helicases, hydrolyzes ATP, whereas you will recall that the T7 enzyme hydrolyzes dTTP. In addition, the hole in the middle of the bacterial enzyme has a diameter that is larger than the diameter of a DNA double helix, and in fact, duplex DNA can slide through the central hole if it is appropriately loaded onto the enzyme. This accounts for an interesting activity of the protein, namely its ability to knock some duplex DNA-binding proteins off DNA (Figure 3.3B). Obviously, the helicase must be loaded appropriately onto DNA in order to catalyze the DNA unwinding at the replication fork and pass only a single strand of DNA through the donut hole. This brings us to another major difference from the T7 helicase. The *E. coli* helicase requires a special protein to get loaded onto DNA for unwinding (Figure 3.3C). When not bound to DNA, these two proteins form a tight complex, which is competent for bringing the

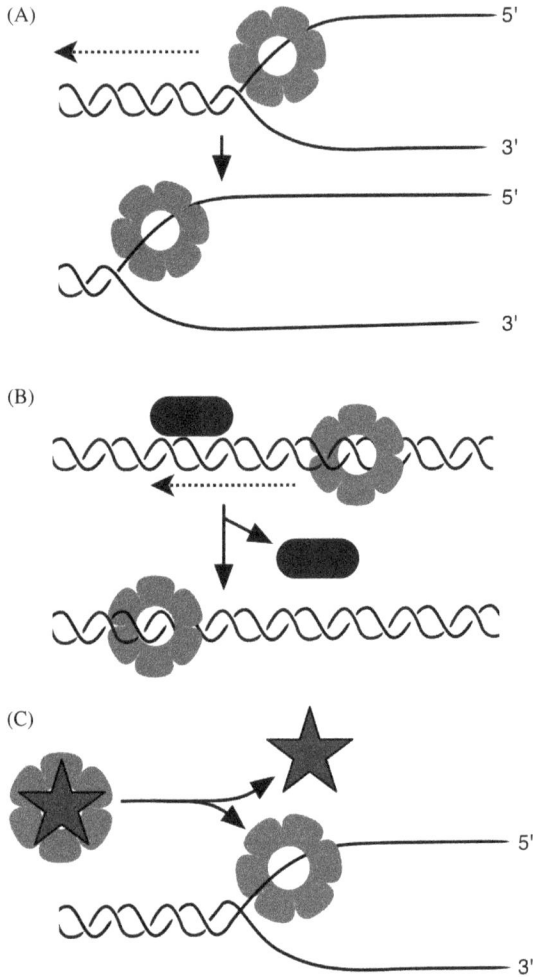

Figure 3.3. Functioning of the *Escherichia coli* replicative helicase. The hexameric replicative helicase unwinds DNA at the replication fork by traveling along the lagging-strand template in the 5′ to 3′ direction (panel A). The central channel in the helicase is large enough to accommodate duplex DNA, and the helicase can track along duplex DNA and displace bound proteins (panel B). Loading of the replicative helicase is carefully controlled by its association with a loading protein, which blocks the central channel when in solution but which helps deliver the helicase to appropriate structures such as forked DNA (panel C).

replicative helicase to the special origin site where DNA replication initiates (see Chapter 5). This loading process is carefully regulated to prevent inappropriate helicase loading at random sites in the genome, which could cause partial replication of only portions of the genome.[2]

An interesting aspect of replication fork dynamics is that the helicase and primase need to travel in opposite directions along the lagging-strand template, even though they are bound to each other (*E. coli*) or are one and the same protein (T7). Recall that the helicase is moving along the lagging-strand template in the 5' to 3' direction, while the primase synthesizes the new strand in the 5' to 3' direction (which means it is traveling 3' to 5' on the lagging-strand template strand). This again requires careful coordination, in that the unwinding catalyzed by helicase needs to pause, while primase synthesizes the short RNA primer. Once the primer is synthesized, the clamp component of holoenzyme is targeted to the RNA primer site as described earlier. This is a key step in the hand-off of the new primer to DNA polymerase for synthesis of the next Okazaki fragment.

During *E. coli* DNA replication, the lagging strand is looped around and the fork behaves according to the trombone model described in the previous chapter. As mentioned earlier, the *E. coli* DNA polymerase holoenzyme contains two copies of the core polymerase, and so the leading- and lagging-strand polymerases are physically coupled to each other during replication. As mentioned earlier, the clamp loader plays a key role in this coupling, with two of its subunits binding the polymerase on the leading and the lagging strand.

While the two polymerases are clearly coordinated with each other, they are apparently not as tightly coupled as in the bacteriophage T7 system. Under at least some conditions, blockage of the lagging-strand polymerase does not cause the leading-strand polymerase to stall. Instead, the leading-strand polymerase and helicase continue onward, generating a longer patch of ssDNA on the lagging-strand template (Figure 3.4A). This in turn can lead to

---

[2] We will discuss an exception in the section below on replication restart.

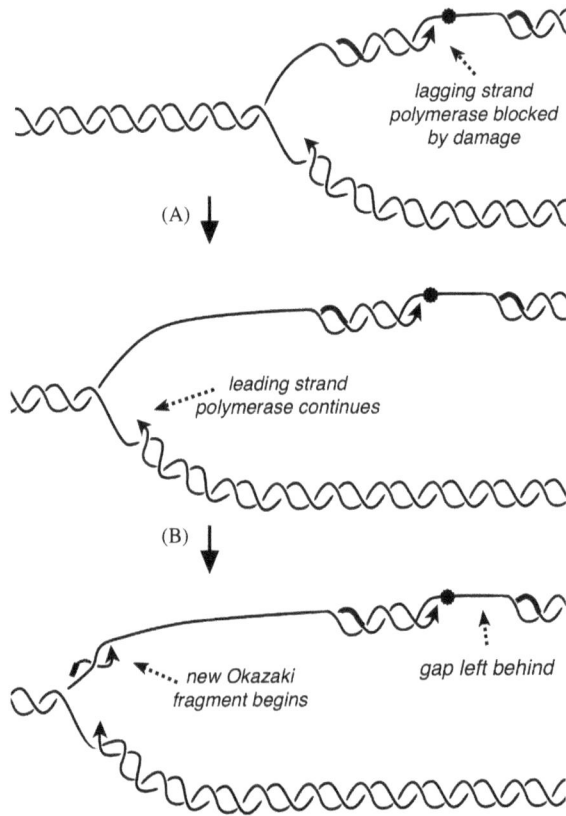

Figure 3.4.   Response of the *Escherichia coli* replication complex to blocking damage on the lagging-strand template. The replication machinery is capable of bypassing blocking damage on the lagging-strand template. The leading-strand polymerase continues onward even though the lagging-strand polymerase is blocked (A). After some delay, the lagging-strand polymerase core complex blocked by damage dissociates from the site of blockage and engages a new RNA primer to resume Okazaki fragment synthesis (B), leaving behind a patch of ssDNA downstream of the blocking lesion.

release of the blocked lagging-strand polymerase so that it can eventually cycle to a new RNA primer and restart Okazaki fragment synthesis (Figure 3.4B). Imagine that a damaged template base initially blocked the lagging-strand polymerase. This series of events would lead to a patch of single-stranded template DNA adjacent to the damaged template base, but the fork would continue onward and replication of the rest of the chromosome could be completed. Some further DNA repair reaction would be needed to deal with the small unreplicated patch (and the blocking lesion), but the cell would be able to complete chromosomal replication and proceed with cell division. We will discuss additional pathways for completing replication in unusual situations in the section on replication restart below.

3.5 The *E. coli* ssDNA-binding protein

The *E. coli* ssDNA-binding protein is required for normal replication and to protect ssDNA that is generated during replication and other cellular processes. The protein is a tetramer that consists of four identical subunits, each with an N-terminal domain that binds ssDNA and a C-terminal domain that binds a variety of cellular proteins. Like the phage T7 ssDNA-binding protein discussed in Chapter 2, the region of the *E. coli* ssDNA-binding protein that binds ssDNA consists of an OB-fold (*o*ligonucleotide/oligosaccharide-*b*inding fold).

The bacterial ssDNA-binding protein plays a crucial role during DNA replication. It rapidly and efficiently binds the single-stranded regions exposed on the lagging strand as replication proceeds (Figure 3.2). This binding has a number of important consequences. First, ssDNA is rather sensitive to chemical modifications and breakage, and ssDNA-binding protein protects from these damages. Second, ssDNA can form secondary structures when two segments happen to have complementary sequences, and these secondary structures can be very serious impediments to replication and compromise DNA stability; ssDNA-binding protein prevents the formation of such secondary structures. Third, the protein is thought to help maintain the proper configuration of ssDNA during the rapid reorganizations that occur in the lagging-strand gymnastics of the trombone model.

While *E. coli* ssDNA-binding protein has all those important functions, it does much more than that. As mentioned earlier, the C-terminus of the protein binds various other proteins and thereby helps to coordinate protein functions for DNA replication. Two key interactions are with primase and with a subunit of the DNA polymerase holoenzyme complex. During the primer handoff mentioned earlier, ssDNA-binding protein is part of the RNA primer–primase complex, and it switches its interaction from primase to the holoenzyme subunit to complete the handoff. ssDNA-binding protein also interacts with a number of proteins involved in repair reactions and replication restart, and we will see in subsequent chapters that the corresponding eukaryotic protein is also central in both DNA replication and repair.

3.6 Housekeeping after the replisome passes — Repairing Okazaki fragments, reducing replicative errors, and recycling clamps from the DNA

While the replisome is an efficient machine, several problems remain to be solved after it passes. The most obvious is that the lagging-strand DNA is not complete, but instead is left with multiple embedded RNA primers and discontinuities at the 5′ ends of these primers. As in the T7 system, the RNA primers need to be excised by a 5′ to 3′ exonuclease, replacement DNA bases need to be inserted, and the final nicks in the DNA need to be sealed. At the beginning of this chapter, we mentioned bacterial DNA polymerase I, which was actually the first DNA polymerase purified and characterized (resulting in the 1959 Nobel Prize for Arthur Kornberg). It

---

1959 Nobel Prize in Physiology or Medicine

This prize was awarded jointly to **Severo Ochoa** and **Arthur Kornberg** for their studies on RNA (Ochoa) and DNA (Kornberg) synthesis; Kornberg isolated DNA polymerase from bacteria and demonstrated its ability to synthesize a complementary strand of DNA using a template strand. *See legend to Figure 2.1 for details and attribution of DNA polymerase structure.*

https://www.nobelprize.org/prizes/medicine/1962/summary/

turns out that this DNA polymerase plays a central role in Okazaki fragment processing. It was initially quite surprising to find that DNA polymerase I contained two exonuclease activities. As with DNA polymerase III, polymerase I has a 3′ to 5′ exonuclease that can edit out mistakes in which the polymerase activity inserted an incorrect base. It also has a 5′ to 3′ exonuclease, and this is the activity that normally removes the RNA primers from Okazaki fragments. Indeed, DNA polymerase I can carry out a concerted reaction called "nick translation," in which it removes bases from in front of the enzyme with the 5′ to 3′ exonuclease and inserts new bases behind with the DNA polymerase activity (Figure 3.5; note that the position of the nick is "translating" along the DNA). Nick translation efficiently removes the short stretch of RNA at the 5′ ends of

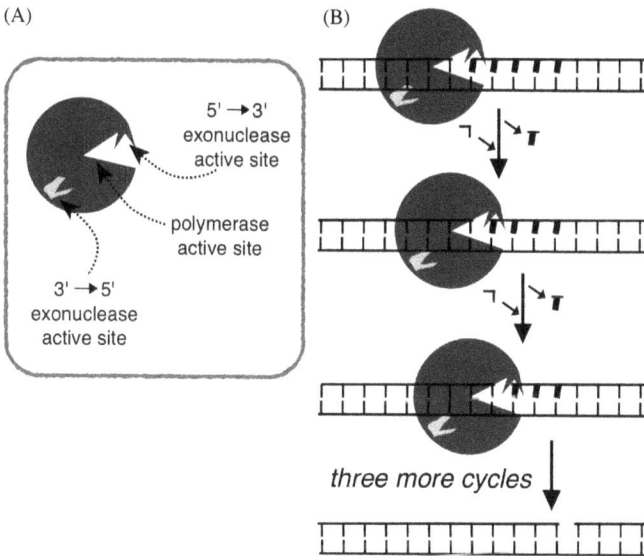

Figure 3.5. Role of DNA polymerase I in final editing of Okazaki fragments. *Escherichia coli* DNA polymerase I has both 5′ to 3′ and 3′ to 5′ exonuclease activities, residing in different domains of the protein (panel A). The coordinated action of the 5′ to 3′ exonuclease and the (5′ to 3′) DNA polymerase is referred to as nick translation (panel B), since the location of the nick in the DNA is moving along the DNA. After the RNA residues are all removed, DNA ligase seals the nick to complete lagging-strand synthesis (not shown).

Okazaki fragments while replacing the RNA bases with DNA. If any bases are misincorporated during this replacement reaction, the incorrect base will very likely be removed by the proofreading 3' to 5' exonuclease activity of polymerase I before completion of the replacement reaction. The 5' to 3' exonuclease activity of DNA polymerase I is in a distinct domain of the protein, which can be released from the enzyme by proteolysis.[3]

Once the RNA segments are replaced with DNA and the nick has migrated accordingly, the nick is sealed by bacterial DNA ligase. *E. coli* DNA ligase also plays critical roles in DNA repair reactions, as we will see later in the book.

The DNA synthesized by either DNA polymerase III or DNA polymerase I has very few remaining errors, thanks to the proofreading activities of these enzymes as described earlier. Nonetheless, without further corrective mechanisms, *E. coli* cells would suffer on the order of one mutation during every round of DNA replication. Measurements of mutation rates show that the frequency is about 100 times lower, demonstrating a very high level of genome stability. The increased fidelity of genome replication is achieved by means of a remarkable and efficient system of post-replication repair called mismatch repair, discussed in Chapter 6.

Another housekeeping issue following DNA replication involves the fate of the sliding clamps. As explained above, a new sliding clamp must be loaded by the replisome for each synthesized Okazaki fragment, which number in the thousands. However, a measurement of protein levels indicates that *E. coli* contains only a few hundred sliding clamps, so clamps must be recycled and reused. In the description of DNA replication above, the polymerase dissociated from the clamp as it cycled to the next RNA primer, but the clamp was left behind encircling the DNA and thus unable to dissociate. Surprisingly, it has been discovered that the clamp loader can also unload clamps, although it does not appear to do so immediately upon completion of Okazaki fragment synthesis. Instead,

---

[3] The resulting enzyme free of 5' to 3' exonuclease activity, called "Klenow enzyme," has been used extensively in biotechnology and DNA manipulation in vitro.

clamps are left behind on the just-synthesized lagging strand as the replisome races away.

This raises some interesting issues. Wouldn't it be more efficient if the replisome itself unloaded clamps immediately for reuse? One compelling possibility is that the clamps serve a further function and are "intentionally" left behind for a period of time. Provocative evidence in support of this possibility is that both DNA polymerase I and DNA ligase can bind to the clamp. Thus, clamps left behind on each Okazaki fragment may help to direct the Okazaki fragment processing steps described earlier. Furthermore, one of the proteins involved in mismatch repair can bind to clamp, raising the possibility that the left-behind clamps also help direct mismatch repair. A good way to think about these results is that the replicative clamp serves as a "toolbelt," capable of bringing multiple tools to bear for different tasks during the overall process of DNA replication. Further experiments are needed to decipher exactly how and when clamps are unloaded, and how extensively these three processes (Okazaki fragment processing, mismatch repair, and clamp unloading) are coordinated with each other.

3.7 Replication restart and other rescue pathways

The above picture of an efficient replication fork that proceeds smoothly around the chromosome from origin to terminus is incomplete. Replication forks can encounter a wide variety of damaged DNA bases, DNA breaks, chemical crosslinks between the strands or between the DNA and a protein, proteins that are bound tightly though noncovalently, and other structures that challenge the process. As we will discuss in later chapters, efficient repair systems greatly reduce the burden of many of these lesions, but these systems are imperfect, leaving lesions in the path of the replisome. Recent research has shown that bacteria have multiple pathways that allow completion of replication in spite of the kinds of DNA lesions mentioned earlier. While the existence and basic mechanisms of these pathways have been uncovered, we have a long way to go to understand many important details, such as which pathways are used for which lesions, how the interplay between the pathways is regulated, and how increased levels of damage alter the efficiency and balance of these pathways. We can also expect to find new pathways as research continues, and there are hints that DNA repair pathways described in later chapters might also be directly triggered by replication fork blockage.

We already discussed one situation, involving damage on the lagging-strand template, in which the replisome is able to essentially bypass the lesion and continue replication (Section 3.4). Surprisingly, recent evidence suggests that a similar pathway can occur with leading-strand damage. Biochemical experiments showed that when a leading-strand polymerase encounters a blocking lesion, primase is able to synthesize a new primer on the leading strand ahead of the blocking lesion and the clamp loader is able to load a new clamp (Figure 3.2B). This series of events primes a new stretch of DNA synthesis on the leading strand, leaving a short gap of ssDNA behind. As with the lagging-strand damage pathway, the gap and blocking lesion still need to be dealt with by a subsequent repair event of some kind, but at least the chromosome can complete its replication.

Recall that *E. coli* has a special protein that loads the replicative helicase at the replication origin and prevents inappropriate loading elsewhere (Section 3.4; also see Chapter 5). The existence of a second helicase-loading pathway in *E. coli* was therefore a long-standing mystery. This pathway, which involves several proteins, was discovered many decades ago as being required for the replication of a bacterial virus called φX174. Surprisingly, however, these proteins were not needed for replication from the bacterial replication origin. Why would *E. coli* encode proteins that are unnecessary for its own replication but allow the replication of a lethal virus? It took a couple of decades to solve this mystery, but the answer revealed that DNA replication is much more robust and flexible than previously thought.

Scientists discovered that *E. coli* uses these proteins to salvage replication when it has been disturbed or blocked in various ways. Depending on the exact combination of proteins involved, four distinct "replication restart" pathways have been defined. Mutations that inactivate the proteins needed for replication restart are nearly lethal to *E. coli*, implying that most rounds of DNA replication must invoke one or more restart pathways. The detailed mechanisms and protein requirements of these pathways are quite complex and beyond the scope of this chapter. Also, the exact situations when the different pathways are used are still being worked out. Some of the pathways seem to provide an alternative to restart replication when either the leading- or lagging-strand polymerase is blocked. It is not yet clear how the choice is made between these replication restart pathways and the re-priming pathways discussed just above (which don't require additional proteins).

Several additional replication-rescue pathways utilize proteins and mechanisms involved in DNA repair and recombination pathways, and these will be described in more detail in subsequent chapters that highlight DNA repair and recombination. A brief overview follows as a teaser for these later chapters.

DNA breaks are very serious lesions that must be repaired for cells to maintain their intact genomes and ultimately to survive, and it is therefore no surprise that cells have mechanisms to repair DNA breaks. A single-strand break or nick would seem to be rather innocuous, because it can be easily repaired by DNA ligase. However, ligase cannot repair a nick unless it has one 5′ phosphate and one 3′ hydroxyl end, and some do not. What happens when a DNA replication fork encounters a nick that cannot be repaired by ligase, or even a sealable nick that just escaped ligation? As the replicative helicase drives through the region of the nick, the arm with the nick is broken off, which seems a rather disastrous event. As we will see in Chapter 11, a special version of a DNA-break-repair pathway can essentially stitch this broken arm back onto its partner and reconstitute a functioning replication fork. This pathway involves reloading the replicative helicase using some of the restart proteins mentioned earlier.

Excision repair pathways replace damaged bases with the correct undamaged base, but these pathways require that the damage is located within a duplex region of DNA (Chapter 10). Again, what happens if the process of replication gets ahead of the process of excision repair of a damaged base? This would be particularly prone to occur when cells suffer a high level of DNA damage from exposure to radiation and/or chemical DNA–damaging agents, overwhelming the excision repair system. When the replication fork reaches such a damaged base, the polymerase is unable to insert a complementary base for certain forms of damage. By then, however, it is too late for excision repair to operate! The damaged base is no longer in a duplex region — the replicative helicase has already unwound this region of the DNA (Figure 3.6, top).

A remarkable kind of gymnastics can come to the rescue — the replication fork can actually back up by a process called replication fork regression (Figure 3.6A). During regression, the two newly synthesized strands anneal with each other and the branch point of the fork is zipped *backward*. Fork regression is driven by special DNA helicases, which have additional roles in DNA repair and/or recombination. Obviously, the process of fork regression must be very carefully regulated or else replication would turn into a hopelessly complicated process. Getting back to the DNA damage that caused the problem in the first place, the process of fork regression has driven the damage back into a duplex region, and thereby allows excision repair to occur successfully and fix the problem (Figure 3.6B). Presumably, excision repair is somehow coupled to this process of regression, but the details are not yet clear. Following excision repair, the fork regression must be reversed (Figure 3.6C), and the normal replication fork restarted, and again one assumes that these are all well-coordinated events.

Fork regression can be used to achieve another remarkable outcome, inserting the correct complementary base opposite a damaged base that itself cannot serve as a template for correct base pairing. This can occur in the situation where the leading-strand polymerase has been blocked by a damaged template base, but lagging-strand replication has proceeded a short distance further on. Fork regression can again extrude the two newly synthesized strands into a (mostly) duplex region (Figure 3.7A). In this case, the 3′ end of the new leading strand, which had been blocked by the damage prior to

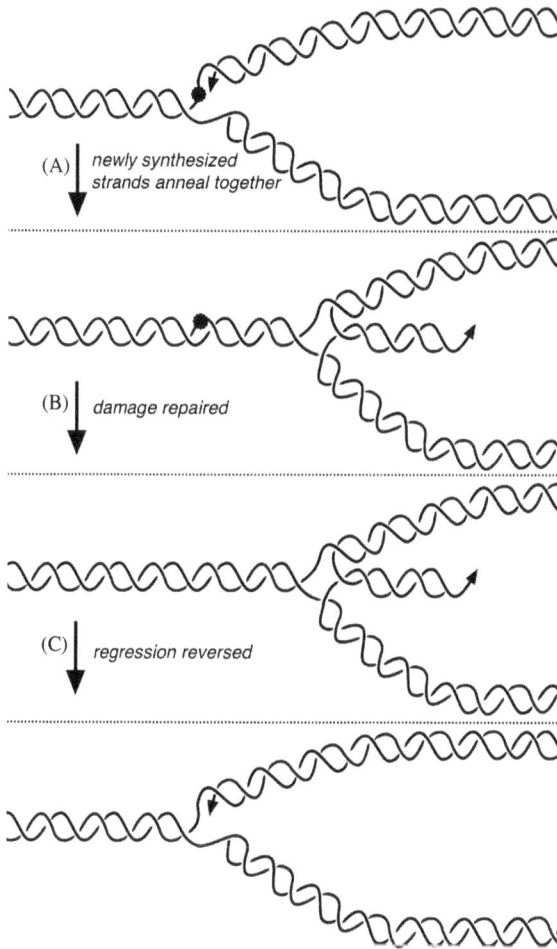

Figure 3.6. Replication fork regression allows repair of blocking template lesions. The two newly synthesized strands at the replication fork are complementary to each other and can thereby anneal to each other and allow the fork to "back up" (step A). Once the lesion is restored into a region of duplex DNA, repair pathways described later in this book can repair the damage (step B). Reversal of the regression restores a normal replication fork structure (step C), and the process of replication can then resume.

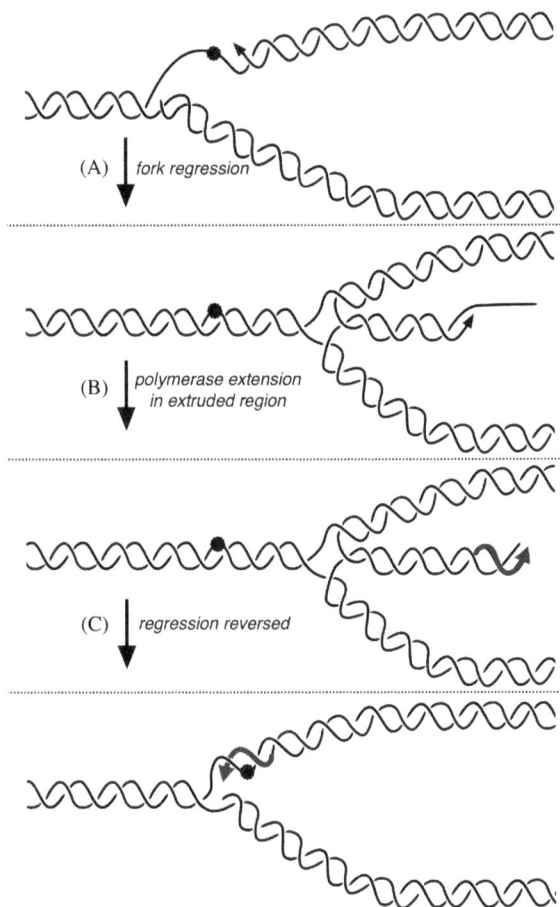

Figure 3.7. Error-free bypass of a blocking lesion via replication fork regression. After blockage of the leading-strand polymerase, replication fork regression places the blocked 3′ end in a duplex with the lagging-strand product (step A). This allows accurate incorporation of several base residues by DNA polymerase (step B). Reversal of regression (step C) then allows resumption of normal DNA synthesis, with the blocking lesion bypassed.

regression, is properly base paired with the new and longer lagging-strand product. Notice that the sequence of the new lagging-strand product would be identical to the sequence of the old leading-strand template, because it is "the complement of the complement."

Thus, the lagging-strand product can serve as an accurate template for leading-strand synthesis! DNA polymerase extension of the 3′ end of the leading strand, even for a few bases, has the effect of circumventing or bypassing the original damage (Figure 3.7B). When the regression process is reversed, the leading-strand product has now been extended past the damage, with the correct base opposite the damage (Figure 3.7C). The fork can restart and complete the replication of the genome. The damaged base remains, but the cell can complete its round of cell division and now has another cell cycle to properly repair the damage, for example, by excision repair.

Importantly, fork regression-driven bypass occurs in an accurate, mutation-free manner. A different and very important form of lesion bypass involves specialized DNA polymerases, and will be covered in Chapter 12. Overall, it is clear that numerous sophisticated pathways have evolved, allowing the completion of bacterial DNA replication even when the DNA is damaged.

3.8 Summary of key points

- The circular *E. coli* chromosome is replicated by two oppositely oriented replisome complexes that each traverse the DNA at a speed of about 1000 base pairs per second.
- The heart of the *E. coli* replisome is the DNA polymerase III complex, which contains the multi-subunit polymerase, the clamp loader and clamp.
- The sliding clamp encircles the DNA and allows DNA polymerase to synthesize long stretches of DNA without falling off the template (i.e., increasing polymerase processivity).
- The five-subunit clamp loader harnesses ATP binding and hydrolysis to crack open the clamp ring structure and load the clamp onto DNA at primer-template junctions.
- The bacterial helicase is a hexamer that travels along the lagging-strand template, associated and interacting with both the bacterial primase and the leading-strand polymerase.

- The *E. coli* replisome can continue past the site of lagging-strand damage, leaving a short single-stranded region containing the damaged base.
- *E. coli* ssDNA-binding protein is a tetramer that protects ssDNA at the replication fork, prevents formation of deleterious secondary structures, and interacts with other proteins in the replisome.
- DNA polymerase I replaces ribonucleotide residues from Okazaki fragments, and DNA ligase seals the remaining nicks.
- *E. coli* and other bacteria have multiple pathways for restarting replication and otherwise rescuing replisomes that have encountered serious problems such as blocking lesions and breaks.

Further Reading

Atkinson, J., & McGlynn, P. (2009). Replication fork reversal and the maintenance of genome stability. *Nucleic Acids Res, 37*(11), 3475–3492.

Hedglin, M., Kumar, R., & Benkovic, S. J. (2013). Replication clamps and clamp loaders. *Cold Spring Harb Perspect Biol, 5*(4), a010165.

Johansson, E., & Dixon, N. (2013). Replicative DNA polymerases. *Cold Spring Harb Perspect Biol, 5*(6), a012799.

Kelch, B. A., Makino, D. L., O'Donnell, M., & Kuriyan, J. (2012). Clamp loader ATPases and the evolution of DNA replication machinery. *BMC Biol, 10*, 34, doi:10.1186/1741-7007-10-34.

Kong, X. P., Onrust, R., O'Donnell, M., & Kuriyan, J. (1992). Three-dimensional structure of the β subunit of *E. coli* DNA polymerase III holoenzyme: a sliding DNA clamp. *Cell, 69*, 425–437.

Langston, L. D., & O'Donnell, M. (2006). DNA replication: Keep moving and don't mind the gap. *Mol Cell, 23*(2), 155–160.

O'Donnell, M., Langston, L., & Stillman, B. (2013). Principles and concepts of DNA replication in bacteria, archaea, and eukarya. *Cold Spring Harb Perspect Biol, 5*(7), a010108.

Stukenberg, P. T., Studwell-Vaughan, P. S., & O'Donnell, M. (1991). Mechanism of the sliding β-clamp of DNA polymerase III holoenzyme. *J Biol Chem, 266*(17), 11328–11334.

Yao, N. Y., & O'Donnell, M. (2008). Replisome dynamics and use of DNA trombone loops to bypass replication blocks. *Mol Biosyst, 4*(11), 1075–1084.

Yao, N. Y., & O'Donnell, M. (2012). The RFC clamp loader: Structure and function. *Subcell Biochem, 62,* 259–279.

Yeeles, J. T., Poli, J., Marians, K. J., & Pasero, P. (2013). Rescuing stalled or damaged replication forks. *Cold Spring Harb Perspect Biol, 5*(5), a012815.

How did they test that?
Does the sliding clamp encircle DNA?

Stukenberg *et al.* (1991) analyzed the mechanism by which *E. coli* sliding clamp stimulates DNA polymerase. One approach was to ask about the nature of clamp binding to DNA. They loaded radioactive clamps onto (nicked) duplex circular DNA, subjected the mixture to various manipulations, and then analyzed the samples with gel-filtration columns. In these columns, large intact protein-DNA complexes elute in the early fractions (around fraction 15), while smaller unbound proteins elute later (fractions 25–35). Previous experiments verified these elution positions. In panel A, radioactive clamp protein is found to be free in solution if the DNA is cleaved with restriction enzyme SmaI prior to the column (filled triangles), but still bound to DNA without restriction enzyme (filled circles).[4] In panel B, clamp was retained on circular DNA even if ATP and the clamp loader (γ complex) were removed; however, clamp again fell off the DNA if the DNA was cleaved with a (different) restriction enzyme. In panel C, clamp remained bound to circular DNA when the DNA nick was sealed with DNA ligase, showing that the clamp does not need a nick to remain bound. Panel D showed that multiple clamps (roughly 20) can be loaded onto DNA; DNA cleavage with SmaI released the loaded clamps. The data in panel E are remarkable. DNA-binding protein EBNA1 was bound in two locations to the loaded clamp–DNA complex, and then the DNA was cleaved in between. Now, the clamp no longer dissociated when the DNA was linearized, showing that another bound protein can block the clamp from sliding off the ends of linearized DNA! (The insert gel shows that the restriction enzyme did indeed cleave the DNA.) These experiments provided strong evidence that clamp encircles DNA, and later crystal structures showed that the clamp is shaped like a donut with a hole that can accommodate duplex DNA. The figure was reproduced from Stukenberg *et al.* (1991), with permission from the American Society of Biochemistry and Molecular Biology; permission conveyed by Copyright Clearance Center, Inc.

---

[4] Only a subset of clamp is loaded onto DNA, explaining the peak in the unbound-protein fractions. The dimeric *Escherichia coli* clamp is named β, explaining the axis label.

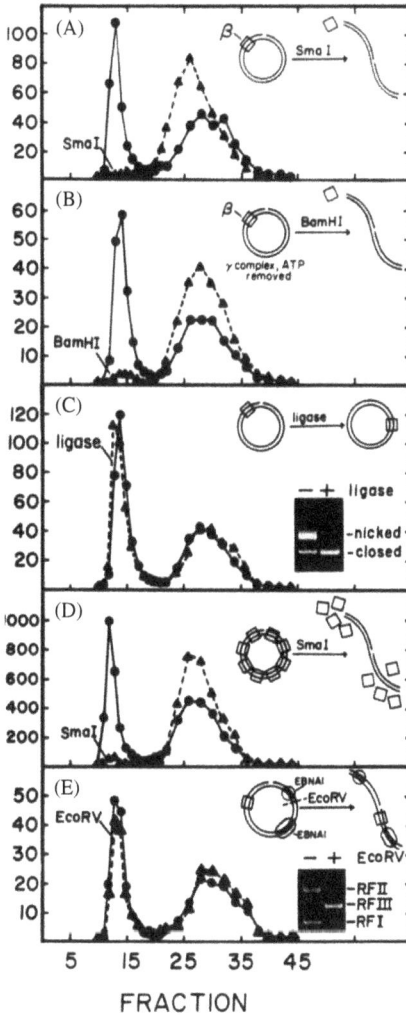

FRACTION

Chapter 4

Eukaryotic DNA replication

Normal diploid human cells carry 46 linear chromosomes: two each of the 22 autosomes and two sex chromosomes. Chromosome numbers in this range are quite typical throughout the animal kingdom, although cells of some animal species have more than 100 chromosomes, and certain plant cells even more (e.g., well over 1000 in a particular fern species). While the number of chromosomes per cell can be large depending on the species, in every case, each and every chromosome must be replicated once and only once per cell cycle. The two replicated copies of each chromosome must also be segregated correctly, one into each of the two daughter cells. Both of these processes have a very low error rate. However, failures do occur, often resulting in non-viable daughter cells or contributing to cancer formation in metazoans.

The machinery that replicates chromosomes in eukaryotic cells is significantly more complex than that of prokaryotes. As we will see, this complexity allows a high degree of flexibility in the replication program and a remarkable degree of regulation to ensure a high probability of correctly replicating each chromosome once and only once. Some of the basic machinery involved in replication is conserved throughout evolution, and proteins involved in replication from bacteria to humans show many commonalities in protein structure and molecular mechanisms. In this Chapter, we will focus on the eukaryotic replisome machine and how it functions to duplicate the chromosomal DNA. In the following Chapter, we will shift our attention to how replication is initiated at replication origins, regulated both at the initiation and later stages, and how replication is completed prior to cell division.

The exact number of proteins involved in eukaryotic DNA replication is still not known and almost certainly varies between disparate species. Nonetheless, the number of different proteins is clearly much higher than in prokaryotes, for at least three major reasons (described in more detail in ensuing sections). First, multimeric proteins that are composed of identical subunits in prokaryotes are instead composed of different but related subunits in eukaryotes (for example, the hexameric replicative helicase). Second, eukaryotes were found to have multiple forms of certain key replication protein complexes, each with specialized functions, in contrast to a singular form in prokaryotes (for example, both clamps and clamp loaders). Third, eukaryotic replication systems must deal with additional complexities such as replicating DNA that contains nucleosomes and preparing their multiple chromosomes for accurate segregation. Many additional proteins, including some that travel with the replication fork, are involved in these processes. The growing list of eukaryotic replication proteins is quite staggering, as indicated by Tables 1 and 2 in the Appendix. In these Chapters, we will again try to minimize the nomenclature to keep the focus on important concepts and functions. A few important and commonly used names will be introduced, particularly when these proteins are also central in other cellular functions such as DNA damage responses and DNA repair.

4.1 Special challenges of replicating multiple linear chromosomes

One of the major challenges of DNA replication in eukaryotes is the large amount of DNA to be replicated. As mentioned above, eukaryotic cells have multiple chromosomes. Each chromosome is believed to contain one double helix of DNA,[1] which can be quite long. For

---

[1] The ends of the linear chromosomes have an unusual structure that will be discussed in Section 5.10.

example, the longest human chromosome, number 1, is just under 250 million base pairs in length, more than 50 times longer than the *E. coli* chromosome. At the same time, the eukaryotic replication machinery travels about 20 times slower than that of bacteria. If human chromosome 1 had only one origin near its center, which initiated bidirectional replication (like in *E. coli*), it would take about a month to replicate the entire chromosome (40 minutes × 50 × 20; recall that chromosomal replication in *E. coli* takes 40 minutes). However, we know that human cells can replicate their DNA and divide in as little as 24 hours. The solution to this riddle is that all known eukaryotic chromosomes contain multiple replication origins that must fire in order for DNA replication to keep up with cell division. Thus, during cellular DNA replication, any given chromosome can have multiple replication complexes, some moving towards each other relative to the parental DNA and some away from each other. Particularly in the next Chapter, we will see how regulation of this origin firing is used to achieve different rates of cell division and to overcome problems, such as replication-fork blockage, that occur during replication.

The slower rate of fork movement in eukaryotes probably relates largely to the complex structure of eukaryotic chromosomes. The DNA is wrapped around nucleosomes in 200-base pair segments, and these nucleosomes are packed into complex higher-order structures. As the DNA is replicated, the replication fork needs to access the duplex DNA wrapped into the nucleosome in order to duplicate the DNA, and then a new nucleosome must be assembled so that each daughter chromosome has the full nucleosome complement. Furthermore, the histones within nucleosomes carry multiple and complex modifications that are particularly important in controlling whether the nearby genes are expressed or not, and these modifications must also be regenerated in the nucleosomes of both daughter DNA molecules. Details about how nucleosomes become duplicated and maintain their proper modifications are the subject of extensive current investigations, but are beyond the scope of this book. It is worth mentioning that a complex set of proteins are needed to manipulate the histones during and immediately after the replication event, and some of these proteins indeed are directly associated with the replication machinery.

As we will discuss in the next Chapter, the multiple origins of eukaryotic chromosomes do not all fire in every cell division. Furthermore, the subset of origins that do fire varies by developmental stage, tissue type, and time during the cell cycle, and DNA damage responses can activate additional origins that would otherwise be inactive. This extreme flexibility in origin usage means that replication forks terminate in very diverse, perhaps random, locations throughout the chromosome. In a large majority of the genome, two replication forks traveling towards each other simply terminate wherever they meet. This process must involve disassembly of the replicative helicase, to prevent replication forks from essentially "passing" each other and replicating DNA that has already been replicated. Two proteins have been implicated in replicative polymerase unloading, and the process is under active investigation.

All known DNA polymerases synthesize DNA in the 5′ to 3′ direction and nearly all require a pre-existing primer to initiate synthesis. These features lead to the "end-replication problem", that is, the need for a special mechanism to allow replication of the 3′ end of the parental strand of a linear DNA. To illustrate the problem, imagine that the RNA primer is synthesized by a primase exactly at the 3′ end of the parental DNA to allow DNA polymerase to begin synthesis (Figure 4.1A). Since RNA primers are erased during Okazaki fragment processing, the new 5′ end will be left with a gap of several unreplicated bases after RNA removal. During the next round of replication, the 5′ end of this daughter chromosome can be replicated up to its 5′ end, but the duplex DNA will be several nucleotides shorter because of the gap that was left behind. Progressive rounds of replication will make the daughter chromosomes shorter and shorter. The problem is worse than this, because all known primase enzymes have some modest sequence specificity. Even if the very end of the chromosome initially has a primase site, this site will be lost with the first shortening of the chromosome, requiring the use of a primase site further down the chromosome. Thus, the shortening will occur much faster than just a few base residues per replication cycle.

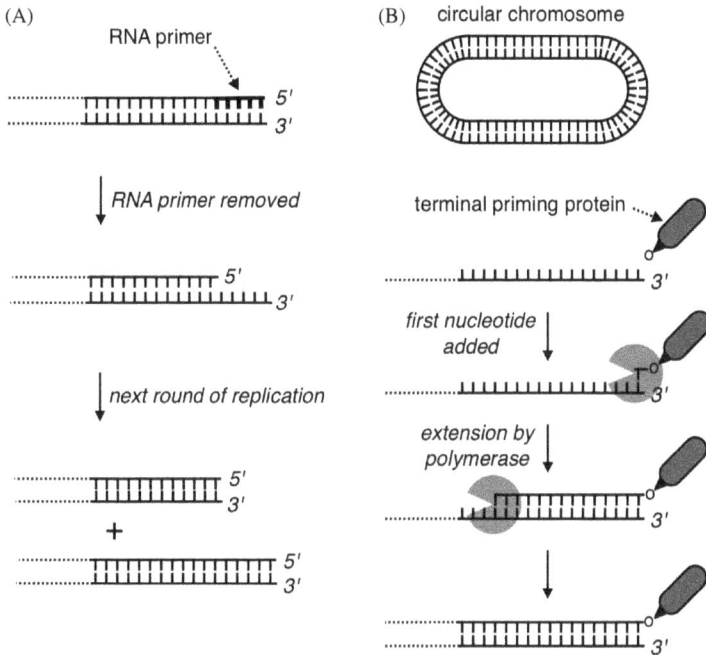

Figure 4.1. The end-replication problem. Replicative DNA polymerases require a pre-existing primer, usually RNA. Even if an RNA primer is synthesized at the precise end of a duplex DNA, removal of the primer after replication and a second round of replication results in shortening of the chromosome end (panel A). Two solutions to the end-replication problem are circular chromosomes (and plasmids) and terminal protein priming (panel B).

Nature has evolved several solutions to the end-replication problem (Figure 4.1B). The first is to avoid ends by using circular DNA, as found in bacteria, some viruses, and mitochondria. Another solution, used by certain viruses including adenovirus, involves the attachment of a special terminal-priming protein to the end of the genome. This terminal protein does not synthesize primers, but rather itself serves as the primer for DNA polymerase. The viral DNA polymerase uses a specific serine-hydroxyl group on the terminal protein to form a phosphodiester bond with the α-phosphate of the first deoxynucleotide to be added. Once the first deoxynucleotide residue is incorporated, the remainder of the strand can be synthesized by the usual 5′ to 3′ reaction of DNA polymerase. The process has additional complexities, and the reader is encouraged to explore this interesting topic with further reading.

The third solution to the end-replication problem, used by virtually all eukaryotic chromosomes, involves the use of special structures called telomeres at the ends of the chromosome. The mechanisms involved in telomere replication will be discussed in the next Chapter, along with the implications of telomere biology with regard to stem cells, aging and cancer.

4.2 How do eukaryotic replication proteins compare to their prokaryotic counterparts?

The possible evolutionary relationships between replication proteins in the different domains of life have been probed by searching for amino-acid-sequence homologies and structural similarities. While one might expect a high level of conservation among proteins that are involved in such a vital process as DNA replication, the results are surprisingly mixed. A subset of replication proteins shows strong enough homology to support their assignment as orthologs evolved from the same ancestor. Others show structural similarities in spite of limited or no evidence of homology at the amino-acid-sequence level. Yet other key replication proteins clearly belong to different protein families, ruling out any close evolutionary relationship. Not surprisingly, even when the proteins are from different families, the basic reactions are usually very similar. Archaeal species that have been studied to date have a set of replication proteins that are much more similar to those from eukaryotes than those from prokaryotes (but with a smaller number of proteins in archaeal replication than in eukaryotic replication).

Near the heart of the replisome, the most conserved protein between prokaryotes and eukaryotes is the clamp-loader complex. Very clear homology exists at the amino-acid-sequence level in the clamp-loader proteins, supporting their assignment as orthologs with a common ancestor. In addition, both complexes are heptameric and form the shape of a horseshoe (Figure 4.2A). Certain

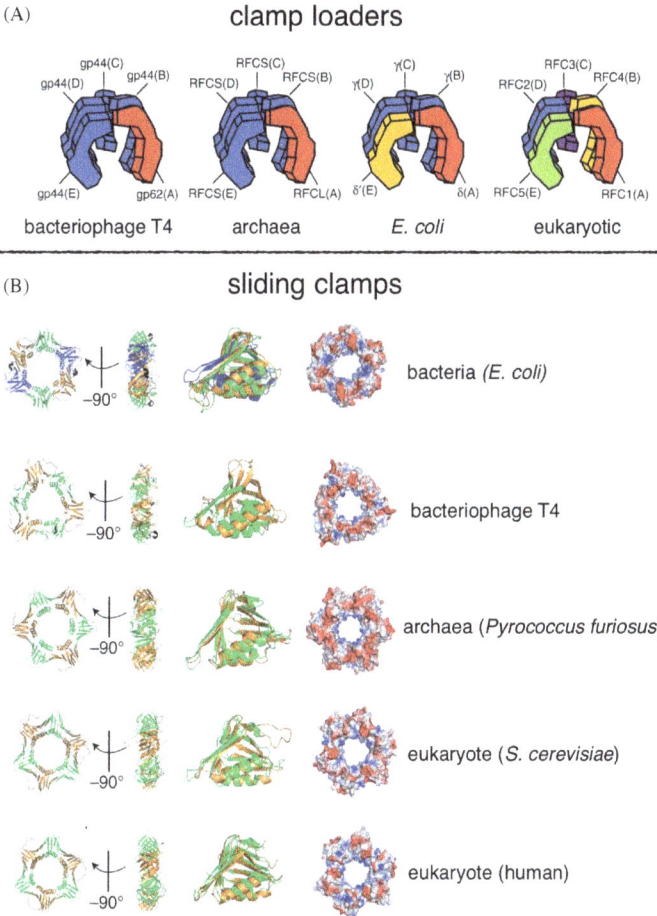

Figure 4.2. The conserved clamp loaders and sliding clamps. Clamp loaders from a bacterial virus, archaea, *E. coli* and eukaryotes are schematized, with protein names indicated (panel A). Crystal structures of sliding clamps from these same organisms are compared in panel B. The two leftward images show the entire clamp with 90° rotation between the two. The next image shows superpositions of the domains that are present six times in each sliding clamp (notice how similar these are between the domains from the various organisms). The rightward images show space-filling models, with positively (blue) and negatively (red) charged regions indicated; note that the central cavity is lined with positively charged residues in each case, allowing DNA (with its negatively charged backbone) to slide freely. The sliding-clamp structures in panel B are as follows: *E. coli*, PDB ID 2POL; bacteriophage T4, PDB ID 1CZD; *Pyrococcus furiosus*, PDB ID 1GE8; *S. cerevisiae*, PDB ID 1PLQ; human, PDB ID 1AXC (see http://www.rcsb.org for all structures). The panels in this figure were reproduced from Hedglin *et al.* (2013), with permission from Cold Spring Harbor Laboratory Press.

topoisomerases (Chapter 7) and enzymes in deoxynucleotide synthesis are also strongly conserved between prokaryotes and eukaryotes.

The clamp itself provides an interesting evolutionary case. The clamps of prokaryotes and eukaryotes show relatively weak amino-acid-sequence homology and are thought to be fairly distant homologs on this basis. The donut-like structures of the prokaryotic and eukaryotic clamps are strikingly similar, each containing six very similar domains that circle around the central hole (Figure 4.2B). Surprisingly, however, the prokaryotic clamp is composed of two identical subunits each with three domains, while the eukaryotic clamp is composed of three identical subunits each with two domains.

Other key proteins in DNA replication have obvious structural similarity but seem to be quite distant from each other in evolution. A case in point involves initiator proteins that bind origins of replication, which will be discussed in Chapter 5. Initiator proteins of both prokaryotes and eukaryotes have a prominent ATPase domain from a particular evolutionary group of ATPase modules, but the prokaryotic and eukaryotic ATPase domains are not particularly close relatives. Considering the initiator proteins of bacteria and that of *S. cerevisiae*, they also share the feature of a C-terminal DNA-binding domain linked to the ATPase domain. These DNA-binding domains are entirely different structures and are not related, although both bind (different) specific sequences within duplex DNA.

A similar story arises with the replicative helicases. Both the prokaryotic and eukaryotic replicative helicases are hexameric enzymes with a hole in the middle for DNA, and so their morphology is similar. However, the ATPase subunits that make up the hexamer are from different families in prokaryotes versus eukaryotes, and so these helicases are clearly not orthologs. Interestingly, in spite of their similar function at the fork, the inherent mechanism of these two helicases differs, as does their pathway of loading (see below).

The two enzymes that actually synthesize nucleic acid at the fork, DNA polymerase and RNA primase, might be expected to be the most highly conserved. They are not. While nearly all DNA polymerases share an architecture resembling a right hand, with the active site in the palm, multiple families of DNA polymerase are found throughout the three domains of life. The main replicative polymerase in bacteria, discussed in Chapter 3, is from an entirely different family than the replicative polymerases of eukaryotes. Likewise, the primases of prokaryotes and eukaryotes are from completely different families and unrelated to each other.

These examples beg the question of how replication machineries evolved from those of our earliest ancestors. There is some speculation that viral DNA-replication machineries may have been co-opted at some point in evolution for cellular replication, to explain the cases where the proteins are not orthologous between the domains of life. In spite of these complex evolutionary relationships, the central reactions in DNA replication are quite highly conserved with few exceptions. Thus, the central steps in eukaryotic replication, described in this Chapter, will be very familiar from the discussion of prokaryotic replication in Chapters 2 and 3.

4.3 MCM complex — the replicative helicase in eukaryotes

As mentioned above, the eukaryotic replicative helicase is a hexameric enzyme with a hole in the middle large enough to accommodate DNA. The enzyme is called the MCM complex, a name that was derived from the phenotype of mutants that first defined the genes encoding the subunits (*Mini-chromosome maintenance*; involved in maintaining small artificial chromosomes in yeast). Like the bacterial and bacteriophage T7 enzymes, the MCM complex tracks along ssDNA by essentially walking along the DNA that passes through the middle; it is not yet known whether the eukaryotic helicase tracks along DNA by the hand-over-hand mechanism used by the T7 helicase (Section 2.4). The eukaryotic MCM complex is not evolutionarily related to the bacterial or T7 enzymes, and key characteristics of the enzymes differ.

The most dramatic difference, which was a big surprise when discovered, is that the MCM complex has the opposite polarity when tracking along ssDNA compared to the bacterial and T7 enzymes.

MCM is a 3′ to 5′ helicase, which means that it tracks along the leading strand of a replication fork rather than the lagging strand. In this fundamental way, the architecture of bacterial and eukaryotic replication differs from each other.

Unlike the homo-hexameric bacterial replicative helicase, the MCM complex is made up of six distinct but related subunits, each the product of a different gene (Figure 4.3). Each of these six subunits is a member of the AAA+ family of ATPase introduced in Chapter 3 in our discussion of clamp loaders. These subunits arrange themselves in a particular order around the ring, and the six different ATPases around the ring play specialized roles in loading, activation and unwinding. There is one particular subunit–subunit interface in the hexamer that cracks open to allow DNA strands to enter or leave, processes that we will discuss in the next Chapter on initiation and termination of replication.

The six-subunit MCM complex by itself is essentially inert for DNA unwinding. Indeed, for several years after its isolation, there was

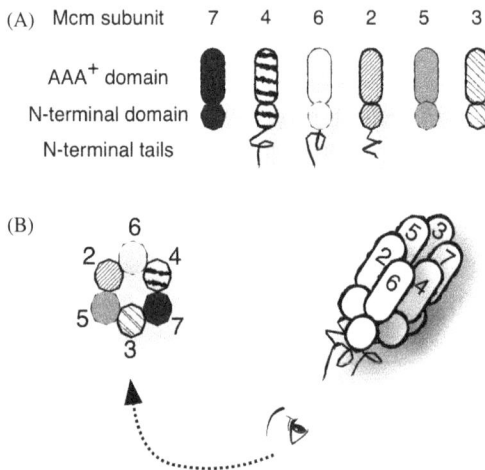

Figure 4.3.   The eukaryotic MCM complex. The six related proteins that make up the MCM complex are depicted side-by-side in panel A, with their names (MCM2 through MCM7) above and the nature of the domains indicated at the left. The six subunits arrange in a specific sequence (panel B). This figure was modified from Bell and Labib (2016).

significant question as to whether MCM was indeed the replicative helicase due to this lack of activity. Direct evidence that MCM is the replicative helicase mounted when it was discovered that the homologous archaeal MCM complex had strong unwinding activity, and also when it was discovered that an aberrant form of eukaryotic MCM complex composed with only three of the six distinct subunits also had helicase activity. Finally, the key discovery was that the helicase activity of the native MCM hexamer was unmasked upon addition of several other proteins. The fully active complex, called the CMG complex, contains a protein called CDC45, the MCM complex, and a 4-protein complex called GINS. All of these proteins are part of the eukaryotic replisome and necessary for ongoing replication at a fork

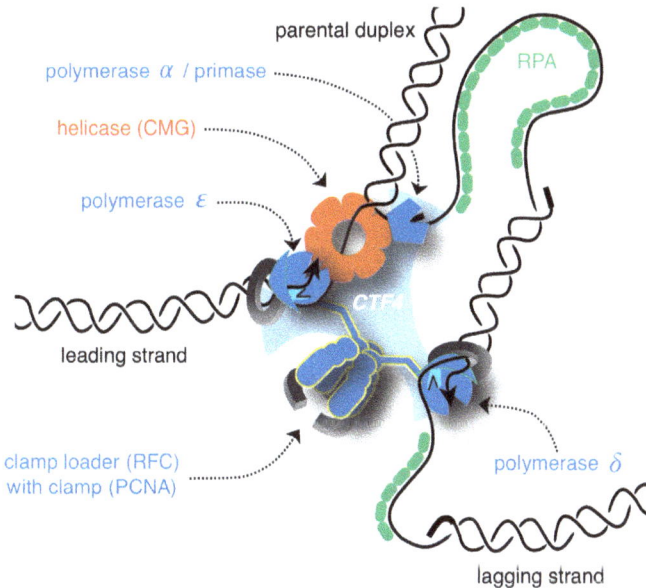

Figure 4.4. The eukaryotic replication fork. The current understanding of eukaryotic DNA replication is consistent with an overall organization as shown. The CTF4 and clamp-loader complex are at the heart of the replisome and are thought to tether leading- and lagging-strand polymerases. The helicase complex (CMG; represented by the orange hexamer) is an 11-subunit complex with the six MCM subunits, the four-subunit GINS complex, and CDC45. The positions of the three eukaryotic replicative DNA polymerases are indicated. Direct evidence for the looping of the lagging strand as shown is currently lacking.

(Figure 4.4). The requirement for additional proteins for MCM activity is very much related to the complex regulation of the helicase, and we will return to this topic when we discuss the loading of MCM at replication origins in the next Chapter.

4.4 Eukaryotic primase is a component of polymerase α

Eukaryotic cells use three different DNA polymerases during normal replication (and several additional polymerases when replicating damaged DNA; see Chapter 12). One of the three replicative polymerases, called polymerase α, is a four-subunit enzyme with the specialized function of providing primers for Okazaki-fragment synthesis. The enzyme also provides the first primer for leading-strand synthesis at replication origins. The smallest subunit of the polymerase α complex has the primase activity and synthesizes RNA primers, which are 8 to 10 nucleotide residues long. The short RNA primers are then handed off to the largest subunit of the enzyme, the actual DNA polymerase, which adds another 10 or so DNA nucleotide residues to form the full primer. It is this RNA-DNA copolymer that is in turn passed to the main replicative polymerase on the lagging strand (see below).

Polymerase α lacks proofreading-exonuclease activity, and therefore makes fairly frequent mistakes during incorporation (compared to a normal replicative polymerase with proofreading activity). As described in Chapter 3 for prokaryotic replication systems, the RNA primers used in Okazaki-fragment synthesis in all cells are replaced during maturation of the lagging strand, and so any errors in the RNA portion are erased during this maturation. It turns out that most of the DNA synthesized in the short primers by polymerase α is also replaced during Okazaki-fragment maturation (see below), limiting the damage caused by polymerase α-induced misincorporation errors.

The activity of the polymerase α/primase complex is coordinated with the overall functioning of the lagging strand. An accessory factor for this enzyme, called CTF4, is required to recruit polymerase α to the replisome as it is assembled at the origin (see Chapter 5). In the intact, functional replisome, this accessory factor links the polymerase α complex to the CMG helicase complex discussed above. Polymerase α may also bind directly to the CMG complex, fortifying the overall interaction. Thus, the unwinding of the parental DNA, which occurs by helicase tracking along the leading strand, is coordinated with the synthesis of Okazaki fragments on the lagging strand through the action of the polymerase α/primase complex.

4.5 Specialized polymerases for the leading and lagging strand

In addition to polymerase α, eukaryotes have two other replicative DNA polymerases called δ and ε. Unlike α, both δ and ε have proofreading activity and therefore synthesize DNA with higher accuracy, consistent with their roles in synthesizing more extensive stretches of DNA than α (see below). All three polymerases belong to the same evolutionary family of polymerases, called the B family. The three mammalian replicative polymerases each have four subunits, the largest of which carries the active site for DNA polymerization. However, the functions of the other three subunits differ between the three enzymes, in part reflecting the different functions of these three polymerase complexes in replication.

Over the years, there has been uncertainty and controversy about the precise roles of polymerase δ and ε with respect to leading-versus lagging strand replication. Indeed, even at the time of this writing, not all scientists agree on the conclusions. Nonetheless, the evidence is now quite strong that ε synthesizes the leading strand while δ synthesizes the lagging strand (with the priming assistance of polymerase α). The evidence includes genetic data using particular mutants of polymerase δ and ε (see "How did they test that?" at the end of this Chapter), biochemical evidence relating to the properties of the enzymes, and physical data in which polymerase ε can be chemically crosslinked to the leading-strand template while polymerase δ can be crosslinked to the lagging-strand template.

The biochemical properties of polymerase ε are well suited for leading-strand replication. Most notably, ε interacts with the CMG helicase complex via multiple contacts, and this interaction increases the processivity of the polymerase. Indeed, a stable complex containing all 15 subunits of both CMG and polymerase ε can be isolated, and is competent for leading-strand synthesis in the test tube. As in the bacterial system, the interaction of polymerase ε with helicase also stimulates the helicase activity of CMG, and so there is mutual stimulation of activities. Polymerase ε lacks the ability to displace the 5′ end of a strand ahead of the enzyme, a process called strand displacement that results in 5′ flaps in the product. As we will discuss shortly, this is a key aspect of lagging- but not leading-strand synthesis.

While the clamp stimulates the processivity of polymerase ε very modestly, the magnitude of this stimulation is very much higher for polymerase δ. A high level of stimulation by clamp is appropriate for lagging-strand synthesis, as we discussed in Chapter 3. Interestingly, one of the four subunits of polymerase ε imparts a high processivity on the enzyme, and this subunit helps the polymerase to encircle the nascent (newly synthesized) stretch of duplex DNA that was just completed. Since the leading-strand polymerase must synthesize a very long stretch of DNA without dissociating, the high processivity imparted by both the CMG helicase complex and the internal processivity subunit of ε is well suited to the job on the leading strand.

In contrast, the biochemical activities of δ are well suited to its job on the lagging strand. Polymerase δ does not detectably interact with the CMG helicase and its processivity is not affected by CMG helicase. Instead, as we will see in the next section, δ has a special relationship with the clamp and clamp loader. As we discussed in the chapter on bacterial DNA replication, such a relationship allows the lagging-strand polymerase to extend Okazaki fragments effectively but desist when the polymerase reaches the 5′ end of the prior Okazaki fragment. In addition, δ can interact with α, suggesting that δ might be recruited to 3′-primer termini on the lagging strand by α. Finally, polymerase δ is capable of strand-displacement synthesis, a general feature of lagging-strand replication in eukaryotic cells. Upon displacing the 5′ end of the RNA primer of the prior Okazaki fragment, δ inserts a short stretch of DNA residues, often just a single residue, to form a flap in the DNA. Flap formation facilitates the proper maturation of Okazaki fragments (see Section 4.8 below).

4.6 The eukaryotic clamp and clamp loader

The eukaryotic clamp and clamp loader were briefly introduced from a structural/evolutionary perspective above (Section 4.2). As we will see, both proteins play key roles both within and outside of DNA replication and their names are worth remembering. The clamp is called PCNA, an abbreviation for *p*roliferating *c*ell *n*uclear *a*ntigen (identified as a nuclear antigen present during S phase in dividing cells). The clamp loader is called *r*eplication *f*actor *C*, or RFC, and was discovered as an essential factor for replicating both animal virus and host DNA.

As mentioned above, the eukaryotic RFC clamp loader is a 5-subunit complex that is homologous and structurally similar to the bacterial clamp loader (see Chapter 3; Figure 4.2A). Both protein complexes contain five subunits that belong to the AAA+ family. In the eukaryotic clamp loader, the five subunits are distinct from each other, encoded by different genes, but are nonetheless homologous (called "paralogs"; proteins within a given species that are descended from a recent common ancestor). There are actually multiple forms of clamp loader in eukaryotes, and these differ from each other in the identity of the largest of the five subunits; the other four subunits are shared. The canonical clamp loader that functions during normal DNA replication is generally called simply RFC, while the multiple alternative clamp loaders with distinct large subunits have other more specialized names.[2]

The PCNA clamp is a homotrimeric ring-shaped processivity factor, composed of six similar subdomains (see Section 4.2; Figure 4.2B). PCNA is loaded by RFC onto primer-template junctions formed by the priming action of the polymerase α/primase complex on the lagging strand. The loading pathway is believed to

---

[2]For example, ELG1-RFC has a special role in unloading PCNA; see Section 4.10 below.

be essentially identical to that of the bacterial clamp-loading system, so the details will not be reiterated here (see Section 3.3). As in the bacterial system, the PCNA clamp is loaded in the orientation in which the polymerase-interacting face is oriented towards the 3′ end of the primer. PCNA then attracts polymerase δ and travels with δ, encircling the template to ensure that the polymerase effectively extends the Okazaki fragment. In parallel to bacterial replication, the δ-PCNA interaction is lost after completion of the Okazaki fragment, allowing δ to recycle and synthesize newer Okazaki fragments.

The clamp loader in the bacterial replisome is a central part of the DNA polymerase holoenzyme complex, linking the polymerases on the leading and lagging strand and positioning the clamp loader very close to the site of primer synthesis (Chapter 3). It is currently unclear how the eukaryotic RFC complex is organized within the eukaryotic replication machinery and whether it plays such a central role. The protein does bind to the eukaryotic ssDNA-binding protein, which coats any ssDNA at the fork (see below), and this could help to localize the loader to the lagging strand. Given the precedent in the bacterial system, it seems likely that additional interactions will be uncovered that integrate the clamp loader more centrally into the eukaryotic replication machinery.

As mentioned above, one face of the PCNA clamp binds to both the clamp loader and the DNA polymerase with which it interacts. Indeed, these two binding events are mutually exclusive. This is called the C-terminal face, since the C termini of all three subunits are on this surface (while the N-termini are on the opposite face of PCNA). As will be described later in this Chapter and in other Chapters, PCNA also interacts with many other key proteins in replication, repair and damage signaling. These proteins also generally bind on the C-terminal face, often by using a common motif called a PIP (*P*CNA-*i*nteracting *p*eptide) box.

4.7 RPA — the eukaryotic ssDNA-binding protein

As in bacterial systems, ssDNA in eukaryotic cells is efficiently coated with a special ssDNA-binding protein. The eukaryotic protein is called RPA, an abbreviation of *r*eplication *p*rotein *A*, being first discovered as a protein essential in the DNA replication of an animal virus. The eukaryotic RPA protein has a different subunit structure than the bacteriophage T7 or *E. coli* ssDNA-binding protein. RPA is a heterotrimer, with three subunits that differ greatly in size (70, 32 and 14 kDa). As with the bacteriophage and bacterial proteins, the protein binds ssDNA by means of OB-folds. However, the largest subunit of RPA has four OB-folds, while each of the two smaller subunits has one.

Again in parallel with the bacterial protein, the RPA protein binds to multiple proteins involved in nucleic-acid metabolism and can drastically alter their activities. As will be evident in later Chapters, RPA plays important roles in various DNA-repair pathways, DNA recombination, and DNA-damage responses. Indeed, a recent review article tabulated more than 35 cellular proteins that have been shown to interact with RPA! These proteins bind to at least three different protein-binding surfaces on the RPA protein. The activities and binding partners of RPA are modulated in complex ways by covalent modifications of RPA protein, including phosphorylation and other modifications.

With regard to DNA replication, RPA enhances the activities of eukaryotic replicative DNA polymerases, increasing both processivity and fidelity. It should be noted that the interactions of ssDNA-binding proteins with replication partners like DNA polymerases are specific — the prokaryotic ssDNA-binding protein enhances the activity of only prokaryotic polymerases while RPA enhances only the eukaryotic polymerases. Given the extensive stretches of ssDNA on the lagging-strand template, RPA would be expected to play a key role in Okazaki fragment synthesis and the dynamics of the lagging strand. In this regard, RPA interacts with both polymerase α/primase and with the clamp loader RFC, reflecting a key role in the switch from primer synthesis to extension of Okazaki fragments.

4.8 Coordination of the fork and the trombone model

It is widely assumed that the lagging strand in the eukaryotic replication fork is looped around to allow replication to proceed in the trombone-like manner described in Chapter 3 for bacterial replication.

Direct evidence for looping of the eukaryotic replication fork is eagerly awaited, as defined biochemical systems of eukaryotic replication have developed rapidly over the last several years.

Looping of the eukaryotic lagging strand is supported by the finding that the CMG helicase complex and one of its associated accessory factors is at the center of a web of protein-protein interactions between leading- and lagging-strand replication (also see Figure 4.4). As mentioned above, the accessory factor CTF4 links the CMG complex with polymerase α on the lagging strand, and the human version of this protein also binds to polymerase δ on the lagging strand. On the leading strand, this accessory factor binds to the CMG helicase complex and to polymerase ε (CMG also binds to ε through yet another intermediary accessory protein). Since CTF4 is a homotrimer, it is easy to imagine this protein physically linking all three polymerases within the functional replication fork. Yeast mutants that are deficient in the CTF4 accessory factor are viable but show heightened levels of genome instability, consistent with a poorly orchestrated replication fork.

In bacteria, the clamp loader and clamp play a central role in organizing the trombone action of lagging-strand replication (see Chapter 3). Given the key role of reiterative clamp loading in Okazaki fragment synthesis, it seems likely that eukaryotic RFC and PCNA are also central in the dynamics of lagging-strand looping, but the details remain to be elucidated.

4.9 Continuing DNA replication past template lesions that block replicative polymerases

Blockage of DNA polymerase due to lagging-strand template DNA damage does not necessarily halt the replication fork, since a new Okazaki fragment can be initiated downstream (see Section 3.4). However, blockage of the leading-strand DNA polymerase is potentially more problematic. In bacterial cells, leading-strand DNA synthesis can resume downstream of a blocking lesion by means of a new priming event catalyzed by the bacterial primase (see Section 3.7). To date, there is no direct evidence that the primase activity of eukaryotic polymerase α plays this role. On the other hand, recent studies have uncovered a remarkable enzyme that is now strongly implicated in re-priming leading-strand synthesis in many eukaryotes. This enzyme is called PrimPol because it has both primase and DNA polymerase activity, dependent on a single active site (unlike the dual active sites in different subunits of the polymerase α complex).

PrimPol is a very unusual DNA polymerase in multiple regards. First, it is the only known DNA polymerase that does not require a pre-existing primer — PrimPol initiates synthesis by linking two nucleoside triphosphates and then polymerizes additional residues onto this dinucleotide by chain extension. This chain-extension only proceeds for a few residues, limiting the role of PrimPol to providing a short primer for other DNA polymerases. Second, PrimPol has a strong preference for dNTPs over rNTPs, and requires template T residues for its priming activity. Third, PrimPol does not share the right-hand-like architecture of all other DNA polymerases. Instead, the enzyme is evolutionarily related to a family of primases from certain eukaryotic viruses. Fourth, PrimPol is present in many eukaryotes, including humans, but absent from some others, including many fungal species and Drosophila.[3]

Current evidence argues that the major role of PrimPol is to reprime leading-strand synthesis after polymerase ε is blocked by template lesions or other impediments. Later in the book, we will consider "translesion" DNA polymerases, which provide an important pathway for replicating through lesions that block replicative polymerases (see Chapter 12). In that context, scientists have much to learn about how PrimPol is coordinated with translesion DNA polymerases, how PrimPol is regulated to prime DNA synthesis only in appropriate circumstances, and which kind of blocking lesions trigger PrimPol priming activity.

4.10 Events after the fork passes

Eukaryotic cells must accomplish the same tasks as prokaryotic cells immediately after DNA replication, including the repair and sealing

---

[3] Perhaps the priming activity of polymerase α substitutes for PrimPol in these species.

of Okazaki fragments on the lagging strand, correcting replicative errors, and recycling clamps from the DNA. We will discuss the process of mismatch repair, which corrects replicative errors, in Chapter 6, and focus on the other two processes here.

The completion of the lagging strand in eukaryotic cells is similar in broad outline to that in prokaryotic cells, but with significant differences in detail. We have already encountered two significant differences in lagging-strand synthesis in eukaryotes, namely the use of a combined RNA-DNA primer synthesized by the polymerase α/primase complex and the generation of short flaps by polymerase δ as it completes the synthesis of each Okazaki fragment. Both of these features of eukaryotic replication are reflected in the processing pathway that completes the lagging strand.

The removal of primers on the eukaryotic lagging strand involves an iterative process with the sequential actions of a nuclease and a DNA polymerase (Figure 4.5). The major nuclease involved in this process is a so-called *flap endo*nuclease (called FEN1), which recognizes the flaps left after completion of Okazaki fragment synthesis (see above) and cleaves the duplex region one-nucleotide residue ahead of the branch point of the flap (Figure 4.5A). Since the flap is often itself one nucleotide residue long, FEN1 often releases 2-nucleotide long products. Polymerase δ then extends the 3′ end of the prior Okazaki fragment until a new flap is formed, again generally a short flap (often a single nucleotide residue). The process repeats itself until all of the RNA primer is removed as well as the beginning segment of DNA at the 5′ end of the Okazaki fragment, the segment that was synthesized by the DNA-polymerase activity of polymerase α (Figure 4.5A). This removes a large majority of the DNA that was synthesized by polymerase α, a DNA polymerase that lacks proofreading exonuclease and therefore makes more frequent mistakes, and replaces it with DNA that is carefully proofread by polymerase δ. The final step in Okazaki-fragment processing is the ligation of the final nick, and presumably this occurs immediately after polymerase δ extends up to the 5′ end of the preceding Okazaki fragment without generating a short flap (Figure 4.5A).

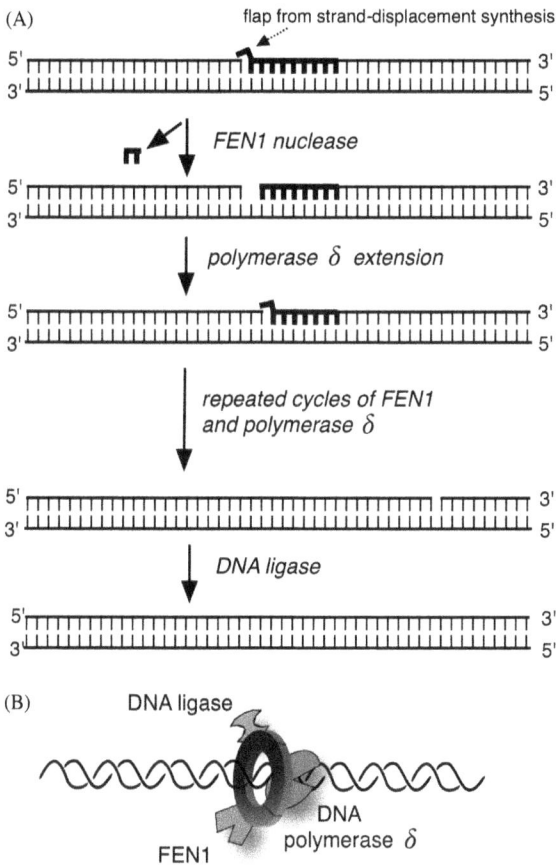

Figure 4.5. Okazaki fragment processing on the lagging strand. Repeated cycles of cleavage by FEN1 and extension by polymerase δ replaces the ribonucleotide residues from the RNA primer, allowing final ligation (panel A). A schematic of the clamp-toolbelt model is shown in panel B, with the three major functions in Okazaki fragment processing bound to the clamp toolbelt.

How are these repeated cycles of cleavage and polymerization, followed in the end by ligation, coordinated? A key feature of this coordination is that the three involved proteins all bind to the clamp PCNA (Figure 4.5B). Recall that a clamp is left behind at the 3′ end of each Okazaki fragment as the fork moves forward — this places the clamp in the perfect position to coordinate the completion and sealing of Okazaki fragments. Furthermore, since PCNA is a trimer, it is very possible that FEN1, polymerase δ and DNA ligase simultaneously bind to form a processing complex with PCNA at the center. Again, as we saw in the previous Chapter, the replication clamp is acting like a toolbelt to coordinate various activities during DNA replication. It should be mentioned that there is a back-up pathway involving a second flap endonuclease, which is important particularly when longer flaps are generated on the lagging strand.

After the primer regions of Okazaki fragments are replaced and the discontinuities are sealed, the clamps encircling the lagging strand must still be removed. As in bacteria, eukaryotic clamp loaders can also unload clamps, although this unloading process is not as well understood as loading. One particular question of interest is how clamp unloading is regulated to occur only after the final processing of Okazaki fragments generates an intact duplex lagging strand. Alternative clamp loaders were introduced above, and one called ELG1-RFC appears to be a specialized RFC involved in unloading clamps after DNA replication. Thus, when the amount of ELG1-RFC is artificially reduced (in yeast or mammalian cells), the PCNA clamp over-accumulates on chromatin, while overexpression of ELG1-RFC reduces the amount of PCNA on chromatin. This alternative clamp loader has also been implicated as having a special role in DNA damage responses.

4.11 Summary of key points

- Eukaryotic genomes contain much more DNA than those of prokaryotes, generally spread between different chromosomes, and yet the replisome itself travels some 20-times slower than a prokaryotic replisome.
- Eukaryotic chromosomes are replicated from multiple replication origins.
- The DNA end-replication problem can be solved by circular DNA, by special proteins attached to the DNA ends, or, as in nearly all eukaryotic chromosomes, by specialized telomeres at the chromosomal ends.

- A subset of replication proteins shows sequence and/or structural conservation between prokaryotes and eukaryotes.
- The eukaryotic replicative helicase (CMG complex) consists of a hetero-hexameric protein called MCM, along with a 4-protein complex called GINS and a protein called CDC45.
- The CMG complex travels along the leading-strand template.
- Eukaryotic polymerase α/primase is a 4-subunit enzyme that provides Okazaki-fragment primers consisting of both RNA and DNA residues.
- Eukaryotic replication employs two different polymerases for extension on the leading (ε) and lagging (δ) strands.
- Eukaryotic clamp loader is a 5-protein complex called RFC (*rep*lication *f*actor *C*), while the clamp is a homotrimeric protein called PCNA (*p*roliferating *c*ell *n*uclear *a*ntigen).
- In addition to the replicative clamp loader, eukaryotic cells have alternative clamp loader complexes, which serve other functions such as removing clamps after DNA replication/Okazaki fragment processing.
- The PCNA clamp is critical for polymerase processivity on the lagging strand and plays other key roles in DNA metabolism described in later chapters.
- The ssDNA-binding protein of eukaryotes (RPA) is a 3-subunit protein with multiple OB-folds for binding DNA.
- The network of interactions that organize the eukaryotic replisome are currently under intensive investigation.
- Blocked leading-strand synthesis in many eukaryotes can be reprimed by the PrimPol enzyme.
- Okazaki-fragment processing in eukaryotes involves a special flap endonuclease (FEN1), extension by polymerase δ, and nick sealing by DNA ligase, all coordinated by the PCNA clamp.

Further Reading

Balakrishnan, L., & Bambara, R. A. (2013). Okazaki fragment metabolism. *Cold Spring Harb Perspect Biol, 5*(2), a010173.

Bell, S. D., & Botchan, M. R. (2013). The minichromosome maintenance replicative helicase. *Cold Spring Harb Perspect Biol, 5*(11), a012807.

Bell, S. P., & Labib, K. (2016). Chromosome duplication in *Saccharomyces cerevisiae*. *Genetics, 203(3)*, 1027–1067.

Burgers, P. M. (2009). Polymerase dynamics at the eukaryotic DNA replication fork. *J Biol Chem, 284*(7), 4041–4045.

Burgers, P. M. J., & Kunkel, T. A. (2017). Eukaryotic DNA replication fork. *Annu Rev Biochem, 86*, 417–438.

Georgescu, R., Yuan, Z., Bai, L., de Luna Almeida Santos, R., Sun, J., Zhang, D., . . . O'Donnell, M. E. (2017). Structure of eukaryotic CMG helicase at a replication fork and implications to replisome architecture and origin initiation. *Proc Natl Acad Sci USA, 114*(5), E697–E706.

Guilliam, T. A., & Doherty, A. J. (2017). PrimPol — Time to Reprime. *Genes (Basel) 8*, 20, doi:10.3390/genes8010020.

Hedglin, M., Kumar, R., & Benkovic, S. J. (2013). Replication clamps and clamp loaders. *Cold Spring Harb Perspect Biol, 5*(4), a010165.

Langston, L. D., Zhang, D., Yurieva, O., Georgescu, R. E., Finkelstein, J., Yao, N. Y., . . . O'Donnell, M. E. (2014). CMG helicase and DNA polymerase epsilon form a functional 15-subunit holoenzyme for eukaryotic leading-strand DNA replication. *Proc Natl Acad Sci USA, 111*(43), 15390–15395.

Leman, A. R., & Noguchi, E. (2013). The replication fork: Understanding the eukaryotic replication machinery and the challenges to genome duplication. *Genes (Basel), 4*(1), 1–32.

Nick McElhinny, S. A., Gordenin, D. A., Stith, C. M., Burgers, P. M., & Kunkel, T. A. (2008). Division of labor at the eukaryotic replication fork. *Mol Cell, 30*(2), 137–144.

Pursell, Z. F., Isoz, I., Lundstrom, E. B., Johansson, E., & Kunkel, T. A. (2007). Yeast DNA polymerase epsilon participates in leading-strand DNA replication. *Science, 317*(5834), 127–130.

Shiomi, Y., & Nishitani, H. (2017). Control of genome integrity by RFC complexes; conductors of PCNA loading onto and unloading from chromatin during DNA replication. *Genes (Basel), 8*(2), doi:10.3390/gcncs8020052.

How did they test that?
Which DNA polymerase synthesizes the leading strand in yeast?

A long-standing controversy involved which DNA polymerase synthesizes which strand (leading or lagging). Pursell *et al.* (2007) provided strong evidence that polymerase ε is involved in leading-strand synthesis in yeast. A related study (Nick McElhinny *et al.*, 2008) provided evidence that polymerase δ is involved in lagging-strand synthesis, together arguing that ε is the major leading-strand polymerase, while δ is the major lagging-strand polymerase. The scientists created a mutant polymerase ε with amino acid substitution M644G, which reduces discrimination against mispaired incoming nucleotides. The mutant enzyme is worse at discriminating certain mispairs than others, for example, it misincorporates T opposite template T about 40-times more often than it misincorporates A opposite template A (tested in vitro). Both of these events create a T to A substitution on one strand (with corresponding A to T on the other strand; see diagram at the top). With the mutant polymerase, a large majority of these mutations are inferred to arise when dTMP mispairs with template T (recall the 40-fold preference), thus revealing the template strand used by mutant ε. The six panels show a stretch of yeast DNA with two sites (positions 686 [**] and 279 [*]) that frequently suffer mutations in strains with the mutant polymerase. In the different panels, this stretch of DNA is located in different orientations with regard to the strong replication origin (black box) called ARS306 or ARS501 (created by engineering the yeast strains before the experiment). These constructs position the two key template T residues on either the leading-strand template (panels A, C, and E) or lagging-strand template (panels B, D, and F). The collection of T's just above or below the DNA sequence reflect the number of mutations detected, with the mutation rates indicated in the tables just below. With the mutant polymerase, nearly all of the mutations at the two hotspots were detected when the T residue was in the leading-strand template, implying that polymerase ε replicates that strand in vivo. The data figure at the bottom was reproduced from Pursell *et al.* (2007), with permission from the American Association for the Advancement of Science; permission conveyed by Copyright Clearance Center, Inc.

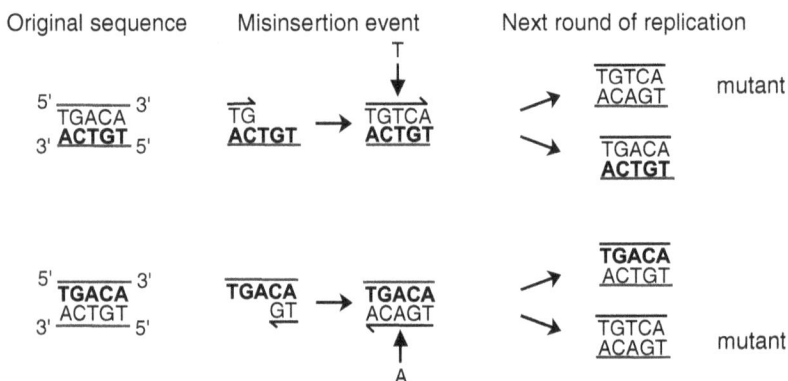

Original sequence   Misinsertion event   Next round of replication

T

5' $\overline{\text{TGACA}}$ 3' $\overrightarrow{\text{TG}}$ $\overline{\text{TGTCA}}$ $\overline{\text{TGTCA}}$ mutant
3' **ACTGT** 5'   **ACTGT** → **ACTGT** $\overline{\text{ACAGT}}$

                                            $\overline{\text{TGACA}}$
                                            ACTGT

5' $\overline{\text{TGACA}}$ 3'   **TGACA** → **TGACA**   **TGACA**
3' $\overline{\text{ACTGT}}$ 5' $\underleftarrow{\text{GT}}$ $\underleftarrow{\text{ACAGT}}$ $\overline{\text{ACTGT}}$

                                            A $\overline{\text{TGTCA}}$
                                            $\overline{\text{ACAGT}}$ mutant

A R-ARS306-OR1			C L-ARS306-OR2			E R-ARS501-OR1		

...AGAAAATTTGC...ATCTGACATTA... (ARS306) / ...TCTTTTAAACG...TAGACTGTAAT...

...TAATGTCAGAT...GCAAATTTTCT... (ARS306) / ...ATTACAGTCTA...CGTTTAAAAGA...

...AGAAAATTTGC...ATCTGACATTA... (ARS501) / ...TCTTTTAAACG...TAGACTGTAAT...

Mutation Rate (x10⁻⁸)	A686**	A279*	Mutation Rate (x10⁻⁸)	A686**	A279*	Mutation Rate (x10⁻⁸)	A686**	A279*
wt	≤0.6	0.6	wt	n.d.	n.d.	wt	≤0.2	≤0.2
pol2-M644G	13	2.2	pol2-M644G	20	5.6	pol2-M644G	14	1.5

B R-ARS306-OR2			D L-ARS306-OR1			F R-ARS501-OR2		

...TAATGTCAGAT...GCAAATTTTCT (ARS306) / ...ATTACAGTCTA...CGTTTAAAAGA

...AGAAAATTTGC...ATCTGACATTA... (ARS306) / ...TCTTTTAAACG...TAGACTGTAAT...

...TAATGTCAGAT...GCAAATTTTCT (ARS501) / ...ATTACAGTCTA...CGTTTAAAAGA

Mutation Rate (x10⁻⁸)	A686**	A279*	Mutation Rate (x10⁻⁸)	A686**	A279*	Mutation Rate (x10⁻⁸)	A686**	A279*
wt	≤0.7	≤0.7	wt	n.d.	n.d.	wt	≤0.1	≤0.1
pol2-M644G	≤0.4	≤0.4	pol2-M644G	≤0.8	≤0.8	pol2-M644G	1.3	≤0.7

Abbreviations:

wt, wild-type; pol2, gene for DNA polymerase ε; nd, not detected

Chapter 5

Replication dynamics — initiating, regulating and terminating cellular DNA replication

The process of DNA replication must be very carefully controlled so that each daughter cell ends up with the same DNA content as its parent cell. In eukaryotic cells with multiple chromosomes, failure in this careful control can result in missing chromosomes or extra copies of chromosomes, a situation called aneuploidy. Depending on the organism and cell type, aneuploid cells can be inviable or severely abnormal, and in higher organisms such as humans, aneuploidy is associated with cancer. As we will see below, the careful control of DNA replication is exerted at the point of initiating DNA replication, and this initiation of replication is tightly coupled to the overall cell-division cycle.

The completion or termination of DNA replication is also a critical process required for normal cell division and growth. In bacterial cells, the two replication forks traveling in opposite directions around the circular chromosome meet in a special region called the terminus, where the colliding complexes need to replicate all intervening DNA completely and disassemble so they don't continue replicating past each other. In eukaryotic cells, there is no special terminus region and opposing replication forks meet at many locations throughout the chromosomes, but again, the replication must be terminated in a careful and accurate manner.

5.1 Defining the bacterial replication origin

The bacterial *ori*gin of chromosomal replication, called *oriC,* was first defined in the model system *E. coli.* Early evidence for a single origin of replication arose in the 1960s from clever genetic studies that essentially measured the copy number of various genes around the circular chromosome. The key to this study was that the copy numbers of these various genes were compared in *E. coli* cultures that were growing rapidly versus non-growing cells of the same strain. Genes in one region of the circular chromosome were found to be present in several-fold higher copy numbers in growing relative to resting cells, whereas the region opposite showed the lowest ratio of copy numbers in growing versus resting cells. Modern molecular techniques that directly measure the number of copies of DNA segments around the chromosome reproduced this gradient of gene copy number in spectacular detail. In this case, the copy numbers of genes around the chromosome were uniform in the resting cells (Figure 5.1A). However, in growing cells, gene copy numbers peaked at one location of the chromosome, were minimum at the opposite side of the chromosome, and showed a more or less continuous decline in both directions going from peak to trough (Figure 5.1B). When nutrients were withdrawn from the growing cells to inhibit further growth, the profile gradually returned to that of the resting cells (as replication gradually ceased). The inference is that the peak region in growing cells contains the replication origin, and that replication proceeds in both directions from the origin toward the opposite end of the circular chromosome.

Why, in growing cells, are genes near the origin present at higher copy numbers than genes near the terminus? At any one moment, the culture contains cells at different stages in their replication; in some, the replication forks have just left the origin, in others, they have traveled halfway to the terminus, and in yet others, the forks are nearly finished replicating the chromosome (Figure 5.1C and D).

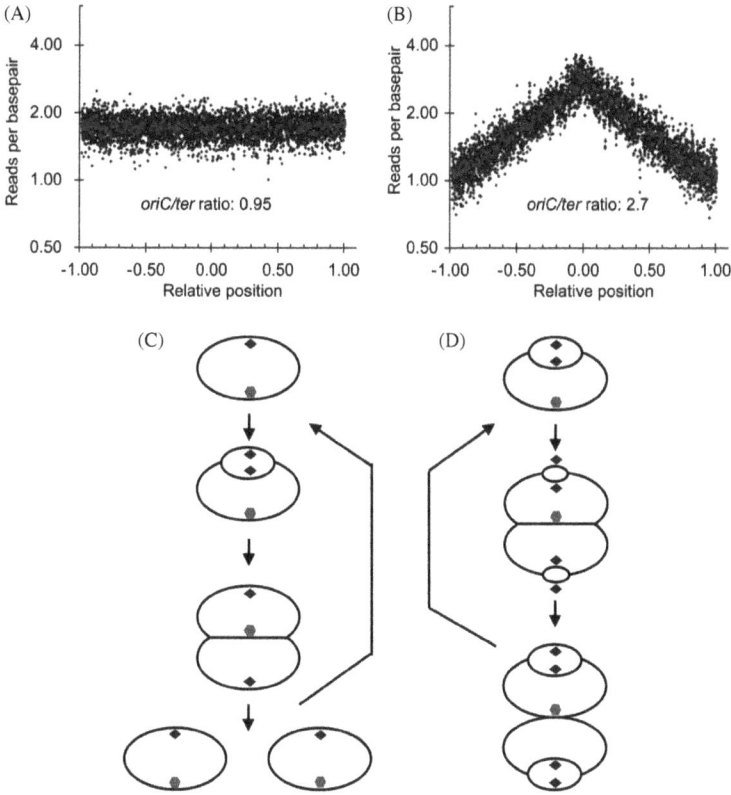

Figure 5.1. Bi-directional replication of the *E. coli* chromosome. The relative abundance of numerous positions on the *E. coli* chromosome was measured by tallying the number of DNA sequence reads in particular DNA samples (panels A and B; see Chapter 15 for discussion of DNA sequencing technologies). The DNA analyzed in panel A was from resting cells (no replication occurring) whereas the DNA analyzed in panel B was from growing cells undergoing replication. The average overall ratio of sequences close to the *oriC* origin is nearly three-times higher than that of sequences close to the terminus in the replicating sample (but close to unitary in the resting cells). The overabundance of DNA from the origin region is explained by the bidirectional replication, as depicted in panels C and D. Cells growing with a doubling time of about 40 minutes or more have a single round of replication at any one time (panel C), while cells growing faster utilize multiple simultaneous rounds of replication (panel D). The diagrams depict a few particular stages in each replication cycle, crudely mimicking the mix of cells at different stages of replication in a growing culture. Counting up the number of copies of origin-proximal DNA (rectangles) and terminus-proximal DNA (hexagons) in these diagrams gives ratios of 7/5 (= 1.4) for panel C and 10/3 (3.33) for panel D. Panels A and B are reproduced from Skovgaard *et al.* (2011), with permission from Cold Spring Harbor Laboratory Press.

The key is that, considering individual cells, genes behind an active replication fork will always be present at twice the copy number than those in front of the fork. When the DNA from all these cells is pooled, the average copy number is highest at the origin and lowest in the terminus region. As already mentioned in Chapter 3, rapidly growing bacterial cells can have two or even three rounds of replication going on at one time, and so the average copy number of genes near the origin can be more than four times as high as those near the terminus. Not surprisingly, genes that are highly transcribed and whose products are needed at high levels in rapidly growing cells tend to be located near the origin, where their copy number will, in a sense, be amplified when the cells are growing rapidly. The most notable examples are the genes that encode ribosomal RNAs.

A second and complementary approach to defining the bacterial replication origin involved testing segments of the chromosome for their ability to support the replication of a bacterial plasmid. Bacterial plasmids themselves have their own replication origins, and if this segment of DNA is cut out, the plasmid cannot propagate when transformed into cells. Plasmid transformation is generally tested with a drug-resistance marker on the plasmid — transformed bacteria become resistant to the drug only if the transformed DNA can replicate in the cells. Scientists ligated fragments of the chromosome to the drug-resistance gene in this kind of experiment and found that one particular segment allowed the plasmids to replicate autonomously. This segment turned out to be the bacterial replication origin. Further experiments showed that the origin DNA segment could be trimmed down to about 250 base pairs without destroying its ability to replicate the plasmid.

DNA sequencing and functional analyses showed that the 250-base pair *oriC* contains specific binding sites for an initiation protein as well as a segment of DNA that is easily unwound, and the function of these will be described in the next section. As the genomes of well over 1000 bacterial species have now been determined, it is clear that the *E. coli* model is typical. These origins invariably contain multiple binding sites for the initiator protein and an easily unwound segment of DNA, although the exact arrangements of these elements and the overall origin size show significant variability between species.

5.2 Initiating DNA replication in bacteria

In all cells, loading of the replicative helicase is a central step in the initiation of DNA replication and is carried out with the help of a special helicase-loading protein. Loading proteins like this are also called "chaperones", because they escort their partner protein to the correct target site. In the case of the bacterial helicase chaperone protein, this target site is the singular replication origin, *oriC*, described above.

The bacterial helicase chaperone protein, like the helicase itself, is a hexamer, and so the chaperone:helicase complex has 12 total subunits arranged as a double hexamer (Figure 5.2). As discussed in Chapter 3, one strand of DNA passes through the central hole of the bacterial replicative helicase as it travels along unwinding the two strands from each other. This central hole is plugged and unavailable in the chaperone:helicase complex, and so the complex is blocked from any unwinding activity.

While the chaperone has the key role of delivering the helicase to *oriC*, the chaperone:helicase complex cannot find *oriC* without the help of another key protein (Figure 5.2). This is the so-called initiator protein, a site-specific binding protein that recognizes the replication origin. In bacteria, the initiator protein binds multiple sites within the *oriC* DNA sequence in a complex and regulated manner. The initiator protein, an ATPase, has very different properties depending on whether it is bound to ATP or ADP. The ADP-bound form can bind some of the sites in *oriC*, and indeed the initiator protein is bound to *oriC* even when initiation is not imminent. However, to trigger DNA replication, *oriC* must be bound by the activated ATP-bound form of the initiator protein. Roughly 10 or more active monomers of initiator protein are found in the activated origin, and as we will see below, the ATP form of the initiator protein binds to DNA sites that are excluded from the ADP form. The key step of loading the replicative helicase can only occur when

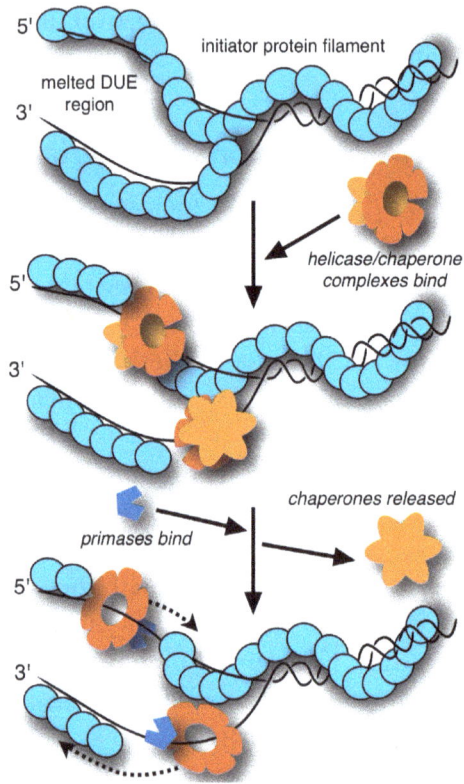

Figure 5.2.   Assembly of the helicase/primase complex onto the *oriC* replication origin. A filament of the initiator protein (blue ovals) assembles on the *oriC* DNA, leading to unwinding of the DUE region of the origin. The complex of replicative helicase and its chaperone (orange/yellow hexamer-shaped figures) is attracted to the unwound DNA/initiator complex, and the chaperone is released as the helicase is loaded around each of the two single strands of origin DNA. As these steps occur, the primase (blue "pac-man") associates with the helicase to form the functional helicase/primase complex. The dotted arrows show the (opposite) directions of travel of the two loaded helicase/primase complexes. This diagram was modified from Mott *et al.* (2008).

oriC is fully bound by these multiple monomers of ATP-bound initiator protein. Several additional proteins can also bind the protein:DNA complex at the origin; these proteins are not strictly required for initiation but they modulate the efficiency and timing of replication initiation and thus participate in the regulation of replication initiation in the cell cycle.

There are several challenges in assembling the helicase onto DNA at the replication origin. The central hole of the helicase must be unplugged from the blockage by the chaperone, the duplex DNA must be locally unwound to provide a single-stranded region to pass through the helicase, and that single-stranded region must be somehow slid into the central hole of the hexameric helicase. What are the requirements for this challenging assembly process?

First, the origin DNA must be properly prepared for the chaperone:helicase complex by the initiator protein. One segment of *oriC* DNA is called a "*D*NA-*u*nwinding *e*lement" or DUE, and is located immediately adjacent to sites where the initiator protein binds regardless of whether it is in the ATP or ADP-bound form. The DUE segment is relatively rich in AT base pairs, which makes it easier to unwind locally into single-stranded segments (the *E. coli* DUE region is about 78% AT while the average of the *E. coli* chromosomal overall is only 49.2%). Two factors promote a localized unwinding of the DUE within the origin. One is the complex of ATP-bound initiator protein within the origin complex — indeed, there is evidence that only the ATP-bound form of the initiator is able to bind to sites in the single-stranded DUE after it is unwound. The second factor that facilitates unwinding of the DUE is a sufficient level of negative supercoiling in the DNA; we will discuss why negative supercoiling favors DNA unwinding in Chapter 7. Thus, an activated *oriC* complex consists of supercoiled DNA with the complex of about 10 to 20 monomers of the initiator protein, mostly or all bound to ATP, with unwound DNA in the DUE segment.

A second requirement for helicase loading is a direct interaction between the helicase and the initiator protein. When the hexameric helicase within the chaperone:helicase complex binds to the initiator protein in a properly assembled origin complex, its binding affinity for the chaperone is lost as the chaperone hydrolyzes its bound ATP into ADP (Figure 5.2). The ADP-bound form of the chaperone in turn has lost its affinity for the helicase and has completed its key role in the initiation process as it departs the scene. There is also evidence that the helicase chaperone can bind directly to the initiator protein, although the importance of this interaction needs to be further investigated.

The third requirement for helicase loading is that the hexameric ring of the helicase must be temporarily cracked to allow a single strand of the DUE to pass through into the central cavity. This step occurs as the helicase is loaded by the chaperone, and the chaperone appears to play an active role in cracking open the helicase hexamer to allow the ssDNA to enter the interior of the helicase hexamer. Once loaded, the proper hexameric ring configuration is re-established so that the helicase cannot dissociate from the DNA. This ring-cracking pathway should sound familiar — it is very similar, at least in outline, to the pathway of clamp loading discussed in Chapter 3. As you recall, clamp loading also involves ATP hydrolysis by the loader protein, reflecting a general role of ATP hydrolysis in complex assembly and protein rearrangement processes.

A fourth protein is likely also involved in the loading pathway and leads directly to subsequent steps in DNA replication. The primase protein has been shown to bind to the chaperone:helicase complex and stimulate the disassembly of the chaperone from the helicase. Presumably, this primase-assembly step is carefully regulated such that the chaperone:helicase disassembly only occurs as the helicase is loaded onto the ssDNA at the DUE. This provides an orderly pathway of assembly in which the primase is now ready to synthesize the first RNA primer to initiate leading-strand synthesis.

As we discussed in Chapter 3, the primase remains associated with the helicase as replication forks progress around the chromosome. As was also discussed, one subunit of the clamp loader (which is part of the DNA polymerase holoenzyme complex) has an affinity for the replicative helicase, and this interaction brings the polymerase holoenzyme complex into the growing replication complex at *oriC*. Recall that the clamp loader also has affinity for the 3′ end of RNA primers on DNA (primer-template junction), and thus once assembled into the growing complex at the origin, a clamp is naturally delivered to the site where actual DNA synthesis begins. Since the clamp loader is part of the DNA polymerase holoenzyme, the polymerase core is also ready to adopt the primer-clamp complex and begin synthesis.

The above description leaves out an important aspect of the initiation process, namely that two helicase complexes must be loaded at *oriC* in order to achieve bidirectional synthesis. One complete hexamer of the helicase is required for each of the two replication forks, and these two hexamers must be loaded on opposite strands of the unwound DUE region (Figure 5.2). The architecture of the protein-DNA complex, with its multiple monomers of initiator protein and unwound DNA at the DUE, presumably contributes to proper loading of these two helicase complexes each in the correct orientation on opposite strands. Details of this interesting aspect of the assembly process remain to be investigated.

While the above description fits *E. coli* and many other bacterial species, it is worth noting that not all bacterial species follow this paradigm for helicase loading at the origin. Some bacterial species have additional loading proteins while others appear to load the helicase with no loading protein other than the initiator protein at the origin. These systems have not been studied in as much detail as the very well-studied *E. coli* paradigm.

5.3 Regulation of origin firing in bacteria

The assembly pathway discussed above already introduced aspects of the regulation of origin firing. As mentioned, accumulation of the ATP-loaded version of the initiator protein at the origin is important for initiation to occur. Conversely, ATP hydrolysis occurs during the initiation process leading to the ADP form of the protein, which is inactive for replication initiation but which can remain bound to the origin. Multiple and complex factors govern the balance of the ATP and ADP bound forms of the initiator protein, and these factors differ significantly in different bacterial species. The key point, however, is that once ATP has been hydrolyzed by the initiator protein, another round of replication is prevented until the system recharges the initiator with ATP for the next round of replication.

Another interesting layer of regulation in *E. coli* requires a special DNA modification that occurs after DNA synthesis. *E. coli* encodes a deoxyadenosine methylase that specifically methylates the

N6 position of the adenine residue in the sequence 5′-GATC-3′ within duplex DNA. GATC sites are spread throughout the chromosome, but are especially concentrated very close to the replication origin. In duplex DNA, every GATC site is base paired with the same sequence, since the reverse complement of GATC is also GATC (5′-GATC-3′/3′-CTAG-5′). Most GATC sites in *E. coli* at any given time have the A in both strands methylated (call it A*). However, when DNA replication passes through this site, each of the two daughter duplexes will temporarily have only one of the strands methylated (Figure 5.3A). One duplex will have the sequence 5′-GA*TC-3′/3′-CTAG-5′ while the other will be 5′-GATC-3′/3′-CTA*G-5′ — these are

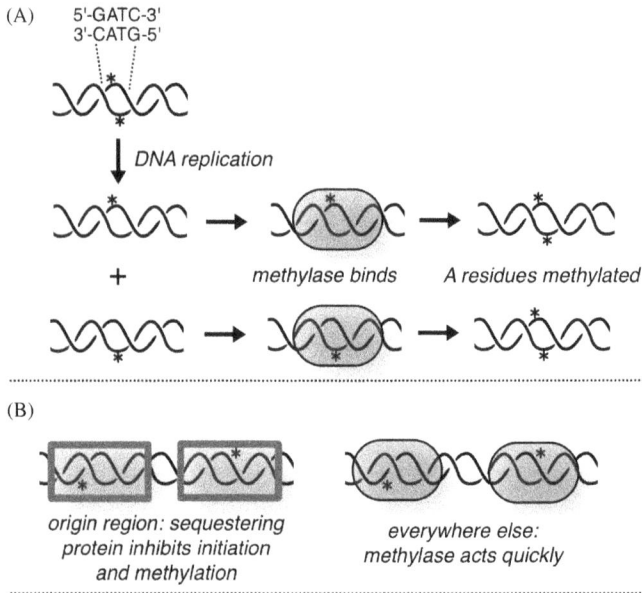

Figure 5.3. Methylation of A residues in GATC sequences of *E. coli* DNA. GATC sites in the *E. coli* chromosome are methylated on the A residue of both strands during most of the cell cycle. Immediately after DNA replication, only the parental strand of each daughter duplex is methylated (panel A). The DNA adenine methylase enzyme recognizes these "hemi-methylated" duplexes and transfers a methyl group to the unmethylated strand to restore the site to fully methylated status. In the *oriC* region, a particular sequestering protein generally binds to the hemi-methylated sites before the methylase has a chance to act, thereby delaying both replication initiation and methylation (panel B).

each called "hemi-methylated" sites. The methylase enzyme binds to hemi-methylated sites and then methylates the unmodified A residue to restore fully methylated DNA, but this takes some time after the replication fork passes.

Getting back to the regulation of bacterial replication, *E. coli* also has a special regulatory protein that competes with the methylase enzyme and binds to hemi-methylated GATC sites near the replication origin (Figure 5.3B). This binding actually sequesters the origin DNA into an inactive form, inhibiting the binding of the initiator protein and thus preventing replication initiation. Over time, the methylase eventually succeeds in methylating the hemi-methylated sites to release the sequestering protein and prepare for the next round of replication. *E. coli* carrying mutations that inactivate the sequestering protein initiate replication too quickly and frequently, demonstrating the importance of this process in careful regulation of initiation within the cell cycle. Methylation of GATC sites also plays a key role in the pathway of mismatch repair, which reduces the frequency of replication errors, as we will see in Chapter 6.

5.4 Termination of replication in *E. coli*

The two replisome complexes assembled at the origin each travel in opposite directions around the circular chromosome until they meet in a region on the other side of the circle. Not surprisingly, this region is called the terminus and has evolved to play an active role in the termination process and subsequent events in cell division.

One key feature of the terminus region is that the two borders of the region contain special sites, called *Ter*, which actively block replication forks (Figure 5.4). A specific termination protein binds to the 23-base pair sequence within these sites, and it is this termination protein that actually blocks the helicase within the replisome. Not only does the termination protein block helicase, it actively promotes the dissociation of the replicative helicase from the stalled replisome. Amazingly, this termination protein acts in a directional fashion, allowing replisomes to pass freely in one direction but blocking replisomes from the other direction. The sites are each

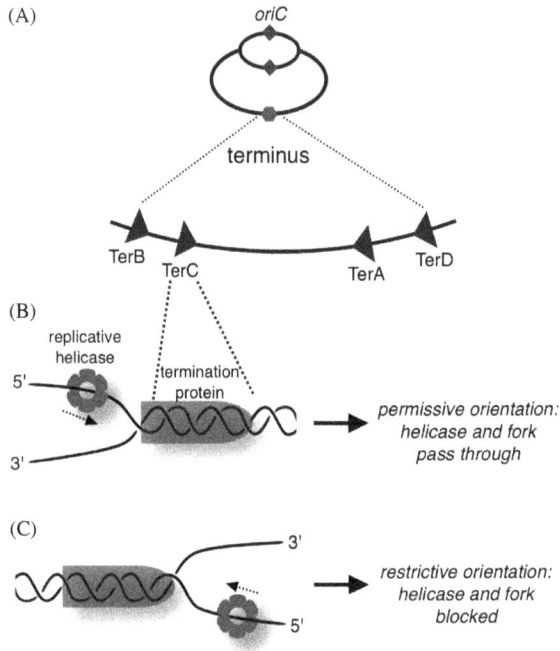

Figure 5.4.   Replication fork trap in the terminus region of the *E. coli* chromosome. The terminus region has four major Ter sites, oriented such that replication forks are allowed into the region but cannot exit from the other side (panel A). At each Ter site, the termination protein binds and functions to block replicative helicase in a direction-specific manner (panels B and C).

oriented such that replisomes can enter the terminus region but cannot exit at the other side. The terminus region contains four such sites, two near each border (Figure 5.4). Both sites near each border are arranged to allow forks to enter the central part of the terminus region, and the presence of two sites apparently provides redundancy to overcome a termination process that is not 100% effective.

The replication machinery must be disassembled when replication terminates upon the collision of two oppositely oriented forks, to prevent parts of the chromosome from being replicated twice in a given cell cycle. As mentioned above, the replicative helicase from the fork that is blocked at a Ter site is disassembled from its encounter with the termination protein. Given the extensive interactions of the helicase with the rest of the replication machinery, loss of the helicase likely cascades into disassembly of the remaining proteins of that replisome. However, the Ter site does not affect the replisome that approaches from the "permissive" side of the site, and so some other mechanism must result in disassembly of this helicase.

Because the terminus region is about 180° around the circle from the origin, you might expect that the two replication forks would meet in this region even without the termination protein mentioned above, and that the cells would still be able to complete replication. Indeed, *E. coli* mutants that have lost the termination protein are viable, so this expectation is reasonable. What then is the advantage(s) of the termination protein and oriented termination sites?

Before discussing the possible advantages, it is important to point out that the viability of a particular mutant does not prove that the system is without any importance, just that it is dispensable for survival under the tested conditions. Cells without the termination system might grow at a slightly slower rate or suffer from cell death at a relatively low frequency due to problems with replication termination or subsequent steps, or they might have particular problems under some alternative growth conditions that are more stressful than standard laboratory growth media.

The overall organization of genes in the chromosome provides one hint about the function of the terminus region. As discussed above, the replisome travels clockwise from *oriC* to the terminus in one half of the chromosome and counterclockwise in the other half. It turns out that the direction of transcription of most genes, particularly those that are highly transcribed (like those for ribosomal RNA), match the direction of replication. This means that the preferred transcriptional direction flips at or near the origin of replication, clearly indicating some kind of selective pressure that relates the direction of transcription to that of replication. Scientists have been able to show directly that replication forks are sometimes blocked when they collide with oppositely oriented transcription complexes, providing a molecular explanation for this selective pressure. These blocked replication forks can lead to fork breakage and/or genetic rearrangements (we will return to this topic in Chapter 11). Regarding the replication-termination region, restricting termination to this region minimizes these problematic collisions of oppositely oriented replisome and transcription complexes. Without the termination system, one of the two forks might pass the termination region before the opposite one arrives (in some fraction of replication cycles); this wayward fork could then collide with a transcription complex after it enters the half of the chromosome from which it was supposed to be restricted.

Additional hints about possible function of the termination system also emerged from studies of the genome, but this time from detailed genomic DNA sequence analyses. Two different systems appear to be involved in coupling the completion of replication to later steps, particularly the process by which the daughter duplexes are segregated into daughter cells. The details are quite complex, but here is a brief overview. First, the *E. coli* terminus region has many copies of a specific DNA sequence that is used by a special protein that holds the two daughter duplexes together for a period of time after replication is complete. This is a process called cohesion, and it assists in completing the cell cycle in part by allowing efficient separation of interlocked chromosomes (see discussion of separating chromosomal catenanes in Chapter 7). Second, *E. coli* has a motor protein that helps drive the terminus region of the chromosome towards the "septum", which is the physical location on the cell envelope where division occurs to create two daughter cells. One end of the protein is localized in the septum region, and by "driving" along the duplex DNA molecule with the other end, the DNA is corralled toward the septum. The trick is that the motor protein loads onto particular sites on the DNA, and these sites are directional with regard to the duplex DNA. The orientation of each site determines which direction the protein drives along the DNA. These sites are spread throughout the chromosome, but a very large majority of them point the motor protein towards the terminus region in the chromosome. At the DNA sequence level, this means that the preferred orientation of the sites is opposite on the two sides of *oriC*, and also on the opposites sides of the terminus region.

The action of this motor protein is important in coordinating chromosome segregation with cell division, and it also plays a complex role in keeping sister chromosomes from getting tangled with each other after replication is completed.

5.5 Location of replication origins in eukaryotes

Eukaryotic cells contain multiple chromosomes composed of linear rather than circular DNA. This of course means that each chromosome needs to have at least one origin of replication, but in fact replication initiates at multiple sites on every eukaryotic chromosome that has been analyzed. The problem is more complicated than that, however, because cell division is much faster during embryogenesis than later in life. For example, embryos of Drosophila (the common fruit fly) show a cell-division cycle of less than 10 minutes during the first two hours of their development. Amazingly, DNA replication occurs in less than 4 minutes during these rapid embryonic divisions. This is about ten-times faster than that of rapidly growing *E. coli*, even though Drosophila has about 40 times as much DNA. This very rapid cell division allows the embryo to reach a stage of roughly 50,000 cells in just 12 hours. In contrast, quickly growing cells in Drosophila adults take roughly 8 hours per cell division.

Drosophila is not an outlier — very rapid cell division is characteristic of early embryogenesis throughout metazoans. You should appreciate two important factors that facilitate this rapid cell division. First, the egg is loaded by the mother with large amounts of the factors required for rapid DNA replication, and so new synthesis of replication proteins is not needed till later in embryogenesis. Second, the mass of the embryo does not grow in proportion to the number of cells, rather the cells get smaller and smaller with each division. This means that the cells do not need to produce many of the other cell components at the rapid rate that they are duplicating their DNA.

How can cells from the same organism replicate so much faster in embryogenesis than later in life? To a good first approximation, this is possible because functioning origin sites are much more prevalent in the embryonic cells than in the adult cells. This already implies some flexibility and malleability in the utilization of origin sites, and this flexibility turns out to be an important characteristic of eukaryotic replication origins.

The search to define eukaryotic replication origins turned out to be a long and arduous adventure, and there are still many unanswered questions in the field. There have been too many key observations over the years to do justice to the field in a summary like this, but a few of the most prominent results are worth mentioning.

The most rapid progress in understanding eukaryotic origins came, not surprisingly, from the study of simple model systems like that of budding yeast. Putative replication origins in *S. cerevisiae* were defined in much the same way as described above for bacteria, as segments of DNA that allowed autonomous replication of a plasmid. In yeast, these were called ARS (*a*utonomous *r*eplication *s*equence) elements. All ARS elements from *S. cerevisiae* contain a common consensus sequence that turns out to be the binding site for the initiator protein (see below). They also contain a region of DNA that is easily unwound, and so on first blush, the ARS elements appear fairly similar to the *oriC* replication origins in bacteria.

In more recent studies, sophisticated molecular methods have been developed to analyze the locations where DNA replication starts in the native chromosomes of eukaryotic cells, including *S. cerevisiae*. These methods showed that many of the ARS elements defined in the plasmid assay did indeed behave as origins in the chromosome. They also showed that, to a first approximation, the same DNA sequence characteristics (such as the common consensus sequence) were required for plasmid and for chromosomal DNA replication.

While most ARS elements defined in the plasmid assay function as origins on the chromosomes of yeast, a subset did not. A common explanation for this lack of function in the chromosome was discovered — these are ARS elements that were located in regions of heterochromatin,[1] which also silences transcription. Indeed, in at least some cases, moving these ARS-element sequences away from

---

[1] Heterochromatin is a tightly packed form of chromosomal DNA that is generally less accessible to many DNA-binding proteins than euchromatin; several forms of heterochromatin have been characterized containing distinct chromosomal proteins (including histones).

the region of heterochromatin allowed them to again function as origins in the chromosome. Additional studies highlighted the importance of the exact positioning of nucleosomes for proper origin function. In summary, a combination of sequence elements and chromatin structure were defined as being important for origin function in *S. cerevisiae*.

Alas, this story turned out to be too simple when other species were analyzed. Even within other budding-yeast species, the sequence composition of origins did not match that of *S. cerevisiae*. Some showed a different consensus sequence or no obvious consensus sequence at all. Moving further from *S. cerevisiae* but still in the yeast family, the origins of some species are characterized by tracts of repeating A/T base pairs, while other species have origins with a high G/C content. Even though these various yeast species have origins that look different, they all use a conserved initiator protein and share most or all of the same proteins involved in replication initiation (see below). The DNA-binding domain of the initiator protein has been shown to be distinct in the two most highly studied yeast species (*S. cerevisiae* and *S. pombe*), explaining at least the difference in consensus sequence elements in the origins of these two species.

Given the variability in origin structure in various yeast species, it was fortuitous that *S. cerevisiae*, with its clear origin consensus sequence, was chosen as the simple model system for studying the process of eukaryotic DNA replication. For many years, scientists struggled unsuccessfully to identify and purify the eukaryotic replication-initiation protein. In a clever and powerful approach, one group decided to use the defined consensus-sequence element of the *S. cerevisiae* origins as a tool to find the initiator protein (see "How did they test that" at the end of this Chapter). They developed an assay that reveals whenever the consensus-sequence element of an ARS DNA fragment was bound specifically by a protein. They then separated an extract of yeast proteins on a chromatographic column, and measured each fraction for its ability to bind that element. They pooled the fractions from the first column that appeared to have such a binding protein, and ran successive columns of different types, as biochemists do. After each column, they found a peak containing protein(s) that binds to the consensus element. After extensive purification, the protein that was binding to the consensus element was nearly pure, and shown to be a six-protein complex that was christened the *origin recognition complex*, or ORC. The six subunits of ORC are relatives of each other, and as introduced in Chapter 4, contain an N-terminal ATPase module linked to a C-terminal DNA binding domain.

Turning to metazoan cells, the nature of replication origins is much less clear but is currently under active investigation in many labs. Decades ago, it was observed that DNA injected into embryonic cells or added to embryonic cell extracts could replicate regardless of sequence — even a bacterial plasmid could replicate! These results provided an early hint that metazoan origins would be difficult to define. Indeed, there are currently no clearly defined consensus-sequence elements that define metazoan replication origins. Furthermore, while metazoans have ORC-protein complexes with the conserved ATPase domains, these proteins do not have much sequence specificity and very likely require other proteins to bind specifically to replication origins. While much remains to be learned about replication initiation in metazoans, there is accumulating evidence that higher-order chromatin structure as well as local nucleosome positioning play important roles in origin recognition by the metazoan ORC protein.

5.6 The overall logic of origin usage in eukaryotes — many are licensed but (relatively) few are fired

Before delving into the detailed mechanism of initiation of eukaryotic replication, it is helpful to consider the overall logic of origin usage and consider how the above-mentioned flexibility is achieved. As a general reminder, the eukaryotic cell cycle is divided into four sequential phases, G1, S, G2 and M (Figure 5.5). All DNA replication is confined to the S (or *synthesis*) phase, and the M phase is when *mitosis* occurs. The G1 phase before S and G2 between S and M are "*gap*" phases between the milestone S and M phases of cell division. The timing of the phases of the cell cycles is orchestrated

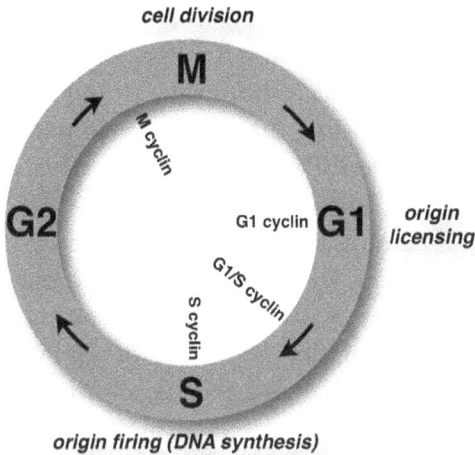

Figure 5.5.   The cell-division cycle of eukaryotic cells. The cell cycle consists of a gap period before DNA synthesis (G1), a period for DNA synthesis (S), another gap period after DNA synthesis (G2), and the period when the cells undergo mitosis (M). The cell cycle is controlled in part by cyclins, which are depicted in a generic fashion inside the circle. The licensing of replication origins occurs during G1 only and origin firing during S phase only.

by a class of proteins called "cyclins", which were originally named based on the finding that their abundance varied in a defined manner across the cell cycle. Cyclins can be categorized according to the phase of the cell cycle when they are active and trigger key cell cycle events (e.g. G1 cyclin, G1/S cyclin, S cyclin, etc).

The most important regulatory constraint of eukaryotic DNA replication is that each segment of DNA should be replicated once and only once per cell cycle, so that each daughter cell has exactly the same DNA complement as its parent. How can cells guarantee that a given segment of DNA is not replicated twice or more during a particular S phase? This problem would be particularly difficult if the assembly of the replication machinery at origins occurred during S phase, when the actual replication occurs. How would the assembly mechanism distinguish between origin segments that have not yet replicated versus those that have, perhaps even passively from a fork that started at a different origin?

Evolution has produced an elegant and nearly fail-safe system to prevent such re-replication during S phase. The key to the system is a two-step process that involves both the G1 and the S phase of the cell cycle (Figure 5.5). During the G1 phase, all potential origin elements undergo a process called "licensing", in which a subset of the replication machinery is assembled in a manner that is poised to replicate in the upcoming S phase. However, licensing is not sufficient to begin replication, and the licensing process is strictly limited to the G1 phase. As the S phase ensues, licensing can no longer occur, but additional factors required for DNA replication are assembled into particular licensed origins, allowing them to become functional in a process called origin "firing". As we will see, not all licensed origins fire, but any time a licensed origin is replicated passively by a fork from a different, distant, origin, the license is erased. In this way, once an origin segment replicates, either by its own action or the action of another origin, it can no longer fire. Because licensing is restricted to the G1 phase, an origin also cannot gain a new license during this particular cell cycle.

Only a subset of licensed origins actually fire during any given S phase (Figure 5.6A). This provides a useful degree of flexibility for the S phase. First, as mentioned above, rapidly growing cells need to use more origins in order to complete replication faster; they can increase the frequency of firing compared to slowly growing cells, with no change in the licensing that occurs in the G1 phase. Second, as the S phase progresses, fewer and fewer licensed origins remain unreplicated (because the license is erased as the DNA replicates), allowing the replication process to, in a sense, focus on the few unreplicated regions of the chromosome (those that still contain a licensed origin).

In a related manner, this system provides flexibility under conditions that cause replication-fork blockage. Because of the multiple origins along a linear chromosome, if two forks from different origins are converging, and one becomes blocked, the other fork can replicate the entire intervening region (Figure 5.6).

What happens if both converging forks become blocked? Repriming of replication by PrimPol (Section 4.9) provides one possible solution, but activation of an intervening origin can also

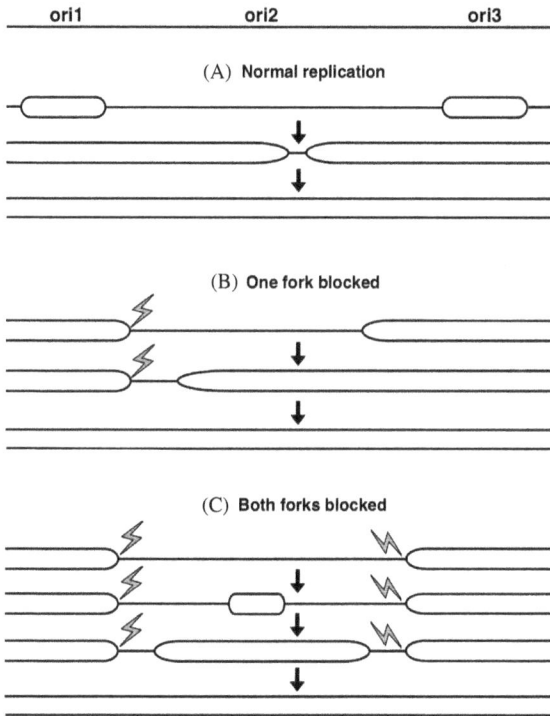

Figure 5.6. Overcoming blocked replication forks in eukaryotic cells. The multiplicity of eukaryotic replication origins and their flexible usage allows cells to complete replication in spite of fork-blockage events. In this hypothetical chromosome, *ori1* and *ori3* are relatively efficient origins, and in most cell cycles, normal replication from these origins suffices to complete replication of the intervening region (panel A). When one fork is blocked, the fork from the opposite origin is sufficient to complete replication of the intervening region (panel B). Even if both forks are blocked, a relatively inefficient origin within the intervening region (*ori2*) can be activated, allowing completion of DNA replication (panel C).

resolve the stalemate. As long as there is another licensed origin in between the two blocked forks, this origin can fire to allow the completion of replication (Figure 5.6C). In this sense, eukaryotic chromosomes are rich with latent licensed origins that are activated only when needed. We will return to this important aspect of regulation when we consider the S-phase checkpoint below (see Section 5.9 and Chapter 13).

5.7 Licensing during the G1 phase

Scientists developed powerful methods to deduce how DNA replication operates during the cell cycle. Some of these methods allowed them to determine which proteins were bound to the origin region at each phase of the cell cycle, and these results were instrumental in our understanding of origin licensing and firing. A result that was initially surprising was that the ORC initiator complex introduced above was bound to the replication origin throughout the cell cycle, not just during the S phase when replication occurs or G1 phase in preparation for DNA replication. The ORC complex can be viewed as a platform, always present, but utilized to load the replication machinery only when it is needed.

These same methods showed a progression of proteins that are bound to origins along with ORC as the G1 and then S phase progress. During G1, a key event in licensing is that the MCM complex is loaded onto the origin by the ORC complex, with the help of two other proteins including one that is an ORC-protein relative (Figure 5.7). These two proteins act essentially as a "clamp loader" for the ring-shaped MCM complex, with the help of the already loaded ORC complex. The loading process has several important features. First, while the eukaryotic MCM complex is a hexamer in solution, loading creates a stable double hexamer in which the two hexamers are associated with each other but facing in opposite directions. We will see below that this allows the establishment of two replication forks going in opposite directions. Second, the MCM complex loaded during G1 phase does not have helicase activity yet. Recall that MCM helicase activity requires five additional proteins that are present in the CMG complex, and these five proteins are assembled onto MCM later in the cell cycle. Third, and this was a big surprise, the G1-loaded MCM complex does not have a single strand of DNA threaded through the hole in the middle of the hexameric ring, but rather has duplex DNA! This certainly fits with the inability of MCM to unwind DNA during G1 phase, because a helicase encircling duplex DNA could only slide back and forth along the duplex. Fourth, for reasons that are still debated, there is an excess of MCM

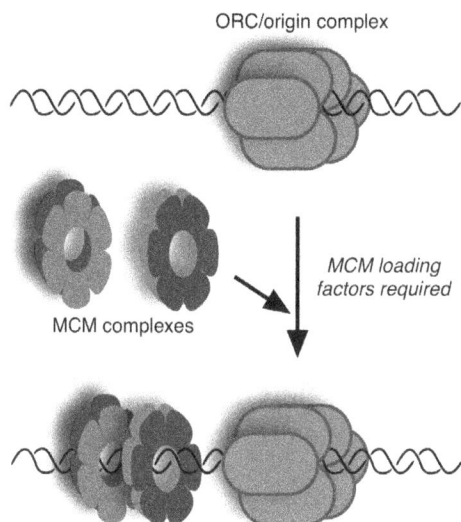

Figure 5.7. Loading the MCM complexes at a eukaryotic origin. The mechanism of localization of ORC/origin complexes varies somewhat between eukaryotic cells (see text), but once assembled, this complex along with additional loading factors allows the loading of MCM complexes. The MCM complexes are depicted with the two faces in contrasting dark and light colors. MCM is loaded around duplex DNA during the G1 phase of the cell cycle.

protein loaded during G1, roughly on the order of 10 MCM double hexamers per bound ORC complex. This implies that each ORC can load multiple MCM complexes during G1.

As described above, the MCM complex in eukaryotes is composed of six distinct but closely related subunits. During the loading process, DNA is passed between the interface of two particular MCM subunits, not randomly through any of the six interfaces. Presumably, the two clamp-loader-type proteins hold the MCM hexamer in an open configuration that allows DNA to pass through the crack in the ring. Once DNA is loaded and the loading proteins dissociate, the MCM complex relaxes to the normal ring-shaped hexamer that totally encircles the DNA molecule.

So, what is the actual "license" which allows an origin to subsequently fire during S phase? It is none other than the loaded MCM complex, which is fully competent to go on to the assembly and firing steps described in the next step. As mentioned above, one key feature of the licensing process is that the license must be erased as DNA is replicated during S phase to prevent any DNA from replicating twice. Indeed, the MCM double hexamers that were loaded during G1 are somehow unloaded as the replication fork passes during S phase. The details are yet to be unraveled, but presumably the active CMG complex in the moving replication fork plays a key role in displacing any MCM double hexamers that it encounters. The CMG complex might be directly involved in some disassembly reaction, or at least push any MCM double hexamers forward into the yet-unreplicated region.

A second key feature of the licensing process is that it is carefully restricted to the G1 phase. This important restriction is accomplished by multiple redundant mechanisms that are triggered at the beginning of S phase and that are focused on the two MCM-loading factors. An inhibitory protein that accumulates in S phase binds to one of the loading factors, the loading factors are ejected from the nucleus, and the loading factors are subjected to proteolytic degradation. These redundant mechanisms ensure that active loading factors are not present during S phase, effectively preventing any further MCM loading (licensing).

The inactivation of MCM-loading factors is triggered by phosphorylation events (and other protein modifications) that occur as S phase begins and progresses. The protein phosphorylation events are particularly critical in driving the progression from G1 to S phase and the transition of licensed origins into origins that are capable of firing. As mentioned above, cyclin proteins are active in specific phases of the cell cycle, and the protein kinases that are responsible for the phosphorylation events that occur as S phase begins are dependent on these cyclins; these kinases are therefore called *c*yclin-*d*ependent *k*inases or CDK's. The inactivation of MCM-loading factors is triggered by CDK activity, and we will see shortly that CDK activity is also critical for assembly of the remaining replication complex onto loaded MCM/ORC complexes and the firing of replication origins.

5.8 Assembly of the complete replication machinery and origin firing

The subunits of the MCM complex are key targets of CDK activity in early S phase. Indeed, MCM phosphorylation is required for the recruitment and activation of the remaining members of the CMG complex (see Chapter 4), a prerequisite for the helicase activity of the complex. The MCM complex is loaded as a double hexamer encircling duplex DNA during G1 phase, but must transition to two single hexamers traveling in opposite directions, with each encircling only one of the two strands of the duplex (Figure 5.8). This dramatic transition is at the heart of initiating bidirectional replication during origin firing, but is still not very well understood.

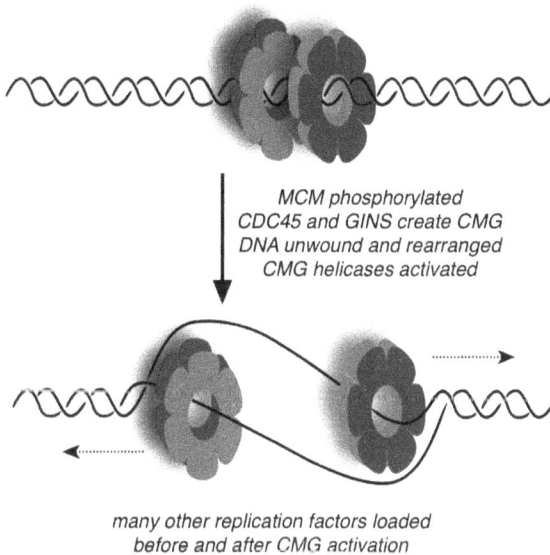

MCM phosphorylated
CDC45 and GINS create CMG
DNA unwound and rearranged
CMG helicases activated

many other replication factors loaded
before and after CMG activation

Figure 5.8. Activation of the CMG complex during S phase. The MCM complexes, initially loaded around duplex DNA, are converted into the active CMG complex by their association with CDC45 and GINS. During the key transition that allows DNA unwinding, one strand of the duplex is passed out of the interior of CMG to leave the opposite strand encircled. The two CMG complexes that become active during replication initiation are oriented in opposite directions and encircle the two different DNA strands, allowing bidirectional replication.

A substantial number of other replication proteins are recruited to the origin complex prior to the initiation of replication, and many of them are also phosphorylated by CDK's. The functions of some of these proteins are still being investigated, but prominent players in this group include the DNA polymerases that catalyze replication and the ssDNA-binding protein RPA. Some of the proteins required for replication are loaded prior to the initiation of DNA unwinding by the CMG complex, while others are loaded only after the key transition to active unwinding by CMG. The overall assembly process seems to be carefully ordered, and scientists are currently studying the details. Recently, DNA replication has been reconstituted in vitro using purified yeast proteins, and this breakthrough will allow careful probing of the assembly process.

5.9 Completing replication: Converging forks and the replication checkpoint

Termination of replication occurs at numerous sites throughout the chromosomes of eukaryotic cells, wherever two opposing replication forks happen to run into each other (see above). This process has been difficult to study in part because it generally occurs at essentially random sites. Clearly, the two opposing CMG complexes must be disassembled, and at least two proteins have been identified that appear to play roles in this disassembly process.

As described above, bacterial chromosomes contain a special termination region, bounded by Ter sites that block replication forks in a unidirectional fashion. While eukaryotic chromosomes do not have comparable termination regions, they do contain a few sites that behave much like bacterial Ter sites. These so-called *r*eplication *f*ork *b*arriers (RFB's) function to prevent replication forks from colliding with transcription complexes within the heavily transcribed genes for ribosomal RNA. The RFB sites are generally located downstream of rRNA genes, preventing replication forks from entering the gene from the downstream side. As in the bacterial case, special proteins bind to the RFB sites and interact with the replication machinery to impose the unidirectional blockade.

Eukaryotic replication forks can be blocked or paused in many other situations, including the presence of tightly bound proteins or DNA lesions in the template, secondary structures in the DNA, and inhibition of polymerases (for example by depletion of dNTPs). Numerous studies have probed how the cell responds to these various blockages and the consequences of improper responses. These are medically important issues, since improper responses can lead to genome rearrangements (e.g. in development of cancer) or to a variety of diseases linked to nucleotide repeats in the DNA (including Huntington's disease, myotonic dystrophy, Friedreich's ataxia, and Fragile-X mental retardation syndrome; also see Chapter 14).

As mentioned above, the flexible nature of origin firing in eukaryotic cells helps to mitigate the problems of fork stalling and blockage. Even when two forks traveling towards each other are both blocked, the intervening segment can still be replicated by activating a normally dormant (but licensed) origin in between the two converging forks (Figure 5.6C). Experimental evidence for the importance of this backup system emerged from studies of mice with artificially low levels of the MCM proteins, which leads to abnormally low levels of licensed origins. These mice show a predisposition to cancer, presumably due to genome instability caused by incomplete replication. As might be expected, reduced levels of MCM also cause hypersensitivity to inhibitors of DNA replication.

The activity of normally dormant origins is subjected to a higher level of regulation that, surprisingly, can both increase and decrease origin firing. To ensure proper completion of various cellular processes, cells have a number of "checkpoints", which monitor the completion of a process and halt the cell-division cycle if the process is incomplete (see Chapter 13). This provides more time to complete the process and also avoids catastrophes that might ensue (such as attempting chromosome separation when replication is not yet completed). Stalled or blocked replication forks trigger one such checkpoint, called the replication (or S phase) checkpoint. The replication checkpoint can be triggered by excess RPA-coated ssDNA at a replication fork and leads to a number of downstream responses. Globally, origin firing is temporarily repressed, which presumably prevents additional problems (i.e. additional blocked forks) from developing. However, in the immediate vicinity of the blocked fork, dormant origins are somehow activated, presumably to attempt completion of replication in the vicinity of the problematic fork (as in Figure 5.6C). It will be fascinating to learn how the replication checkpoint can distinguish between distant and nearby origins!

5.10 Telomeres and their replication

The invariant 5′ to 3′ directionality of DNA polymerases demands some special mechanism to replicate DNA ends, as described in Section 4.1. The ends of eukaryotic chromosomes have a special structure, called a telomere, which promotes this end replication. In addition, telomeres protect chromosome ends from degradation or recombination events, which are normally triggered by DNA ends (such as those generated by a DSB). As we will see, telomeres are also noteworthy because they play a crucial role in human disease including cancer and are connected to the aging process.

The sequence and structure of telomeres is quite remarkable. At the DNA sequence level, the chromosome ends contain a short repetitive sequence, only 6 to 8 bases in most species, which is repeated tens or hundreds of times (again depending on species) at each chromosomal end. Vertebrates have the 6-base sequence 5′-TTAGGG-3′ on the strand with the 3′ end of the chromosome (the complement 5′-CCCTAA-3′ is on the other strand). As expected from the polarity of DNA polymerases, the 3′ end cannot be completely replicated by the replisome complex, and thus vertebrate telomeres contain a single-strand extension with a number of 5′-TTAGGG-3′ repeats (Figure 5.9A).

When the actual structure of chromosome ends is carefully analyzed, it turns out that this single-stranded 3′ end is usually paired in an interesting manner. The 3′ end circles around and essentially

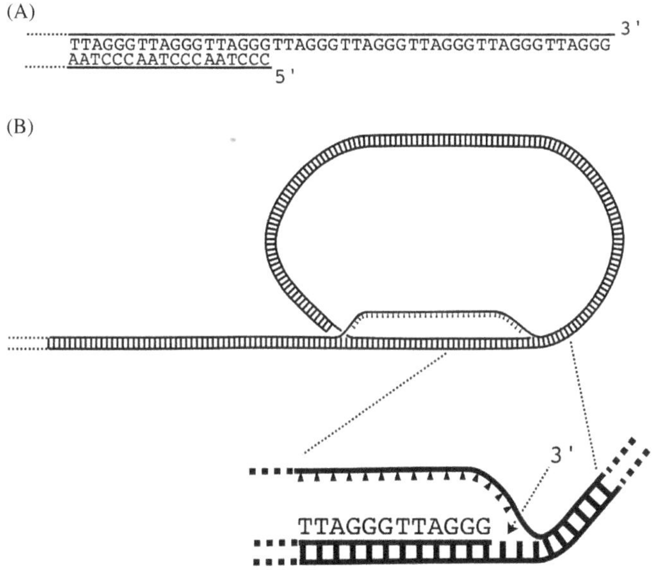

(A)

(B)

Figure 5.9. T-loop configuration of DNA at telomeres. The 3′ single-stranded ends of eukaryotic chromosomes are generally "tucked into" an upstream region of the same chromosome. Because the end region has extensive repeats of the same sequence, the 3′ end is complementary to the opposite strand at many positions in this upstream region, allowing base pairing to stabilize the t-loop structure. A related structure, called a D-loop, will be discussed in the context of homologous recombination in Chapter 11.

"tucks into" the duplex some distance back to form a looped structure, called the t loop (*t*elomere loop) (Figure 5.9B). The invasion of this 3′ end occurs within a more internally located region of the terminal repeats, and so normal base pairing (with the 5′-CCCTAA-3′ complement) is involved in stabilizing the t-loop structure. We will encounter this kind of "strand invasion" again later in the book when we discuss the process of DSB repair (Chapter 11).

The remarkable property of the t-loop structure is that is disguises the end of the chromosome, ensuring that the cell does not mistake the natural chromosome end for a detrimental DSB, which the cell would attempt to repair. Each t-loop must be undone temporarily when DNA replication occurs, but rapidly reforms after replication to protect the chromosomal end. The t-loop is a carefully programmed structure that depends on a complex of bound proteins[2] that helps protect the chromosomal ends. Telomeres also generally contain repeats of a different sequence that are more distant from the chromosomal end, called sub-telomeric repeats, and these are also the platform for binding of important proteins involved in telomeric function.

In addition to their important structural role, the terminal repeats of telomeres also play a critical role in promoting telomere replication. The enzyme telomerase, discovered in the 1980's, was found to add new repeats at these 3' ends by recognizing telomeric repeats and extending their 3' end (Figure 5.10). Telomerase contains an RNA component with an extended version of the sequence that is complementary to the repeat; in vertebrates the RNA carries the sequence 5'-CCCUAACCC-3'. This portion of the RNA can base pair with the 3' end of the chromosome in such a way that the 5' end of the RNA sequence is single stranded. The other component of telomerase is a specialized polymerase that can use an RNA template to make DNA; this class of enzyme is called reverse transcriptase (also prominent in many RNA viruses including HIV). The telomerase enzymatic activity can thereby extend the 3' end with copies of the same repeat, by using the RNA within the enzyme as the template for these repeats. The process can work iteratively to add multiple repeats during a given cell cycle.

The process is not yet complete, however. After the 3' end is extended, generally by multiple repeats, the conventional DNA replication machinery fills in most of the 5' end, which has become

---

2009 Nobel Prize in Physiology or Medicine

This prize was awarded to **Elizabeth H. Blackburn**, **Carol W. Greider** and **Jack W. Szostak** for their studies on the structure of telomeres and discovery of telomerase, which explain how chromosome ends are protected and copied during DNA replication.

https://www.nobelprize.org/prizes/medicine/2009/summary/

---

[2] Called "shelterin"

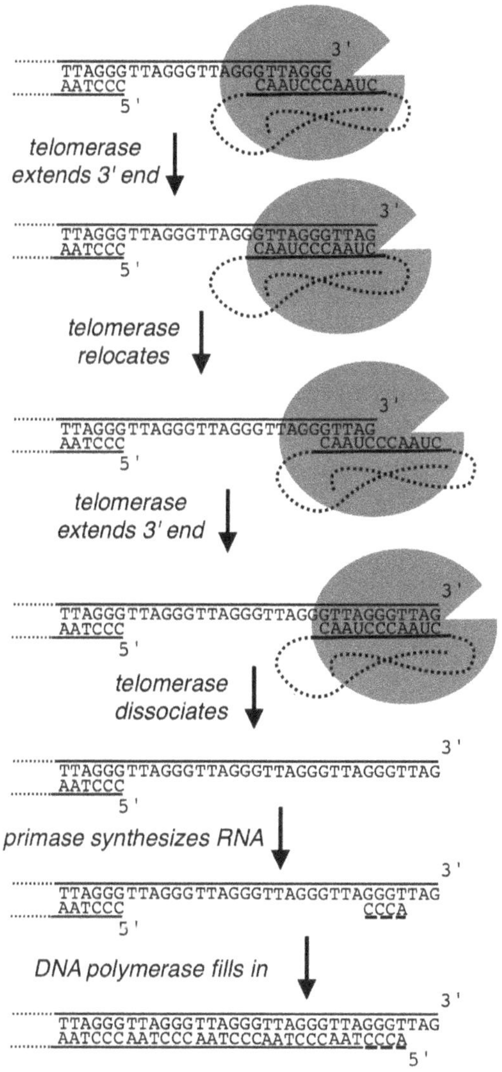

Figure 5.10. Mechanism of replication of telomeric DNA. Telomerase is depicted as the grey "pac-man". The telomerase RNA molecule is depicted as the mostly dotted line associated with the pac-man; the solid line segment of the RNA represents the critical RNA region that allows base pairing and templating. Extension by telomerase is arbitrarily shown to begin using the 3′G of the GGG segment of the repeat, but other positions within the repeat can also be extended when they constitute the 3′ end and there is sufficient base pairing to the telomerase RNA. The short RNA primer synthesized by primase is the four-nucleotide dashed segment in the two molecules at the bottom of the figure.

much shorter than the 3′ end (Figure 5.10). An RNA primer is synthesized by primase, using a distal copy of the TGGG sequence on the 3′ strand as template. As in normal DNA replication, DNA polymerase then utilizes the RNA primer to synthesize DNA in the direction towards the interior of the chromosome. Because the chromosome ends normally contain 3′ single-stranded extensions of the repeat, erasure of the RNA primer is not consequential.

Telomere and telomerase have received extensive coverage in the mainstream media due to their connections to aging, cancer and other diseases. It was rather surprising when researchers found that many vertebrate primary cells lack telomerase, and that telomeres generally get shorter as vertebrate organisms age. Importantly, ectopic expression of telomerase is sufficient to allow various primary human cells to become immortal in culture (meaning they can replicate and undergo cell division indefinitely). Nonetheless, it is important to note that certain vertebrate cell types do normally contain telomerase and maintain long telomeres, not surprisingly including germ cells and stem cells.

The repression of telomerase activity in somatic cells of adults appears to limit cancer formation — once telomeres are sufficiently eroded, cells can no longer propagate and often undergo cell death. How then do some cells become cancerous, a state in which they propagate indefinitely? It turns out that about 90% of cancer cells have reactivated telomerase by some mechanism, thereby short-circuiting the protective circuit just described. The remaining small fraction of cancers do not show reactivation of telomerase, but these turn out to have activated an alternative mechanism of telomere replication (*a*lternative *l*engthening of *t*elomeres, or "ALT"), essentially the exception that proves the rule. A number of other diseases have also been linked to telomeres and telomerase, for example hematopoietic diseases related to bone-marrow-stem cell function.

Books have been written about the implications of these and other related findings with respect to the possible prevention of aging, cancer incidence and treatment, and lifestyle choices that might influence telomere length. Some of these topics are quite controversial, so keep a critical and open mind as you delve into these topics further!

5.11 Summary of key points

- Replication initiation is carefully regulated in all cells to maintain proper chromosome copy number (ploidy).
- Termination is also carefully orchestrated to prevent DNA from replicating twice in a cell cycle and to assist in the process of chromosome separation and segregation into daughter cells.
- The bacterial replication origin, *oriC*, was defined as the highest copy number region in rapidly growing cells and as a segment that allows autonomous replication of plasmids; replication from *oriC* is bidirectional.
- The bacterial initiator protein binds in multiple copies to *oriC*, unwinds a region of the origin, and serves as a binding platform for the chaperone:helicase complex.
- As the replicative helicase is loaded around ssDNA at *oriC*, the chaperone departs, aided in part by primase binding to the helicase.
- Replication termination occurs in the chromosome terminus region, opposite *oriC*, aided by Ter sites that flank the terminus region and block replicative helicase in a directional manner.
- The directionality of DNA replication has evolved to avoid head-on collisions between DNA replication and transcription (particularly of heavily transcribed genes).
- Replication origin usage in eukaryotic cells is flexible, with more origins utilized in rapidly growing (e.g. embryonic) cells.
- Yeast replication origins, ARS elements, were defined as segments that allow autonomous replication of plasmids; replication from ARS elements is bidirectional.
- The yeast initiator protein, ORC, was first purified based on its binding to the most conserved sequence within ARS elements.
- Replication origins in metazoan cells are still poorly defined, but appear to be dependent more on chromatin structure than on conserved DNA sequence elements.

- Eukaryotic origins are licensed in G1 phase and a subset of licensed origins fire during S phase; the licensing system provides flexible origin availability but ensures that no DNA is replicated twice during a single S phase.
- Licensing involves the loading of the MCM helicase complex around duplex DNA; correspondingly, MCM helicase activity is not activated by licensing.
- Origin firing in S phase involves activation of the MCM helicase activity as it transitions to a fully active CMG complex encircling single strands of ARS DNA.
- Eukaryotic replication initiation and progression is regulated by numerous covalent modifications of key proteins, particularly involving phosphorylation by cyclin-dependent kinases.
- Eukaryotic chromosomes also contain replication-fork-blockage sites (RFBs) that reduce replication-transcription collisions, but replication termination occurs throughout the genome and does not generally depend on RFBs.
- Chromosome ends contain special structures called telomeres, which protect the ends and promote end replication via a process that involves telomerase (a reverse transcriptase).

Further Reading

Bell, S. P., & Kaguni, J. M. (2013). Helicase loading at chromosomal origins of replication. *Cold Spring Harb Perspect Biol, 5*(6), a010124.

Bell, S. P., & Labib, K. (2016). Chromosome duplication in Saccharomyces cerevisiae. *Genetics, 203*(3), 1027–1067.

Bell, S. P., & Stillman, B. (1992). ATP-dependent recognition of eukaryotic origins of DNA replication by a multiprotein complex. *Nature, 357*(6374), 128–134.

Chodavarapu, S., & Kaguni, J. M. (2016). Replication initiation in bacteria. *Enzymes, 39*, 1–30.

Costa, A., Hood, I. V., & Berger, J. M. (2013). Mechanisms for initiating cellular DNA replication. *Annu Rev Biochem, 82*, 25–54.

Coster, G., & Diffley, J. F. X. (2017). Bidirectional eukaryotic DNA replication is established by quasi-symmetrical helicase loading. *Science, 357*(6348), 314–318.

Duderstadt, K. E., & Berger, J. M. (2013). A structural framework for replication origin opening by AAA+ initiation factors. *Curr Opin Struct Biol,* *23*(1), 144–153.

Duderstadt, K. E., Reyes-Lamothe, R., van Oijen, A. M., & Sherratt, D. J. (2014). Replication-fork dynamics. *Cold Spring Harb Perspect Biol,* *6*(1), a010157.

Fragkos, M., Ganier, O., Coulombe, P., & Mechali, M. (2015). DNA replication origin activation in space and time. *Nat Rev Mol Cell Biol,* *16*(6), 360–374.

Ilves, I., Petojevic, T., Pesavento, J. J., & Botchan, M. R. (2010). Activation of the MCM2-7 helicase by association with Cdc45 and GINS proteins. *Mol Cell,* *37*(2), 247–258.

Leonard, A. C., & Mechali, M. (2013). DNA replication origins. *Cold Spring Harb Perspect Biol,* *5*(10), a010116.

McIntosh, D., & Blow, J. J. (2012). Dormant origins, the licensing checkpoint, and the response to replicative stresses. *Cold Spring Harb Perspect Biol,* *4*(10), a012955.

Mott, M. L., Erzberger, J. P., Coons, M. M., & Berger, J. M. (2008). Structural synergy and molecular crosstalk between bacterial helicase loaders and replication initiators. *Cell,* *135*(4), 623–634.

Pfeiffer, V., & Lingner, J. (2013). Replication of telomeres and the regulation of telomerase. *Cold Spring Harb Perspect Biol,* *5*(5), a010405.

Remus, D., Beuron, F., Tolun, G., Griffith, J. D., Morris, E. P., & Diffley, J. F. (2009). Concerted loading of Mcm2-7 double hexamers around DNA during DNA replication origin licensing. *Cell,* *139*(4), 719–730.

Reyes-Lamothe, R., Nicolas, E., & Sherratt, D. J. (2012). Chromosome replication and segregation in bacteria. *Annu Rev Genet,* *46*, 121–143.

Rhind, N., & Gilbert, D. M. (2013). DNA replication timing. *Cold Spring Harb Perspect Biol,* *5*(8), a010132.

Siddiqui, K., On, K. F., & Diffley, J. F. X. (2013). Regulating DNA replication in eukarya. *Cold Spring Harb Perspect Biol,* *5*(9), a012930.

Skarstad, K., & Katayama, T. (2013). Regulating DNA replication in bacteria. *Cold Spring Harb Perspect Biol,* *5*(4), a012922.

Skovgaard, O., Bak, M., Lobner-Olesen, A., & Tommerup, N. (2011). Genome-wide detection of chromosomal rearrangements, indels, and mutations in circular chromosomes by short read sequencing. *Genome Res,* *21*(8), 1388–1393.

How did they test that?
Isolation of ORC, the eukaryotic replication-initiation protein

The eukaryotic replication-initiation proteins and the genes that encode them were difficult to identify and remained mysterious for some 20 years after the gene for the bacterial initiator protein (DnaA) was identified and 10 years after that bacterial protein had been purified. Bell and Stillman (1992) used a novel and powerful strategy that allowed them to purify the eukaryotic replication-initiation proteins without any initial knowledge of the encoding genes, by simply searching for proteins that bind specifically to a defined *Saccharomyces cerevisiae* replication origin. They fractionated yeast extracts over a series of different chromatography columns and other purification methods (diagram on left). The various fractions isolated from each of these purification steps were assayed for their ability to bind to the origin-DNA sequence by a method called DNA footprinting. In this method, a radioactively labeled origin DNA fragment is incubated with the protein fraction of interest and then subjected to a brief treatment with a relatively randomly acting nuclease. The DNA fragments are then separated by high-resolution gel electrophoresis and visualized by autoradiography. As seen in most of the lanes of the lower figure on the right, the nuclease generated many fragments throughout the gel when the origin was unoccupied by protein (same as a control reaction with no added protein). However, in fractions 16 through 20, numerous bands in the region of the origin (critical origin region at boxes labeled A and B1) were eliminated and other bands were either increased or created anew (so-called DNase hypersensitive sites); the specific binding of the protein caused these alterations. During the purification procedure, these kinds of fractions would be pooled and used for the next purification step, as diagrammed on the left, leading to preparations that were more and more pure over subsequent steps. The figure shown at the bottom of the right panel presents the footprinting results with fractions from the final purification step, and the gel at the top of the right panel is a polyacrylamide gel showing the sizes of the proteins in each fraction corresponding to the footprint below (top gel is stained with a generic protein-binding stain). These proteins were named the *o*rigin *r*ecognition *c*omplex, or ORC, which consists of six different proteins, ORC1 through ORC6. The data figures on the right were reproduced from Bell and Stillman (1992), with permission from Springer Nature; permission conveyed by Copyright Clearance Center, Inc.

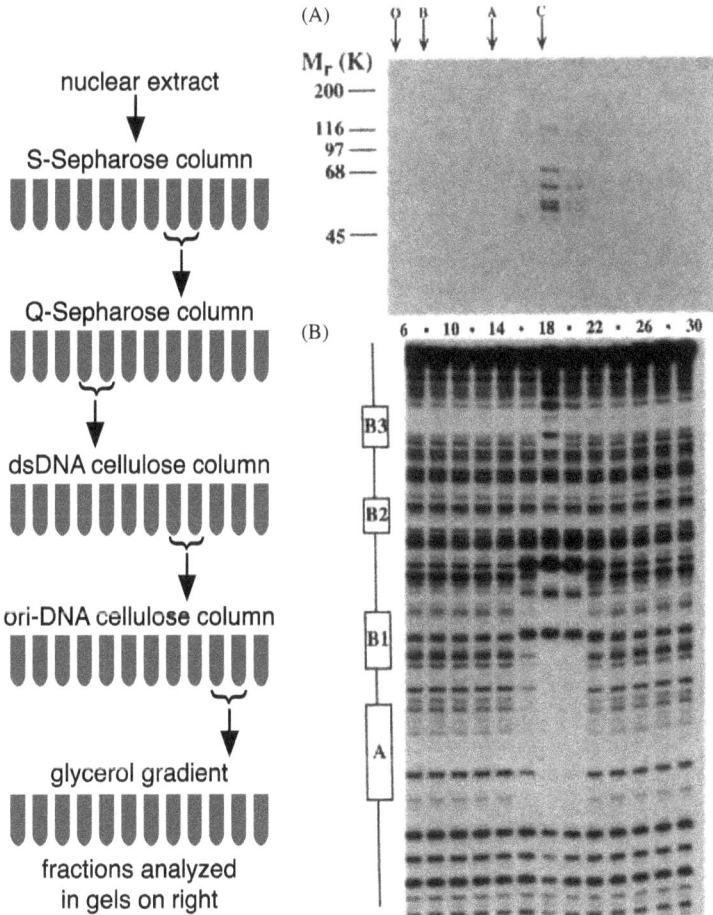

Chapter 6

Postreplication repair of mismatches and ribonucleotides

As described in previous chapters, the replication machinery is extremely accurate. Replicative polymerases have a high fidelity for base insertion as well as potent proofreading activities that remove most incorrect bases that have been incorporated. Nonetheless, a fraction of these incorrect bases remain in the newly synthesized DNA after the replication fork has passed. Such replication errors are estimated to occur about once every 10 million base pairs replicated in each replication cycle.

Virtually all cells have an error-correction mechanism that reduces this error rate further, correcting more than 99% of the errors left behind by the replication machinery. An incorrectly inserted base in DNA will cause an abnormal base pair, also called a DNA mismatch. Accordingly, the error-correction mechanism is called *mismatch repair* (MMR). As we will see, MMR operates in various situations, and the version that operates on replication errors is often called post-replicative MMR.

Replicative DNA polymerases are also very accurate in selecting deoxyribonucleotide precursors over ribonucleotide precursors, but again they do occasionally make mistakes and insert ribonucleotide residues. The rate of incorporation of ribonucleotide residues is augmented by the fact that ribonucleotide pools are much higher than deoxyribonucleotide pools, thus DNA polymerases are fighting against an excess of the wrong kind of precursors. In the closing section of this chapter, we will consider another post-replicative repair process, *ri*bonucleotide *e*xcision *r*epair (RER), which removes these ribonucleotide residues from duplex DNA.

6.1 The overall function and logic of post-replicative MMR

When considering a DNA mismatch caused by misincorporation during replication, let's say a G:T mispair, how can you know whether that position should be G:C or A:T? Both G:C and A:T occur normally throughout the DNA molecule, and there is no way to tell which is correct without some additional information. Thus, if a form of MMR was to act on replication errors without some additional information, it would not improve the accuracy of DNA replication because it would change the mismatch to the incorrect base pair roughly half the time on average.

The critical additional piece of information that allows MMR to greatly improve the accuracy of DNA replication is the identity of the strand in the helix that was synthesized in the most recent round of replication. The base in the mismatch that is on the newly synthesized strand is reckoned to be the mistake. As we will see shortly, evolution has generated two different ways to identify the newly synthesized strand, one that operates in *Escherichia coli* and certain other bacterial species, and another that operates in eukaryotes and many other bacterial species that do not follow the *E. coli* model.

Many pathways of DNA repair, including MMR, take advantage of the redundancy of information in the DNA duplex to accurately remove damaged or incorrectly inserted bases. We will consider the detailed mechanism below, but the outline of the MMR process is that a segment of the newly synthesized strand is removed by nuclease action, and then DNA polymerase resynthesizes that strand using the intact partner strand as the template. The relatively high accuracy of replicative DNA polymerases ensures that the new patch of synthesized DNA is nearly always free of errors. The net result is that the mismatch has been effectively replaced with a proper

Watson–Crick base pair, the same base pair as that in the original parental DNA, thereby avoiding a mutational event.

6.2 Methyl-directed MMR in *E. coli*

Much of the early research on MMR utilized the *E. coli* model system, and so the *E. coli* MMR system has been intensively studied and has served as the paradigm for all other studies of MMR. Beginning in the 1970s, scientists were able to demonstrate that DNA containing a mismatch, when transformed into *E. coli*, is subjected to a repair reaction that corrects the mismatch. In addition, genetic studies identified mutations in four *E. coli* genes that increased the overall mutation rate by about 100-fold; when these mutants were transformed with mismatched DNA, no repair was detected. The protein products of these four genes are required for MMR. Because mutations in these genes increase the *mut*ation frequency, the genes were named *mut* (followed by a letter)—three of the four were *mutH*, *mutL*, and *mutS*; the fourth gene had previously been identified as *uvrD* (mutations in this gene had already been found to cause sensitivity to *UV* radiation).

Another key discovery that helped unravel the pathway of MMR in *E. coli* was the finding that a certain form of DNA methylation provides the information about which strand is newly synthesized. *E. coli* and some other bacteria encode a *DNA a*denine methyltransferase (methylase), called Dam, which adds a methyl group to the N^6 position of the A residue within the sequence 5′-GATC-3′. Like the recognition sites of many restriction enzymes and their modification partners, this sequence has reverse complementarity, that is, the complementary strand also reads 5′-GATC-3′ (in the opposite direction, i.e., 5′-GATC-3′/3′-CTAG-5′). The enzyme can methylate both of the A residues in the duplex site, and is very efficient at methylating the unmethylated A of a site that already has one strand methylated (hemi-methylated site) (see Section 5.3).

The *E. coli* genome has many thousands of Dam recognition sites scattered around the circle, and these sites are fully methylated most of the time. However, because methylation occurs after DNA replication (and not on the nucleotide precursor), each Dam recognition site is transiently hemi-methylated immediately after DNA replication (see Chapter 5; Figure 5.3). Importantly, the newly synthesized strand is always the one without the methylated A, and this is the basis of strand recognition by the MMR system. Soon after replication (about 2 minutes on average), the Dam enzyme methylates the unmethylated A of each hemi-methylated site. At that point, the strand signal for MMR is lost. Thus, the MMR system has to act very soon after DNA replication to take advantage of the transient signal for the newly synthesized strand (Figure 6.1).

Some of the early evidence for this strand discrimination system was provided by experiments similar to those mentioned earlier. Duplex DNA was constructed with a mismatch at one site and with Dam methylation (at the GATC sites) on only the Watson or the Crick strand, and was then transformed into *E. coli*. The mismatch was always corrected on the strand without the methylation, using the strand with methylation as the template for DNA replication. Furthermore, when neither strand was methylated, repair occurred but showed little strand preference, and when both strands were methylated, repair was blocked. Genetic evidence also showed that Dam methylation was important in improving replication fidelity in *E. coli* — mutations that inactivated the Dam methyltransferase caused a large increase in overall mutation frequency, just like the MMR mutants introduced above.

The molecular mechanism of MMR was ultimately unraveled with elegant biochemical experiments that purified the key components and reproduced the reaction in vitro (Figure 6.2; see "How did they test that" at the end of this chapter). The MutS protein was found to recognize and bind to a variety of DNA mismatches. MutS acts as a dimer, with only one of the two subunits actually contacting the mismatch. Binding of the MutS dimer to the mismatch induces a dramatic bend in the duplex DNA, and the MutL protein is then able to bind to the complex in an ATP-dependent fashion. The role of the MutH protein is to cleave the unmethylated strand at a nearby GATC site, but this activity is suppressed unless MutH binds

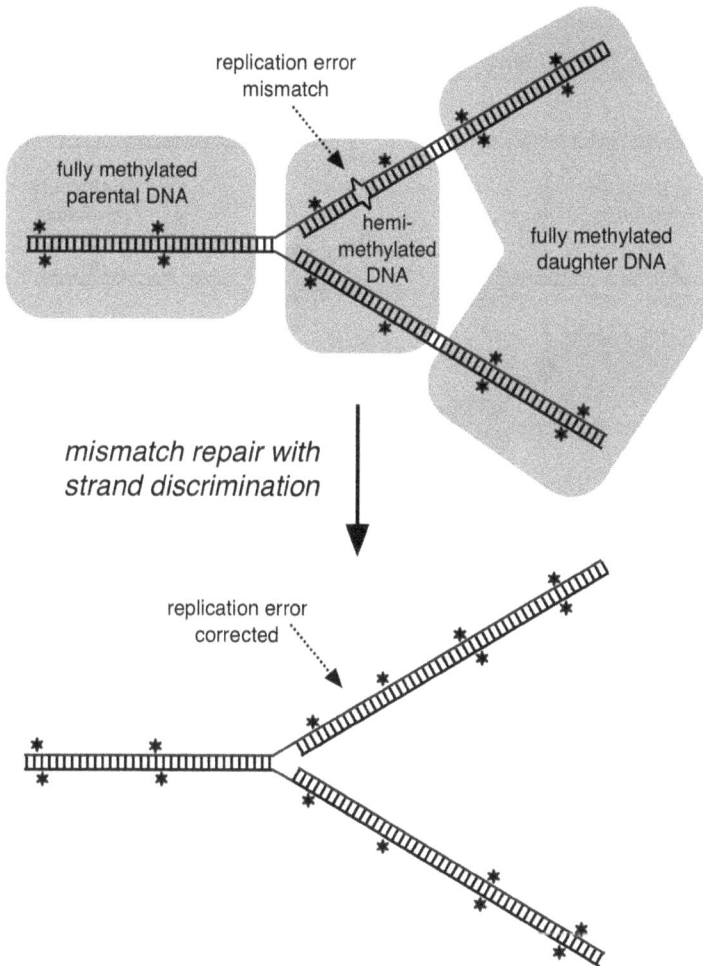

Figure 6.1. Methylation status and strand discrimination for mismatch repair in *Escherichia coli*. As previously depicted in Figure 5.3A, DNA is transiently hemi-methylated immediately after DNA replication in *E. coli*. This hemi-methylated state allows a period of time in which mismatch repair can correct replication errors (mismatched segment with two opposing triangles) in a strand-specific manner, preferentially replacing the newly synthesized strand. This figure was modified from Modrich (2016).

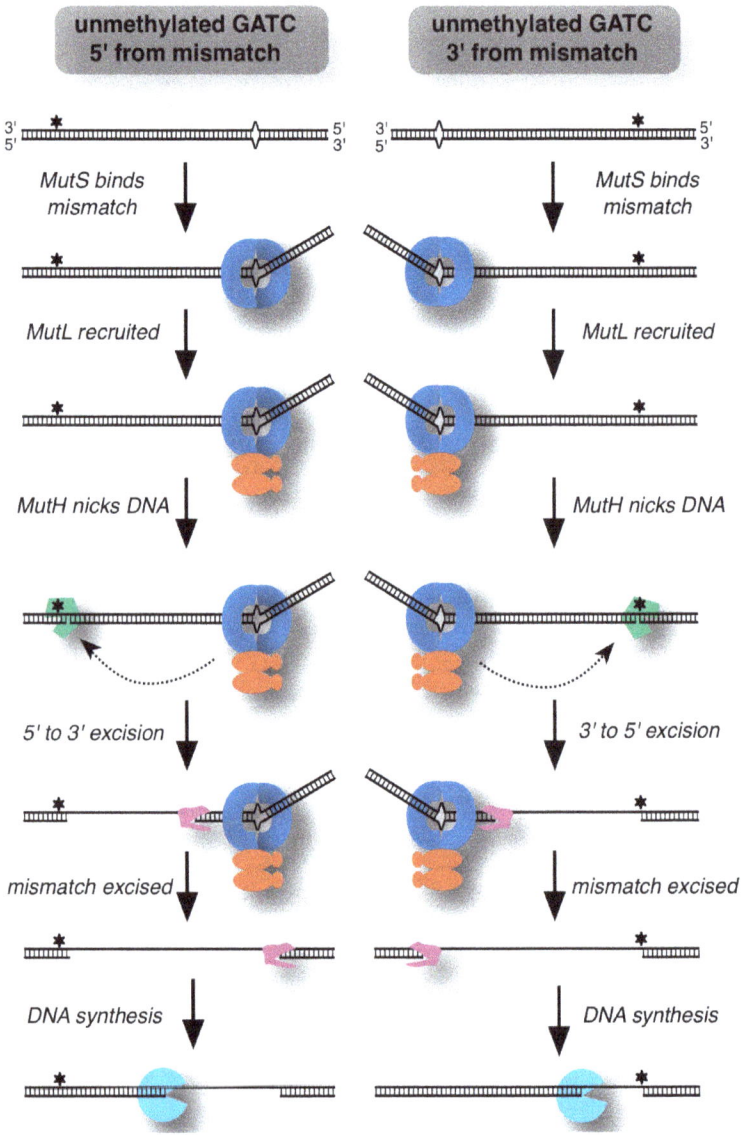

Figure 6.2. Mechanisms of methyl-directed mismatch repair in *Escherichia coli.* Two closely related pathways of mismatch repair are depicted, differing in the direction of the hemimethylated GATC site from the mismatch. MutS is in dark blue, MutL in orange, MutH in green, exonucleases in pink, and DNA polymerase in light blue. Modified from figures of Modrich (2016).

2015 Nobel Prize in Chemistry

This prize was awarded to **Tomas Lindahl, Paul Modrich**, and **Aziz Sancar** for elucidating mechanisms that cells use to repair damaged DNA and maintain genome fidelity, via the pathways of base excision repair (Lindahl), DNA mismatch repair (Modrich), and nucleotide excision repair (Sancar).

https://www.nobelprize.org/prizes/chemistry/2015/summary/

---

to the MutS:MutL complex. This restricts the DNA cleavage to DNA regions that actually require repair.

Once the nearby GATC site is nicked, the segment of DNA from the nick to the mismatch (and a bit beyond) is destroyed by exonuclease action (Figure 6.2). The UvrD protein is a DNA helicase involved in multiple repair pathways. The mismatch:MutS:MutL complex loads UvrD at the nicked GATC site in the orientation that allows the helicase to travel toward the mismatch. As UvrD unwinds the intervening DNA, one of several exonucleases degrade the displaced ssDNA to create a large gap in the duplex. Four different exonucleases can participate in this strand erosion step; each has a particular directionality and so different exonucleases are involved depending on whether the nicked GATC site is in the 5' or 3' direction from the mismatch (Figure 6.2). After the single strand is degraded away, the replicative DNA polymerase holoenzyme complex restores the missing DNA strand, with help from the ssDNA-binding protein, the replicative sliding clamp, and DNA ligase to seal the final nick after replication is complete. As described earlier, the repaired patch is very likely to have no errors, given the low error rate of the replicative DNA polymerase (roughly 1 in 10,000,000 base pairs replicated) and the relatively small size of the patch (on the order of 1000 bases or less).

6.3 Eukaryotic MMR

MMR in eukaryotic cells is, in many respects, similar to the *E. coli* MMR described earlier. Both systems use a signal to identify the newly synthesized strand, both systems erode a stretch of ssDNA from that signal to the mismatch region, both systems can act in either direction from the signal to the mismatch, and the eukaryotic system uses proteins that are homologous to the *E. coli* MutS and MutL proteins.

The value of a bacterial model system was very evident in the elucidation of mammalian MMR; indeed, this is a remarkable system to highlight the value of basic research. The biochemical assay that was developed to study *E. coli* MMR was used, with minor modification, to reconstitute mammalian MMR in vitro, first with crude extracts and later with purified proteins. In addition, as will be described in more detail below, the properties of bacterial cells deficient in MMR were remarkably similar to a particular class of human cancer cells, and this led to the demonstration that these cancer cells had a deficiency in MMR. Finally, defective human genes responsible for the high frequency of this class of cancers were identified more easily and quickly due to the clear homology of their protein products to the *E. coli* MutS and MutL proteins.

In spite of the similarities just mentioned, eukaryotic MMR differs from that in *E. coli* in the key aspect of strand discrimination. Eukaryotic cells do not employ DNA methylation as a strand signal, but rather rely on DNA nicks and gaps to identify the newly synthesized strand. It is easy to see how nicks and gaps would identify the newly synthesized strand on the lagging strand. Relatively short Okazaki fragments are the immediate product of synthesis on the lagging strand, providing a strand discontinuity every couple hundred base pairs, always on the newly synthesized strand. How then is the newly synthesized strand recognized on the leading strand side of the fork? To answer this question, we need to look more closely at the detailed mechanism of the pathway.

Recall that in the *E. coli* system, the MutH protein nicks the newly synthesized strand, that is, the one that lacks Dam methylation at a nearby Dam recognition site. Eukaryotic cells do not have a

MutH homolog, which is not surprising given that they also lack Dam methylation. However, the MutL homolog in eukaryotes does have a nicking activity, which is lacking in the *E. coli* MutL protein. This MutL homolog nicking activity turns out to be critical in directing eukaryotic MMR to the newly synthesized strand. However, the MutL homolog cannot provide strand direction on its own. The PCNA protein, the sliding clamp in DNA replication, is the critical partner that orients the MutL homolog to the correct strand. Recall that PCNA is a trimer that encircles the DNA, and also that the two faces of PCNA (up and down the DNA helix) are distinct from each other (see Chapter 4). Like DNA polymerase, the MutL homolog binds to only one face of the PCNA clamp, and this orientation restricts the nicking activity to only one of the two DNA strands (Figure 6.3). For this system to work, loading of PCNA by RFC must be uniquely oriented along the duplex. Indeed, RFC loads PCNA at the 3′ terminus of a single-strand/double-strand junction in a unique orientation (which is also critical to orient DNA polymerase correctly for replication; see Section 4.6).

Going back to MMR on the leading strand of a replication fork, RFC is thought to repeatedly load PCNA at the 3′ end of the growing leading strand as replication proceeds, providing excess PCNA for use in the MMR process. This PCNA is always oriented in the same direction, which is the direction that allows the MutL homolog to cleave the newly synthesized strand behind the replication fork (Figure 6.3). In addition to requiring loaded PCNA, the MutL homolog endonuclease activity also requires the MutS homolog dimer bound to a mismatch. Indeed, this requirement limits nicking by MutL homolog to the vicinity of the mismatch and prevents inadvertent nicking throughout the rest of the genome or in regions of newly synthesized DNA that have no mismatches.

Current areas of research in eukaryotic MMR include the precise nature of the excision and resynthesis reactions. The eukaryotic exonuclease called ExoI, which degrades single strands in the 5′ to 3′ direction, plays an important role in excision of the newly synthesized strand from the MutL homolog-induced nick, in a reaction stimulated by the MutS homolog. This would allow MMR from a

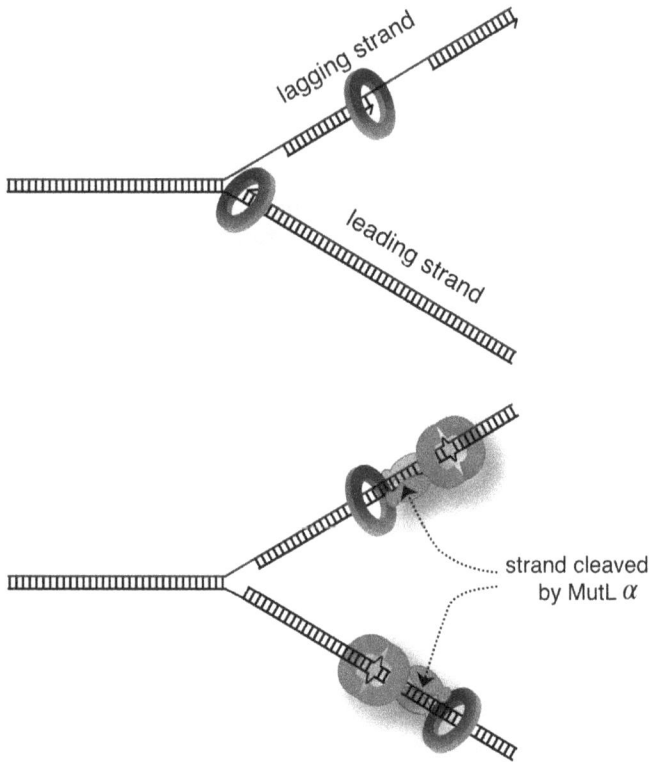

Figure 6.3. Strand-specific nicking by the MutLα protein. PCNA (donut-shaped protein encircling the DNA; present in both panels) loads the MutLα protein (MutL homolog) in opposite directions on the leading and lagging strands because the MutLα protein binds to only one of the two faces of PCNA. When the PCNA-MutLα complex associates with a MutS homolog complex bound to a mismatch (dimeric protein on the DNA segment with oppositely oriented triangles), the nuclease activity of MutLα is activated. Given the orientation of the MutLα protein on the DNA, its cleavage specificity is confined to only one strand, which corresponds to the newly synthesized strand.

nick that is located on the 5′ side of the mismatch, likely using DNA polymerase δ to replace the degraded strand. There is evidence that polymerase δ also provides a pathway for MMR when the nick is on the 3′ side of the mismatch. In this case, the polymerase synthesizes a new strand as it displaces the old strand containing the mismatch, and the displaced strand is cleaved off later.

Up until this point, we have been referring to the eukaryotic MMR proteins simply as MutS homologs and MutL homologs. The next complexity to discuss is the fact that eukaryotic cells have multiple homologs of MutS and MutL. First, in contrast to the MutS and MutL homodimers utilized in the *E. coli* system, eukaryotic cells utilize heterodimers in each case. This has functional consequence, for example, only one of the two MutL homolog subunits has endonuclease activity (which relates to the strand specificity of the nicking activity). Second, eukaryotic cells use distinct forms of each heterodimer in various cellular processes. For example, the primary form of MutS homolog in post-replication repair is the heterodimer MSH2/MSH6 (also called MutSα; MSH stands for *MutS homolog*), which acts on simple mismatches and small insertion/deletion mispairs (mismatched loops of 1–3 bases). Meanwhile, the MutS homolog MSH2/MSH3 (MutSβ) has a specialized role in repairing larger insertion/deletion mispairs (mismatched loops of 2–10 bases; note the overlapping specificity). Yet another heterodimer, MSH4/MSH5, plays an important role during meiotic recombination. Clearly, the detailed functions of the individual subunits of these important MMR proteins have become specialized during evolution to most efficiently operate with different kinds of mismatches and in partnership with various replication and recombination proteins.

Near the beginning of this chapter, it was noted that only certain bacterial species follow the *E. coli* model of methyl-directed MMR. In *E. coli* and the other species that follow this model, the MutL protein lacks the endonuclease active site that allows MMR in the eukaryotic system. However, these bacterial species employ the MutH protein, absent in eukaryotes, which induces the nicking that triggers MMR. Conversely, the bulk of bacterial species do not use Dam methylation, and these species more closely follow the eukaryotic model. Thus, these species lack MutH protein but harbor a MutL protein with nicking activity, just like the eukaryotic homolog. The endonuclease active sites of the eukaryotic and these bacterial MutL proteins are homologous to each other. Ironically, even though *E. coli* provided the key model system that helped unravel the MMR pathway, the methyl-directed MMR pathway is, in a sense, the exception rather than the rule.

6.4 Additional functions of MMR

The repair of DNA mismatches was first inferred in the analysis of meiotic recombination more than 50 years ago. In some lower eukaryotes such as the budding yeast *Saccharomyces cerevisiae*, a diploid cell can enter meiosis and produce haploid products that are fully viable. Meiosis includes a pre-meiotic round of DNA replication, followed by recombination between homologous chromosomes, and then segregation of the chromosomes into four haploid products. (Some yeast species also trigger a post-meiotic round of DNA replication, producing eight products.) Using genetic markers, these four (or eight) products from a single meiosis can be analyzed to infer the events that occurred during the meiotic recombination event. Without going into the gory details, the meiotic products from a heterozygous diploid sometimes displayed an unequal number of haploids with the two different genotypes. It appeared as if one genetic marker had been "converted" into the other during meiosis, and this process was thereby called gene conversion. Detailed study of this kind of gene conversion led to a model in which a DNA heteroduplex was formed at the site of the genetic marker, that is, a segment of duplex DNA consisting of one strand from each of the two parent chromosomes in the diploid (we will return to this process in Chapter 11). The actual gene-conversion event was inferred to result from repair of the mismatch that had been generated at the site of the genetic marker. MMR seems to be an integral part of meiosis — as indicated earlier, eukaryotic cells even have specialized versions of the MMR proteins that operate during meiosis. Furthermore, mutations in these proteins can cause sterility.

MMR also plays a critical role in the stability of DNA containing repeating sequences. This important role relates to the fact that repeating sequences provide a particular challenge for accurate replication by DNA polymerases. A simple run of the same base pair over and over, called a homonucleotide run, leads to errors in which the run gains or loses one base pair. For example, at one particular locus, a T:A base pair was lost from a run of 14 contiguous T:A base pairs at a frequency approaching 1 in a 1000 replication cycles (in an MMR-deficient yeast strain). This frequency is orders of magnitude higher than the frequency of mutation from any simple base:base mismatch error (such as mutations from a G:T mismatch). The difficulty posed by these repeating-mononucleotide runs is that the duplex can essentially slip, that is, one strand can shift register with the other in the forward or backward direction. This is often called strand slippage, and results in one (or more) nucleotide bulging out of the helix (Figure 6.4). If DNA polymerase

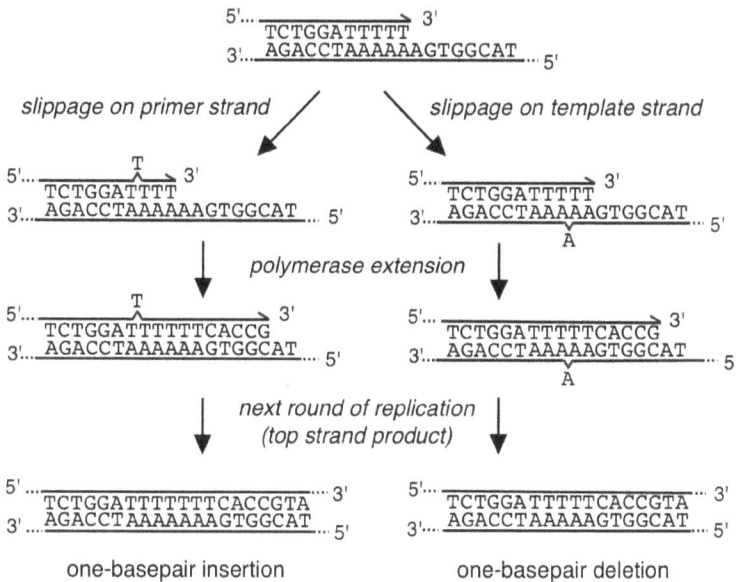

Figure 6.4. DNA slippage during replication causes insertion and deletion mutations. Strand slippage can occur on either the primer or template strand within an elongating DNA polymerase complex. If the polymerase continues unabated, the slipped structure is maintained in the immediate product. This slipped-strand structure can be repaired, for example, by mismatch repair, prior to the next round of DNA replication. However, if this repair does not occur, an insertion or deletion mutation will be created in one of the two daughter products (depicted in the last step).

extends the 3′ end of such a strand-slippage structure past the region of the homonucleotide run, one of the progeny cells will have suffered a one-base-pair insertion (bulge in primer strand) or deletion mutation (bulge in template strand) compared to its progenitor cells (again assuming that the mistake is not corrected by MMR).

The frequency of generation of these replication errors from slipped-strand structures varies in two important ways. First, the longer the repeat, the more frequent the errors. In the above example of the T:A homonucleotide run, the 14-repeat run generated mutations at a rate that was several-hundred-fold higher than a run of 7 T:A base pairs. Second, homonucleotide repeats generate errors more frequently than repeats of a more complex nature, including di- and tri-nucleotide repeats (e.g., a dinucleotide repeat sequence like 5′-CACACACACACA-3′/3′-GTGTGTGTGTGT-5′).

The slipped-strand structures generated by replication errors, such as that in Figure 6.4, can be corrected by MMR in a very similar manner as that of base:base mismatches. This is a key function of MMR, and a failure in this aspect of MMR may contribute to certain degenerative diseases that involve trinucleotide repeats (also see Chapter 14). Mechanistically, the process is analogous to repair of base:base mismatches. The use of strand-specific signals (methylation of Dam sites in *E. coli* or strand-specific nicks in eukaryotic cells) ensures that the MMR system corrects the newly synthesized strand, the one that contains the recent DNA polymerase error. As mentioned earlier, two different eukaryotic MutS homolog proteins have specialized function in correcting mismatched bulges depending on the size of the bulge. The standard heterodimer MutSα (MSH2/MSH6), which corrects base:base mismatches, also corrects bulges of 1–3 nucleotides (such as those in Figure 6.4), while the variant MutSβ (MSH2/MSH3) recognizes bulges of 2–10 bases.

6.5 MMR defects in cancer

Beginning in the 1970s, studies in model systems such as *E. coli* and *S. cerevisiae* revealed that MMR-deficient mutants showed an abnormally high frequency of insertion and deletion mutations in simple repeating sequences (sometimes referred to as "microsatellite instability"). Indeed, these were some of the early studies that led to the conclusions in the previous section about the role of MMR in repairing slipped-strand structures. The connection of MMR to cancer began to emerge with studies of a class of human cancer cells that were also found to exhibit microsatellite instability. These cells were from patients with the most prevalent hereditary form of colorectal cancer predisposition, called Lynch syndrome (previously called *h*ereditary *n*on*p*olyposis *c*olon *c*ancer; HNPCC). These patients suffered increased frequencies of a variety of other cancers in addition to colorectal, and so their genetic predisposition reflected some general problem in DNA metabolism.

Based on the similar phenotypes of the cancer cells to the MMR-deficient strains in *E. coli* and yeast, scientists tested whether the cancer cells could perform MMR and found them to be defective. Furthermore, the causative mutations in families with Lynch syndrome were soon traced to genes that encode the MutS and MutL homologs, directly tying the cancer-predisposition syndrome to defects in MMR. We will return to the cancer predisposition due to MMR defects in Chapter 14 (Section 14.5).

6.6 Postreplicative repair of incorporated ribonucleotide residues

Due to the 2′-OH group on its sugar, RNA is much more fragile than DNA — as much as 100,000 times more susceptible to spontaneous hydrolysis under physiological conditions. This explains why evolution adopted the more stable DNA molecule as the repository of genetic information in virtually all organisms on Earth (with the exception of certain viruses that use RNA). However, RNA is obviously synthesized at very high amounts in the cell, serving as messenger, transfer, and ribosomal RNAs. Accordingly, the pools of ribonucleoside triphosphates in the cell are large, estimated to be on the order of 100-times larger than deoxynucleoside triphosphate pools in eukaryotic cells. This contributes to an important problem in DNA replication. Even though replicative DNA polymerases are quite good at discriminating against RNA nucleotides, they are not perfect in this regard and they are fighting this 100-fold excess of ribonucleotides. The net result is that replicative DNA polymerases sometimes incorporate ribonucleotides into genomic DNA. Current estimates are that ribonucleotides are incorporated on the order of once for every 1000 bases incorporated during eukaryotic genomic replication. This translates into more than 10,000 ribonucleotide residues incorporated into the genome of a yeast (*S. cerevisiae*) cell during a round of replication, and more than a million such residues incorporated during replication of a human cell. Incorporated ribonucleotides are therefore the most common perturbation in genomic DNA structure during normal growth conditions.

In addition to making DNA much more fragile, incorporated ribonucleotide residues cause structural perturbations in the DNA duplex, disturbing the interaction of many proteins with DNA. There is evidence that one such disturbance involves inhibition/ alteration of nucleosome formation. In addition, a ribonucleotide residue incorporated in the template DNA strand (i.e., from a previous round of replication) acts to stall DNA polymerase as it tries to extend the opposite strand. Finally, and perhaps most importantly, incorporated ribonucleotide residues can lead to mutations via pathways that involve single- and double-strand breaks and the cellular enzyme DNA topoisomerase I.

As you might expect from the dangerous nature of incorporated ribonucleotides, normal cells have an efficient repair pathway for removing the vast majority of these residues.[1] The pathway is called RER, and initiates with an enzyme that specifically cleaves DNA at a ribonucleotide residue (Figure 6.5). The enzyme is RNase H2, which recognizes the altered structure of DNA and cleaves one strand immediately on the 5′ side of the incorporated ribonucleotide residue. Next, polymerase δ (or less efficiently, polymerase ε) carries out strand-displacement synthesis to replace the region with

---

[1] There is evidence for less efficient backup pathways in both prokaryotes and eukaryotes.

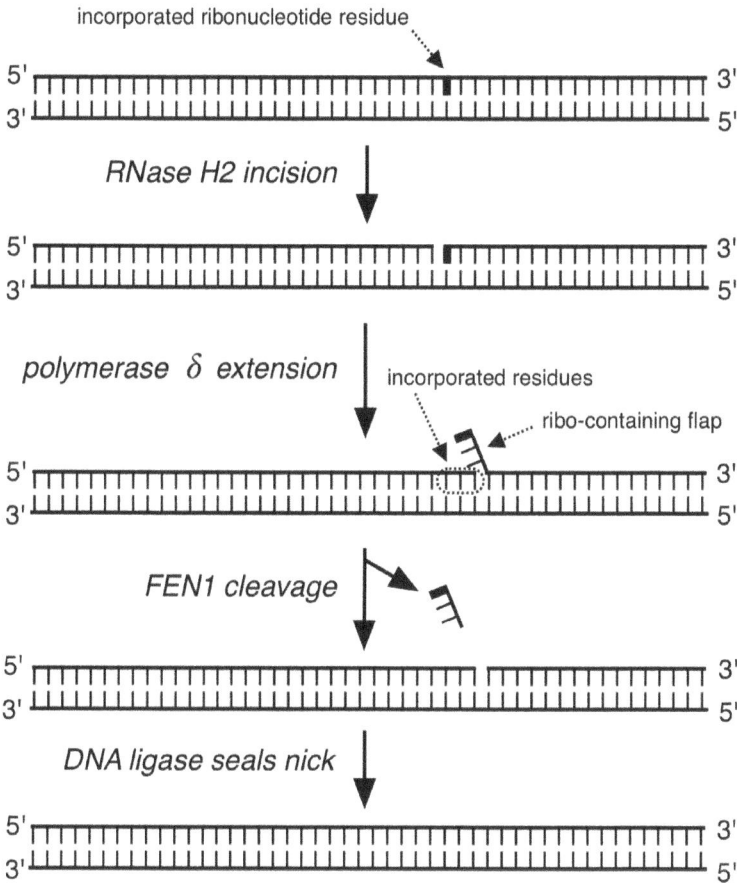

Figure 6.5. Repair of incorporated ribonucleotide residues after DNA replication. The basic steps in ribonucleotide repair are depicted.

the ribonucleotide residue as the ribonucleotide-containing strand is pushed off the duplex into a flap. The flap is then cleaved by the same flap endonuclease involved in Okazaki fragment processing (FEN1), and the resulting nick is sealed by DNA ligase. Importantly, one of the subunits of RNase H2 binds to the sliding clamp PCNA and thus localizes to the replication fork. This provides a direct coupling between DNA replication and repair of the incorporated ribonucleotide residues, and ensures that their lifetime in the genomic duplex is very short.

RER is a critical pathway, and mutations in this pathway have severe consequences in mammals. Mice that are constructed to contain knockout mutations that inactivate RNase H2 show an embryonic-lethal phenotype. Before dying, cells from the embryos show more than a million ribonucleotide residues per genomic DNA equivalent, and display a variety of abnormalities including genome instability and hyper-induced DNA damage responses. In humans, mutations in genes that encode the subunits of RNase H2 are involved in three different inherited diseases (*A*icardi-*G*outières *s*yndrome or AGS, *l*upus *e*rythematosus or LE, and *a*taxia with *o*culomotor *a*praxia *1* or AOA1). These diseases have different characteristics involving autoimmune responses, neurodegenerative problems, and symptoms that mimic congenital viral infection. The mechanisms by which perturbations in RNase H2 lead to these varied diseases are currently under investigation.

6.7 Summary of key points

- MMR corrects more than 99% of replicative errors.
- In *E. coli* and certain other bacterial species, the newly synthesized strand is targeted during MMR by means of hemi-methylated GATC sites that direct MutH-induced nicking of the unmethylated strand.
- In the *E. coli* system, MutS protein recognizes mismatches and then recruits MutL, which activates the strand-specific nicking by MutH.
- A combination of helicase and nuclease action destroys the strand between the mismatch and the nicking site, and DNA polymerase replaces the destroyed strand with new DNA.
- Eukaryotic MMR is similar to that in *E. coli* and uses MutS and MutL proteins that are homologous to the bacterial versions.
- Nonetheless, eukaryotic MMR does not employ a MutH-like protein nor DNA methylation status.
- Instead, the eukaryotic MutL protein nicks the newly synthesized strand, using the directional signal provided by its oriented loading on DNA by the PCNA clamp.

- Eukaryotic MutS and MutL proteins are heterodimers, and multiple forms of the eukaryotic MutS protein have specialized function in mismatch recognition.
- MMR also occurs on recombination intermediates during meiosis and plays a key role in preventing mutations within repeating DNA sequences.
- MMR deficiency causes cancer predisposition.
- Ribonucleotides are occasionally incorporated during DNA replication, promoted in part by the much larger pools of ribonucleotide precursors.
- The process of RER removes incorporated ribonucleotides, beginning with specific cleavage at the residue by a specialized RNase H enzyme.
- FLAP endonuclease, DNA polymerase δ, and DNA ligase complete the process of RER.
- RER deficiency causes embryonic lethality in mice and is responsible for certain inherited human diseases.

Further Reading

Heinen, C. D. (2016). Mismatch repair defects and Lynch syndrome: The role of the basic scientist in the battle against cancer. *DNA Repair (Amst), 38*, 127–134.

Jiricny, J. (2013). Postreplicative mismatch repair. *Cold Spring Harb Perspect Biol, 5*(4), a012633.

Kunkel, T. A., & Erie, D. A. (2015). Eukaryotic mismatch repair in relation to DNA replication. *Annu Rev Genet, 19*, 291–313.

Lahue, R. S., Au, K. G., & Modrich, P. (1989). DNA mismatch correction in a defined system. *Science, 245*(4914), 160–164.

Modrich, P. (2016). Mechanisms in *E. coli* and human mismatch repair (Nobel Lecture). *Angew Chem Int Ed Engl, 55*(30), 8490–8501.

Vaisman, A., McDonald, J. P., Huston, D., Kuban, W., Liu, L., Van Houten, B., & Woodgate, R. (2013). Removal of misincorporated ribonucleotides from prokaryotic genomes: An unexpected role for nucleotide excision repair. *PLOS Genet, 9*(11), e1003878.

Williams, J. S., Lujan, S. A., & Kunkel, T. A. (2016). Processing ribonucleotides incorporated during eukaryotic DNA replication. *Nat Rev Mol Cell Biol, 17*(6), 350–363.

How did they test that?
Reconstitution of methyl-directed mismatch repair in vitro

Mismatch repair is challenging to study because it is an infrequent event in DNA replication, occurring only after rare replicative errors. Lahue, Au and Modrich (1989) reported the first in vitro reconstitution of mismatch repair using purified proteins from *Escherichia coli.* The key to their studies was the assay system, which allowed restriction enzyme cleavage to report on the fate of the mismatched site. The scientists constructed a 6.44-kilobase-pair duplex circle with very special properties by annealing two ssDNA circles derived from a bacterial virus. The resulting duplex circle had normal base pairing throughout with the exception of a single G:T mismatch, located within the overlapping recognition sites of two restriction enzymes (HindIII [H] and XhoI [X]). The G:T mismatch prevents cleavage by either restriction enzyme. Mismatch repair on the V strand creates a functional site for only X, while repair on the C strand creates a functional site for only H. Mismatch repair is detected by cleaving the reaction products with a combination of ClaI (cuts opposite the mismatch site) and either X or H. The substrate generally is hemi-methylated at its lone d(GATC) site, which should direct the mismatch repair reaction to the unmethylated strand (according to prior genetic studies). All reactions contained the mismatch repair proteins, which had been purified from crude extracts largely by using this same assay. The experiment shown here tests the importance of the methylation status of the d(GATC) site. In the first pair of reactions, the d(GATC) site is replaced with an (unmethylated) d(GATT) site, and no mismatch repair occurs (no doubly cleaved DNA products; lanes 1 and 2). Likewise, a substrate with both strands methylated at the d(GATC) sequence is inert (lanes 3 and 4). When the methyl group is on the C strand, repair is directed to the V strand (cleaved by X, not H; lanes 5 and 6), but when it is on the V strand, repair is directed to the C strand (cleaved by H, not X; lanes 7 and 8). Finally, unmethylated DNA with a d(GATC) site is repaired at reduced efficiency but not in a strand-specific manner (lanes 9 and 10). The two panels in this figure were reproduced from Lahue *et al.* (1989), with permission from the American Association for the Advancement of Science; permission conveyed by Copyright Clearance Center, Inc.

V 5'-AAGCTTTCGAG Hind III
C 3'-TTCGAGAGCTC Xho I

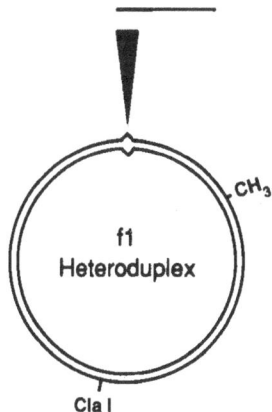

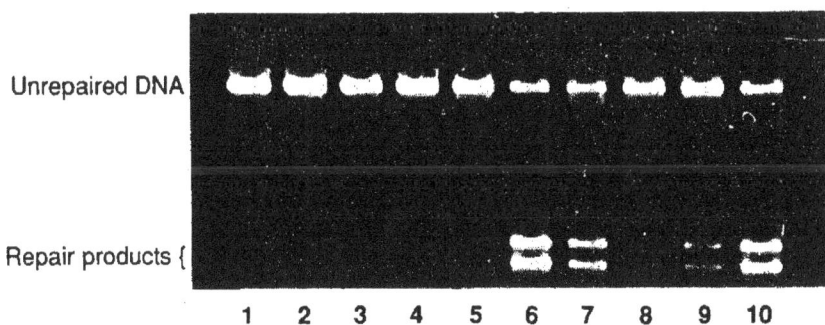

Chapter 7

DNA topology and the enzymes that alter it

7.1 The superhelical structure of duplex DNA

A discussion of the structure of plasmid DNA provides a useful introduction to the higher-order structure of DNA. Plasmids are small duplex DNA molecules, often in the range of a few-thousand base pairs long, which are frequently found in bacteria. Plasmids can carry a wide variety of genes, notably antibiotic-resistance genes, and must be able to replicate as the cells divide.

When plasmid DNA is isolated from bacterial cells and cleaved with a restriction enzyme that has one site in the plasmid, the resulting linear DNA has the expected duplex structure with the two complementary strands wound around each other about once every 10.5 base pairs. DNA ligase can then be added such that the two broken ends ligate back to each other, reforming a circle. However, this reformed circle is structurally distinct from the starting or "native" circle before it was cleaved (Figure 7.1). The native duplex circle winds around itself multiple times, a state referred to as supercoiling or superhelicity. The resealed circle does not show this additional winding, and is thereby referred to as a "relaxed" circle. You can convince yourself that a simple winding can result in supercoiling by removing your belt, if you are wearing one, and twisting one end a few times while holding the other end firm. When you

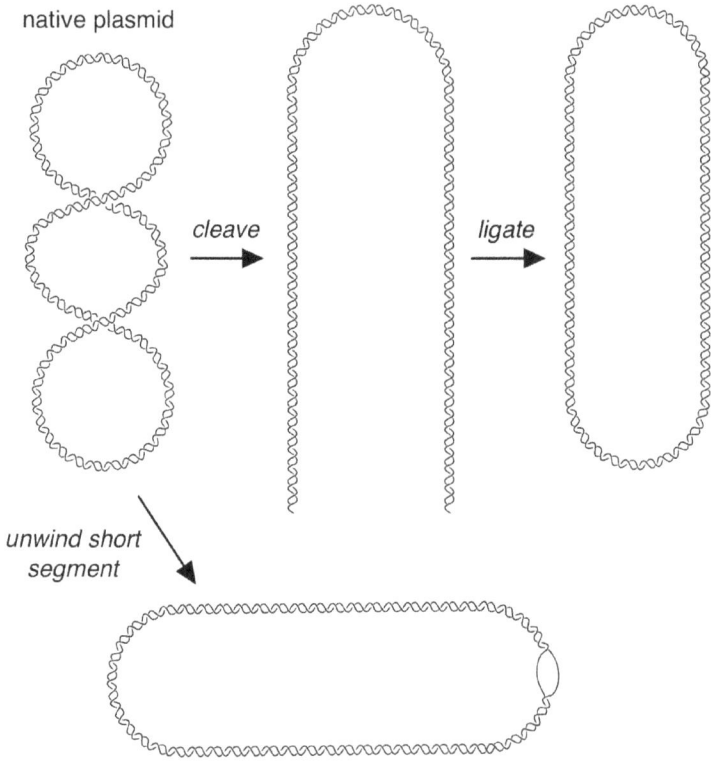

Figure 7.1. Underwinding and unwinding of duplex DNA. Plasmid DNA is isolated from bacterial cells in a negatively supercoiled state (top left). This underwinding can be interconverted to unwinding of a short duplex segment (bottom) with no breakage or resealing of phosphodiester bonds. When negatively supercoiled DNA is cleaved with a restriction enzyme and resealed, fully relaxed DNA with no supercoiling results (top right). Note that most plasmid DNA is much longer, on the order of thousands of base pairs with tens of negative supercoils. Also note that this depiction is idealized — a collection of DNA molecules generally has a distribution of topological states within the individual molecules.

buckle the belt back together, you will see that it forms a circle that is wound around itself.

Duplex DNA is already a helix with the two complementary strands coiled or wound around each other, and so the additional level of winding or coiling is naturally called supercoiling or super-helicity. A critical point is that the winding of the two strands around each other and the supercoiling of the duplex can be interconverted. If you disrupt the base pairing of an appropriate length of the native plasmid DNA and thereby eliminate the corresponding number of turns of the Watson–Crick double helix in that region, you in turn eliminate the supercoiling of the circle so that it looks more like the relaxed circle (Figure 7.1, bottom). This means that the native circle as isolated from the cell has a deficiency of helical turns compared to relaxed DNA, and the sign of the supercoiling is thereby said to be negative; the DNA is "underwound." It is also possible to generate positively supercoiled DNA, which has an excess of winding, and indeed positively supercoiled DNA can also be found in nature (although negative supercoiling is far more common).

The coiling of DNA can be considered in a stricter, mathematical sense, with the concept called the "linking number" or Lk. If a constrained DNA molecule like a circle is forced to lie in a plane, the total number of times the strands wind around each other equals the linking number. The linking number, however, has two components. First, the two strands wind around each other in the Watson–Crick helix about once every 10.5 base pairs, and this is called "twist" or Tw. Second, the double helix winds around itself, which also translates to the strands winding around each other, and this is called "writhe" or Wr. Mathematically, the total linking number Lk is the sum of these two, and so $Lk = Tw + Wr$. Linking number cannot be changed without breaking the DNA constraints, which for the circle, would mean breaking and resealing the DNA (with some kind of rotation in between, like you did with your belt).[1]

Operationally speaking, it is generally more useful to consider not the total linking number, but the linking-number difference compared with otherwise identical DNA that is totally relaxed. Again, one can express this as $\Delta Lk - \Delta Tw + \Delta Wr$. Considering a 4000-base pair long native plasmid isolated from bacterial cells, the ΔLk value is found to be on the order of −20. Notice that the mathematical value is negative, reflecting again that this is negatively

---

[1] Notice that no DNA backbone was broken during the interconversion labeled "unwind short segment" in Figure 7.1; the twist decreased by two coincidentally with a reduction in (negative) writhe of two (from −2 to 0).

supercoiled DNA. It can also be useful to express the linking-number difference in a form that is independent of DNA length, so that molecules of different length can be appropriately compared. This reflects a value called σ, which is equal to ΔLk divided by the Lk value of the same molecule in its relaxed state ($\sigma = \Delta Lk \div Lk^0$; for the 4000-base pair circle, $\sigma \cong -0.05$).[2]

As implied above, the linking number only makes sense for DNA that is constrained, and we have stuck to the example of intact circular DNA as being constrained. With linear DNA free in solution, the ends can rotate freely and so no net supercoiling can possibly accumulate; the DNA is unconstrained. However, if two ends of a linear DNA are held to a surface and not allowed to freely rotate, the concept of supercoiling and the mathematical treatments can be extended to linear DNA. In the very long linear chromosomes of eukaryotic cells, the DNA is held tightly at multiple locations in a complex-looped architecture, and hence it is perfectly sensible to discuss the superhelical state of this DNA. As with bacterial DNA, this eukaryotic DNA is negatively supercoiled. However, we will see below that the negative supercoiling of prokaryotic and eukaryotic DNAs arises from quite distinct mechanisms.

Negative supercoiling of duplex DNA has numerous important consequences. Any process that requires melting (unwinding) of a stretch of base pairs will be favored by negative supercoiling, in that the unwinding relieves some of the energetic cost of the supercoiling of the duplex around itself. The initiation of replication or transcription requires localized DNA unwinding, and so these processes are favored by negative supercoiling. Supercoiling (either negative or positive) also has the effect of condensing DNA to some extent.

7.2 The helical and superhelical dilemma of replicating DNA

As mentioned in Chapter 1, the double-helical structure of DNA immediately posed a puzzling question about DNA replication,

---

[2] Lk^0 for this plasmid is about $4000 \div 10.5 = 381$; σ therefore equals $(-20 \div 381)$ or -0.052.

particularly when it was later discovered that some DNA molecules are circular. Consider the 4000-base-pair plasmid mentioned earlier. The two parental strands are wound around each other nearly 400 times. Let us first imagine that the replication machinery rotates rapidly as it replicates the DNA, one rotation for every 10.5 or so base pairs (i.e., it follows the winding of the parental DNA) (Figure 7.2A). We will also assume that the two parental strands are never broken during the process (even transient breakage could

(A) Replication machinery rotates to follow parental helix

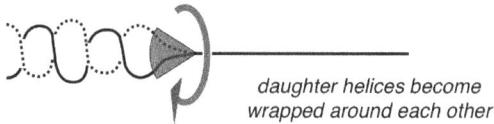

daughter helices become wrapped around each other

(B) Replication machinery fixed; parental helix rotates

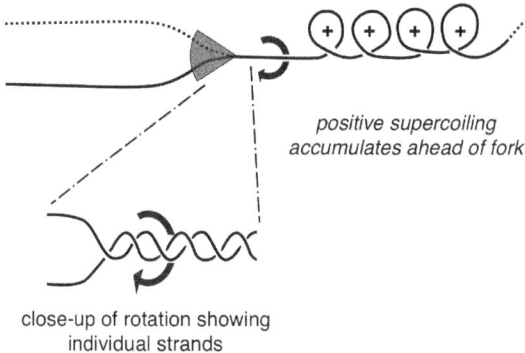

positive supercoiling accumulates ahead of fork

close-up of rotation showing individual strands

Figure 7.2. Two topological possibilities for DNA replication. The process of DNA replication must deal with the fact that the two strands in the duplex wind around each other about once every 10.5 base pairs. One possible solution is that the replication machinery spins rapidly to match the winding of the two strands of parental DNA around each other (panel A). This has the advantage of not disturbing the duplex DNA out in front of the machinery, but the disadvantage of expelling the two daughter helices in a complex form in which they are extensively wrapped around each other. Another possibility is that the replication machinery is fixed (panel B). In this case, the two daughter helices are expelled in a simple, nonintertwined manner. However, the parental helix becomes overwound (positive superhelicity) ahead of the replication machinery because this helix must rotate to unwind as it enters the replication machinery, and the rotation must be counteracted.

allow rotation). As the two daughter duplexes exit the replication machinery, they would need to be wrapped around each other once every 10.5 base pairs — the two daughter duplexes would end up hopelessly tangled, writhing around each other nearly 400 times (Figure 7.2A). Segregating such tangled circles would be nearly impossible. Yet, the circular chromosome of *Escherichia coli* is 1000 times longer still!

You might imagine instead that the replication machinery is fixed in the cell and the DNA passes through this machine as it replicates (Figure 7.2B). Now, the two daughter duplexes would exit the replication machinery in a nice, separated fashion without tangling. However, to achieve this separation, the parental DNA in front of the replication machinery would need to rotate rapidly, once for every 10.5 or so base pairs replicated, to allow the parental DNA to feed into the stationary replication machinery. This rotation would rapidly lead to the accumulation of excess coiling in front of the replication machine, which in this case would constitute positive supercoiling (Figure 7.2B). The magnitude of this supercoiling would be crushing when you imagine replicating DNA that is 4000-base pairs long, let alone chromosome-length DNA that is 1000-times longer. Without some compensating mechanism, replication would rapidly come screeching to a halt due to the excess positive super-coiling ahead of the fork, which makes further unwinding at the fork much more difficult. This kind of fork blockage due to accumulation of positive supercoiling has indeed been observed during in vitro DNA replication under appropriate conditions.

7.3 DNA topoisomerases to the rescue

The topological problems in DNA replication described earlier are solved, in large measure, by a collection of enzymes called DNA topoisomerases (which also have other crucial roles in the cell). As their name implies, DNA topoisomerases have the ability to alter the topology of DNA. From the above discussion of DNA topology, you might infer that DNA topoisomerases must break and reseal DNA in order to perform their function, and you would be correct.

There are two major categories of DNA topoisomerases, type I topoisomerases, which transiently break one strand of duplex DNA (Figure 7.3), and type II topoisomerases, which transiently break both strands of DNA (Figure 7.4). One remarkable aspect of DNA topoisomerases is that they do not create free DNA

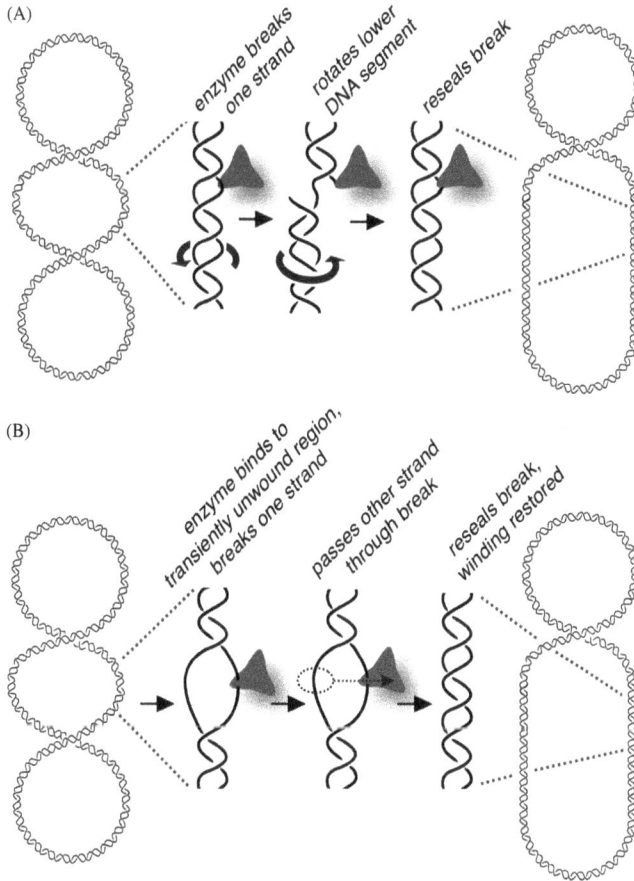

Figure 7.3. Reaction mechanism of type I DNA topoisomerases. Type I topoisomerases transiently break one DNA strand at a time. One group of these enzymes cleave the strand within a duplex region and allow rotation around the break, removing one superhelical turn for each rotation of the helix (panel A). The second group of type I topoisomerases act by a strand-passage mechanism within a partially denatured region of DNA: they pass one single strand through a transient break in the other single strand (panel B). The net result is the loss of a single superhelical turn.

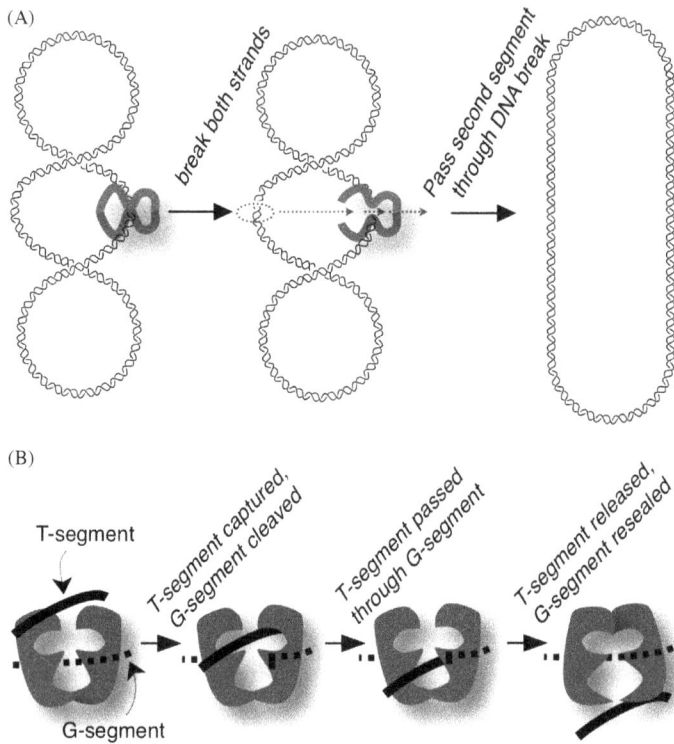

Figure 7.4. Reaction mechanism of type II DNA topoisomerases. Type II DNA topoisomerases transiently break both strands, promoting the passage of another duplex DNA segment through the transient break (panel A). This results in removal (or introduction) of two superhelical turns at a time. The roles of the G-segment (*g*ate) and T-segment (*t*ransport) are shown in more detail in panel B. The cartoon of the enzyme depicts three protein regions where the two topoisomerase subunits (left and right) contact each other, as well as the two large cavities where the T-segment resides just before and just after passage through the G-segment. The reader is strongly encouraged to reproduce the results of the type I and type II mechanisms with a wire (such as USB wire) or belt to convince herself/himself of the numerical change in superhelical turns.

breaks, like nucleases do, but rather transiently break the phosphodiester bond in DNA as they form a protein–DNA covalent bond at the break. This ensures that a topoisomerase cannot "lose its grip" on the broken DNA, which could be disastrous. In addition, the covalent protein–DNA bond is a high-energy bond, like the phosphodiester bond that was broken, and this energy is important in reforming the intact phosphodiester bond at the end of the topoisomerase reaction. Indeed, some (but not all) topoisomerases do not even require an energy source such as ATP for their reaction. All known topoisomerases use the hydroxyl group of an active site tyrosine or serine to carry out the breakage and resealing reaction. The type II topoisomerases are dimeric enzymes with two active sites, so that both strands of DNA become covalently attached to the enzyme (one on each side of the DSB).

Simply breaking and resealing DNA by itself does not change the DNA topology. Instead, the winding of the two strands around each other has to be altered while the DNA is transiently broken. One way of doing this is for the duplex on one side of the break to swivel around itself one or more times before resealing, which is the mechanism used by many type I topoisomerases (Figure 7.3A).

A second way of altering DNA topology is to transiently break one strand and then physically pass the other strand through the break. This mechanism is used by certain type I topoisomerases that act on a region of DNA that is transiently unwound, essentially a single-strand bubble region (Figure 7.3B). The net effect is topologically equivalent to rotating the duplex once, and in both cases, the linking number is changed one step at a time.

As mentioned earlier, the type II topoisomerases transiently break both strands of the duplex. In order to alter DNA topology, these enzymes pass another duplex segment of DNA through the transient break (Figure 7.4). It is useful to define the two segments of duplex DNA — the transient DNA break serves as a gate, and is thereby called the "G-segment," while the second segment of DNA is transported through the gate and is thereby called the "T-segment." Notice that the T-segment is transported through both the DNA gate as well as the enzyme itself (Figure 7.4B). A remarkable aspect of this reaction mechanism is that the type II topoisomerases change the DNA-linking number in steps of two rather than one (Figure 7.4A). Experimental demonstration of a linking-number change of two provided some of the earliest evidence for the mechanism of the type II topoisomerases (see "How did they test that" at the end of this chapter).

In our earlier discussion, we saw two potential topological problems inherent in the process of DNA replication, tangling of the daughter duplexes around each other, and accumulation of positive supercoiling ahead of the replication fork. Most (but not all) DNA topoisomerases are able to remove (or "relax") positive supercoiling, and indeed various topoisomerases have been shown to function ahead of the replication fork to assist DNA replication. In addition, type II topoisomerases can readily solve the tangling of the two duplex daughter molecules behind the fork. As mentioned earlier, these enzymes make a transient DSB in one segment of DNA and pass another duplex segment through the break. Perhaps surprisingly, the other segment of duplex DNA need not be from the same DNA helix — one daughter duplex can be transiently broken, while the second daughter gets passed through the break, as shown in Figure 7.5A. Furthermore, after complete replication of a circle, the two daughter duplexes are likely to end up interlocked at least a few times. This is because a few Watson–Crick helical turns will be unwound during the very last phase of replication and result in daughter interlocking (Figure 7.5B). Even in this situation, type II topoisomerases are able to resolve the interlocks, passing one daughter duplex through the other until the two circles are no longer tangled.

All cells have multiple DNA topoisomerases, often two or more of each of the two types. The precise roles of each of the four or more topoisomerases in a given cell are often complex and overlapping, and it is beyond the scope of this book to delve into all the complexities. However, it is worth noting that when the appropriate topoisomerase (or combination of topoisomerases) is inhibited (by drugs or by mutation), DNA replication fails. In some situations, the failure is seen as a cessation of replication mid-stream, as the release of positive supercoiling is prevented. In other situations, when replication is able to complete, the failure is exhibited during cell division, with two daughter chromosomes being unable to separate from each other due to DNA tangling.

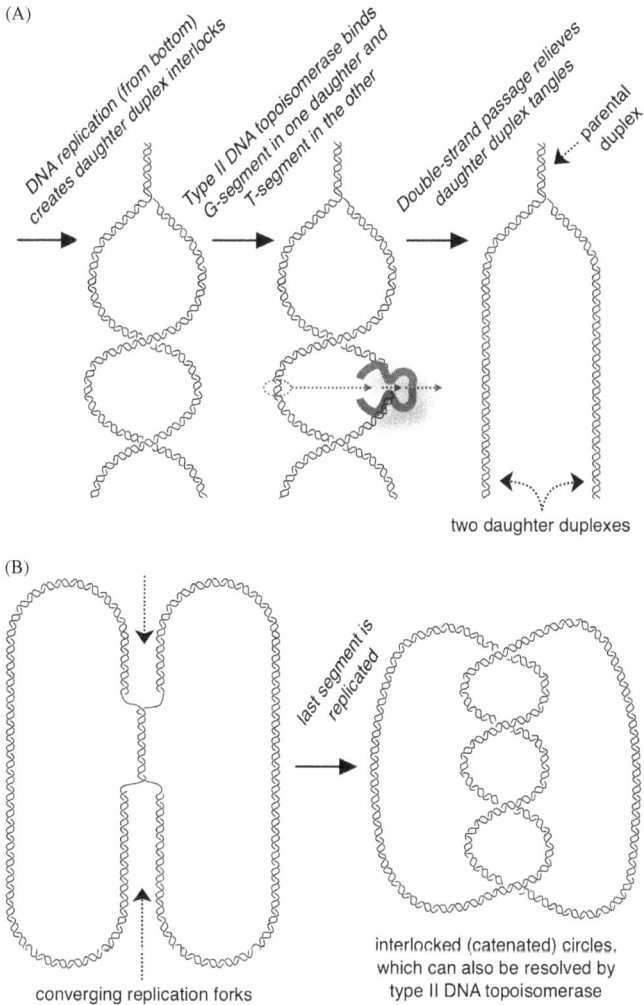

(A)

DNA replication (from bottom) creates daughter duplex interlocks

Type II DNA topoisomerase binds G-segment in one daughter and T-segment in the other

Double-strand passage relieves daughter duplex tangles

parental duplex

two daughter duplexes

(B)

last segment is replicated

converging replication forks

interlocked (catenated) circles, which can also be resolved by type II DNA topoisomerase

Figure 7.5. Topological issues in the final stages of DNA replication. Particularly during the late stages of DNA replication, the two daughter helices can become intertwined with each other due to a failure in removal of positive superhelicity ahead of the fork. (This kind of intertwining is sometimes called catenation, particularly when it applies to the intertwining of two complete circles like the ones at bottom right). Type II DNA topoisomerases are capable of disentangling the two daughter helices by passing one intact daughter helix through a transient double-strand break in the other (panel A). In the very final stages of replication, two converging replication forks readily generate catenated products (panel B). This is particularly evident when you consider that the area of duplex DNA ahead of the forks becomes too small for a type II DNA topoisomerase to bind and remove positive superhelical turns.

7.4 The sources of negative supercoiling inside cells

As already mentioned, DNA from both prokaryotes and eukaryotes is found to be negatively supercoiled when it is extracted from cells (assuming the constraints on the DNA are maintained). However, the sources of negative supercoiling are distinct in the two kingdoms.

A specialized type II topoisomerase called DNA gyrase introduces most (but not all) of the negative supercoiling in bacteria. Like other type II topoisomerases, DNA gyrase makes a transient DSB in the G-segment. However, gyrase has the unique ability to orient the T-segment that is passed through the gate in such a way that net negative supercoiling is introduced. This is accomplished by having the bound DNA wrap around a domain of DNA gyrase in preparation for strand passage, as illustrated in Figure 7.6. The introduction of negative supercoiling by DNA gyrase requires ATP hydrolysis, as

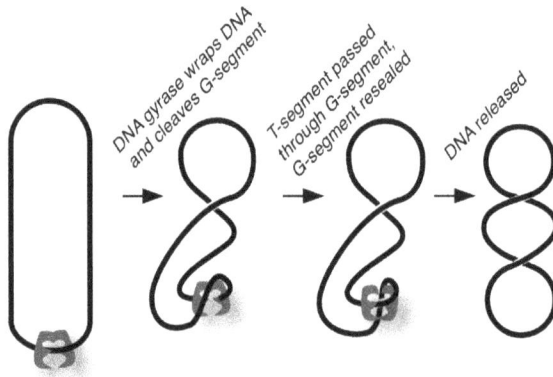

Figure 7.6. Active DNA supercoiling by DNA gyrase. Bacterial DNA gyrase actively introduces negative superhelical turns, two at a time. This reaction is accomplished by a wrapping of a segment of the duplex DNA around the enzyme, followed by a specifically oriented strand passage. Note that when DNA gyrase wraps a segment of the DNA, the remainder of the DNA has to twist around to form a compensating "local" superhelical turn (top of the circle after the first arrow). This local turn is essentially captured into a global negative superhelical turn as gyrase completes its duplex-break resealing; the second global superhelical turn is the one that was constrained in the DNA bound to the enzyme.

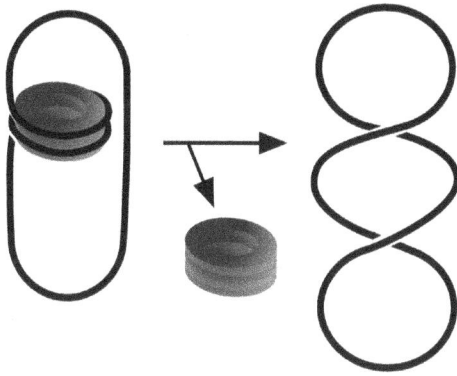

Figure 7.7.   Topological equivalence of DNA wrapping and negative superhelicity. In this diagram, duplex DNA is wrapped twice around a protein such as a nucleosome complex (left). Upon dissociation of the protein, the underwinding manifests as negative superhelicity.

might be expected because negatively supercoiled DNA is in a state of higher energy than relaxed (nonsupercoiled) DNA.

A portion of the negative supercoiling in DNA isolated from bacterial cells can be attributed to a more passive source. Certain DNA-binding proteins bind DNA in a coiled fashion that essentially constrains negative supercoiling in the DNA. When released from the protein, the DNA is negatively supercoiled just like DNA that has been treated with DNA gyrase (Figure 7.7). Several bacterial proteins are responsible for such passive supercoiling, and they are sometimes referred to as histone-like proteins (see below). The bacterial histone-like proteins are involved in a wide variety of cellular processes, beyond just contributing to chromatin structure.

As suggested by the above discussion of bacterial histone-like proteins, the eukaryotic histone proteins and the nucleosomes they form are responsible for nearly all the supercoiling in eukaryotic cells (just as in Figure 7.7). Furthermore, a large portion of eukaryotic DNA is bound within nucleosomes throughout the cell cycle. Nucleosomes consist of a stretch of DNA wrapped around an octamer of histones — two each of four histones called H2A, H2B, H3, and H4. About 140 base pairs of DNA is wrapped roughly twice around the histone octamer, always in the same orientation (which constrains negative supercoiling). In between nucleosome cores, a specialized histone called H1 is bound to so-called spacer DNA.

Eukaryotic histones and the histone-like proteins in bacteria are similar in overall composition, but do not show any particular amino acid homology with each other. All are fairly small, basic proteins that are rich in basic amino acids (lysine and arginine) that can bind to the negatively charged phosphodiester backbone. These proteins play myriad roles in chromosome function and structure, in addition to their structural role in constraining negative supercoiling in the chromosomal DNA.

7.5 Subversion of topoisomerases as a potent tool in chemotherapy

A number of important chemotherapeutic agents target DNA topoisomerases, usually via a mechanism that involves subversion of the normal reaction cycle (Table 7.1). Recall that every topoisomerase makes transient breaks in DNA and becomes covalently attached to

Table 7.1. Chemotherapeutic agents that inhibit DNA topoisomerases.

Category	Enzyme	Drug
Antibacterial	Bacterial DNA gyrase (DNA topoisomerase IV in some species)	Quinolones and fluoroquinolones: nalidixic acid, ciprofloxacin, gemifloxacin, levofloxacin, moxifloxacin, norfloxacin, and ofloxacin
Anticancer	Mammalian type I topoisomerase	Camptothecin, topotecan, irinotecan (CPT-11)
Anticancer	Mammalian type II topoisomerases	Doxorubicin (adriamycin), etoposide, teniposide, daunorubicin (daunomycin), mitoxantrone, ellipticine, amsacrine
Anticancer	Mammalian type II topoisomerases	ICRF-187, ICRF-193

Note: This table lists many of the known antibacterial and anticancer agents that inhibit DNA topoisomerases; many of these compounds are in current clinical usage. All of the compounds in the first three rows stabilize the cleaved topoisomerase-DNA complex, either enhancing the forward cleavage reaction or inhibiting the resealing reaction. The ICRF compounds in the last row act by a different mechanism, acting on the ATPase domain of the enzyme.

the broken DNA during the intermediate phase of the reaction cycle when DNA is passed through the break. A unifying feature of (most of) these chemotherapeutic agents is that they trap or stabilize the reaction intermediate in which the topoisomerase is covalently attached to broken DNA. At the structural level, scientists have found that these inhibitors bind right at the enzyme active site, where the DNA is broken (Figure 7.8). If you look closely, you will see that the (green or yellow) inhibitors are actually localized within the DNA molecule exactly at the site of the DNA break, thereby physically interfering with the resealing reaction. In addition to contacting DNA, the inhibitors also interact with nearby amino acid residues in the enzyme, accounting for the fact that any one topoisomerase inhibitor targets only certain topoisomerases.[3]

The stabilization of this reaction intermediate by the inhibitors causes a destructive effect on cells that is equivalent to a particularly nasty form of DNA damage, and this DNA damage is more important for cell killing than is the simple inhibition of enzyme activity. In a very real sense, these drugs turn what is normally a critical productive enzyme into a cellular poison by subverting the reaction mechanism in this unique manner.

The first example involves the antibacterial quinolones and more powerful fluoroquinolones, which target bacterial DNA gyrase (and in some bacteria, the type II topoisomerase called topoisomerase IV). This group includes the commonly used ciprofloxacin and a number of other related compounds that are prescribed and are named with the "–floxacin" extension (such as norfloxacin and gemifloxacin). Ciprofloxacin itself is one of the most commonly used antibiotics, with a market share that exceeded a billion US dollars annually while it was under patent.

In the realm of anticancer agents, the mammalian type II topoisomerase was shown to be the target of a number of anticancer drugs, including some that were in clinical use before their target was

---

[3] Correspondingly, mutations that alter these residues can cause resistance to the drugs, contributing to the problem of drug resistance in antibacterial and anticancer therapy.

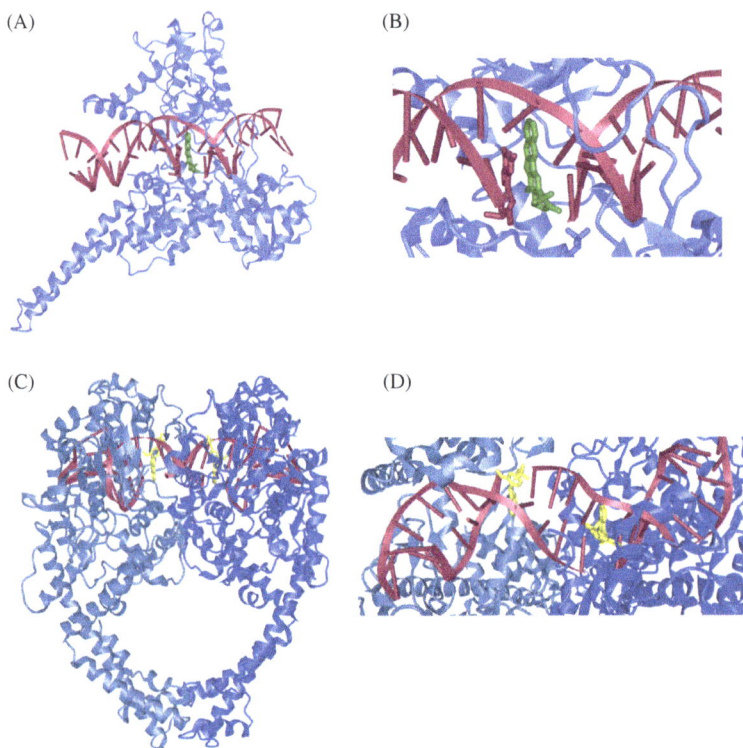

Figure 7.8. Disruption of topoisomerase reactions by chemotherapeutic drugs. The three-dimensional structures of topoisomerases in complex with DNA and chemotherapeutic drugs are shown. The structure in panels A (full view) and B (close-up including inhibitor) is that of a complex of human topoisomerase I (blue), camptothecin (green), and DNA (red), from the RCSB PDB (www.rcsb.org) of PDB ID 1T8I (Staker *et al.*, 2005). The structure in panels C (full view) and D (close-up including inhibitor) is that of a complex of human topoisomerase II β (two subunits in different shades of blue), amsacrine (yellow), and DNA (red), from the RCSB PDB (www.rcsb.org) of PDB ID 4G0U (Wu *et al.*, 2013). The close-up in panel B is from the same orientation as in panel A, while the close-up in panel D is looking down on the complex from the viewpoint of panel C. In both complexes, the inhibitor is bound within the DNA, precisely at the site(s) where DNA cleavage occurred (note two molecules of amsacrine, one at each cleaved phosphodiester bond). These structural images were generated using the web-based visualization suite iCn3D ("I see in 3D", version 2.7.15), developed by Wang *et al.* (2019), using the above-stated coordinate IDs from the PDB.

identified. Drugs targeting mammalian type II topoisomerase include doxorubicin, ellipticine, amsacrine, daunorubicin, genistein, etoposide, and teniposide (a complex with amsacrine is shown in Figure 7.8C and 7.8D). Mammalian type I topoisomerase was found to be the target of a natural compound isolated from the Chinese tree Camptotheca (Happy Tree), extracts of which had been used in traditional Chinese medicine as a cancer treatment. The original compound is called camptothecin, and its more powerful derivatives such as CPT-11 and topotecan are clinically useful against colorectal cancer and certain ovarian cancers (and likely other cancers; a complex with camptothecin is shown in Figure 7.8A and 7.8B).

7.6 Summary of key points

- Nearly all DNA in cells is in a negatively supercoiled state, with a deficiency of turns of the two strands around each other.
- Without some compensating mechanism, the process of DNA replication would result in extensive tangling of the daughter duplexes and/or extensive positive supercoiling of parental DNA ahead of the fork.
- Type I DNA topoisomerases break one strand of the duplex, whereas type II DNA topoisomerases break both strands.
- While the DNA is transiently broken, topoisomerases either swivel the two strands of the duplex around each other or pass an intact (single or double) strand through the transient break.
- DNA topoisomerases assist DNA replication by untangling daughter duplexes and by removing positive supercoiling in the parental DNA ahead of the fork.
- Much of the negative supercoiling in bacteria results from an active process involving a type II topoisomerase named DNA gyrase.
- Passive coiling of the helix around proteins such as histones constrains negative supercoiling, accounting for the negative supercoiling in eukaryotic cells (and a subset of that in bacterial cells).
- Several important chemotherapeutic drugs (antibacterial and anticancer) target DNA topoisomerases, usually by stabilizing the transient intermediate between the enzyme and broken DNA.

Further Reading

Brown, P. O., & Cozzarelli, N. R. (1979). A sign inversion mechanism for enzymatic supercoiling of DNA. *Science, 206*(4422), 1081–1083.

Collin, F., Karkare, S., & Maxwell, A. (2011). Exploiting bacterial DNA gyrase as a drug target: Current state and perspectives. *Appl Microbiol Biotechnol, 92*(3), 479–497.

Liu, L. F., Liu, C. C., & Alberts, B. M. (1980). Type II DNA topoisomerases: Enzymes that can unknot a topologically knotted DNA molecule via a reversible double-strand break. *Cell, 19*(3), 697–707.

Nitiss, J. L. (2009). Targeting DNA topoisomerase II in cancer chemotherapy. *Nat Rev Cancer, 9*, 338, doi:10.1038/nrc2607.

Nöllmann, M., Crisona, N. J., & Arimondo, P. B. (2007). Thirty years of *Escherichia coli* DNA gyrase: From in vivo function to single-molecule mechanism. *Biochimie, 89*(4), 490–499.

Pommier, Y. (2006). Topoisomerase I inhibitors: Camptothecins and beyond. *Nat Rev Cancer, 6*, 789, doi:10.1038/nrc1977.

Schoeffler, A. J., & Berger, J. M. (2008). DNA topoisomerases: Harnessing and constraining energy to govern chromosome topology. *Q Rev Biophys, 41*(1), 41–101.

Staker, B. L., Feese, M. D., Cushman, M., Pommier, Y., Zembower, D., Stewart, L., & Burgin, A. B. (2005). Structures of three classes of anticancer agents bound to the human topoisomerase I-DNA covalent complex. *J Med Chem, 48*(7), 2336–2345.

Vos, S. M., Tretter, E. M., Schmidt, B. H., & Berger, J. M. (2011). All tangled up: How cells direct, manage and exploit topoisomerase function. *Nat Rev Mol Cell Biol, 12*, 827, doi:10.1038/nrm3228.

Wang, J. C. (2009). *Untangling the Double Helix*. Cold Spring Harbor, NY: Cold Spring Harbor Press.

Wang, J., Youkharibache, P., Zhang, D., Lanczycki, C. J., Geer, R. C., Madej, T., Phan, L., Ward, M., Lu, S., Marchler, G. H., Wang, Y., Bryant, S. H., Geer, L. Y., & Marchler-Bauer, A. (2019). iCn3D, a web-based 3D viewer for sharing 1D/2D/3D representations of biomolecular structures. *Bioinformatics.* doi:10.1093/bioinformatics/btz502

Wu, C. C., Li, Y. C., Wang, Y. R., & Chan, N. L. (2013). On the structural basis and design guidelines for type II topoisomerase-targeting anticancer drugs. *Nucleic Acids Res. 41*(22), 10630–10640.

How did they test that?
Type II DNA topoisomerases change linking number in steps of two

Studies published in 1979 and 1980 showed dramatic evidence that certain DNA topoisomerases alter DNA topology by means of transient DNA DSBs. DNA topology can only change when either one or both strands are transiently broken, and all prior studies focused on transient single-stranded breaks by type I topoisomerases. Brown and Cozzarelli (1979) predicted that the topological result of the reaction would reflect the reaction mechanism: transient breaks in one strand would alter the linking number in a unitary fashion, while transient DSBs would alter the linking number in steps of two (top figure). Small DNA circles with distinct linking numbers (distinct topoisomers) can be separated from each other in an agarose gel. The scientists purified a unique topoisomer by running such an agarose gel, cutting out the band of interest, and eluting the DNA. They used this topoisomer as a substrate in DNA gyrase reactions. Lanes d, g, and l show a collection of all relevant topoisomers from an unpurified DNA sample, while lanes a and b show the purified topoisomer ($\Delta Lk = 0$) that was used as a substrate (a very small amount of contaminating linear DNA runs just faster than the circular topoisomer). Samples from a time course with DNA gyrase are in lanes c, e, f, h, i, and j. By the later time points, the products have accumulated so much negative supercoiling that they cannot be resolved in this gel. However, in the first three time points, you can clearly see that the only topoisomers formed have a ΔLk of -2 and -4, with a notable absence of -1 and -3 (compare to all topoisomers in lanes d, g, and l). DNA gyrase carries out the reverse reaction in the presence of a certain inhibitor. When the highly supercoiled product (lane j) was incubated with DNA gyrase and the inhibitor, again only the -2 and -4 topoisomers were generated (lane k). Liu, Liu, and Alberts (1980) obtained comparable results showing changes of linking number in steps of two with another topoisomerase, and were the first to call these enzymes type II DNA topoisomerases.

The top diagram was modified from Brown and Cozzarelli (1979). The data figure at the bottom was reproduced from Brown and Cozzarelli (1979), with permission from the American Association for the Advancement of Science; permission conveyed by Copyright Clearance Center, Inc.

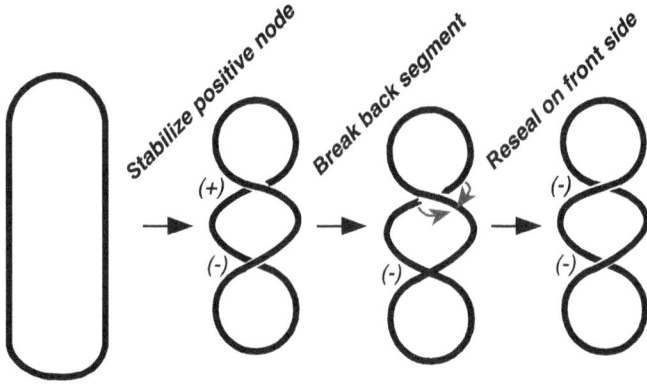

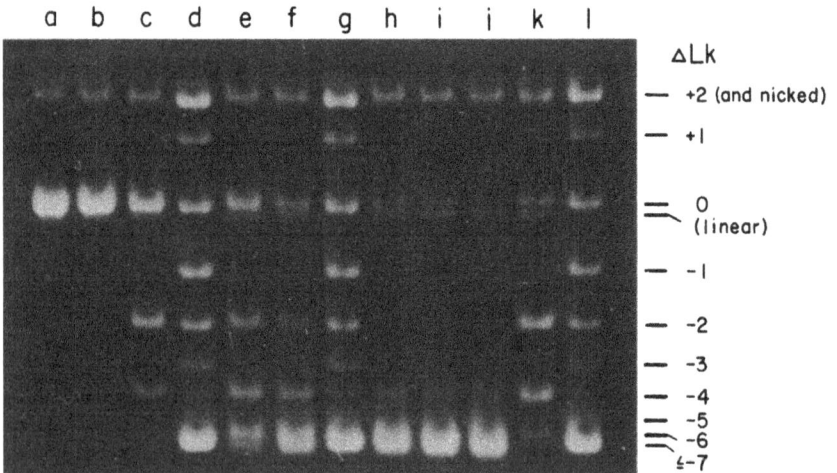

Chapter 8

DNA damage — a persistent threat to the genome

8.1 Introduction

Although DNA faithfully carries genetic information from generation to generation over the eons, it is vulnerable to damage. Individual bases get corrupted, strands break, and crosslinks are formed between the two strands or between DNA and cellular proteins. As described earlier in this book, the double-stranded nature of genomic DNA provides a redundancy to the genetic information that is extremely useful for repairing many forms of damage (Chapters 1 and 10). The diploid nature of most eukaryotic cells provides further redundancy that is useful in two major ways. First, if one of the copies of a particular gene is defective (from mutation or DNA damage), the other copy often provides sufficient function for the organism to thrive. Second, if one of the chromosomes suffers a break, the other can serve as a template to allow the accurate repair of the break (see Chapter 11).

In this chapter, we will consider the myriad forms of DNA damage that have been detected and studied. Studies on DNA damage are much too extensive to review systematically here. In just one slice of the DNA-damage field, scientists have detected about 100 different forms of oxidative DNA damage! Instead, this chapter will provide an overview that focuses on the most common and highly studied forms of DNA damage, and how these forms of damage are generated in cells.

8.2 Damage, repair, and mutation

When thinking about the consequences of DNA damage, it is useful to categorize damage as involving either one or two strands of the DNA molecule. Damage to one strand, generally involving some chemical alteration of one of the incorporated bases, can often be repaired in a very accurate manner by using the information in the complementary strand as a template for the replacement of the damaged base (see Chapter 10). Two-strand DNA damage can be much more serious, since templating by the complementary strand is often impossible. Prominent examples of two-strand damage include a *d*ouble-*s*trand *b*reak (DSB) and an interstrand crosslink that chemically attaches the two strands to each other.

As we will see shortly, individual bases are subject to chemical reactions including oxidative damage, hydrolytic attack, and alkylation of particular residues. The deoxyribose sugar and even the phosphodiester backbone can also be damaged. For example, the phosphodiester backbone can be disrupted with strand breaks, and sometimes these breaks are more complex with residual fragments of the incorporated nucleotide left at the frayed end. Some of the most insidious forms of damage involve the chemical crosslinking of a large organic molecule or even a full-size protein to the DNA molecule. We will encounter examples of all these myriad forms of damage in this chapter.

Before going into further detail about the forms of damage, it is important to place DNA damage in context with repair and tolerance mechanisms and with the process of mutation. As we will discover in the next four chapters, many forms of DNA damage are efficiently repaired by a multitude of repair enzymes. In model systems, such as *Escherichia coli* and *Saccharomyces cerevisiae*, mutants that are defective in particular repair enzymes often show extreme hypersensitivity to the relevant DNA-damaging agent. Certain repair mutants even show aberrant growth under "normal" growth conditions due to their inability to deal with spontaneous DNA damage. Such repair-deficient mutants have played critical roles in the ability of scientists to study repair pathways. The mutants alerted scientists to the existence of the pathway, and then were central in allowing the purification of the relevant repair proteins and the determination of molecular mechanisms of repair.

For particular forms of damage or under conditions with very extensive DNA damage, cells also have damage-tolerance pathways that allow the completion of DNA replication and cell division without actually repairing the DNA damage (Chapter 12). Furthermore, evolution has provided all cells with complex and effective pathways that are induced by heightened levels of DNA damage (Chapter 13). These pathways often lead to the induction of higher levels of DNA repair enzymes and damage-tolerance proteins, as well as alterations in other aspects of cellular physiology to improve survival of the cell or multicellular organism.

It is critically important to understand the distinction between DNA damage and mutation. As we have been discussing, DNA damage involves some kind of chemical alteration to the structure of DNA, resulting in an aberrant base, sugar, or backbone structure. Mutation, on the other hand, is an inheritable change in the sequence of DNA. The chemical structure of a mutant DNA molecule is perfectly normal, it just has a different DNA sequence than

---

1946 Nobel Prize in Physiology or Medicine

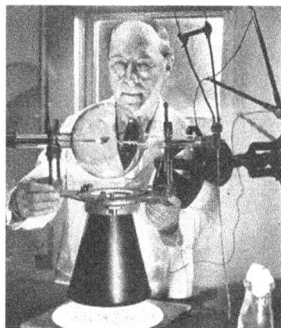

This prize was awarded to **Hermann J. Muller** who discovered that X-ray irradiation of fruit flies causes mutation. *Photo of Dr Muller at his X-ray machine, courtesy of the Genetics Society of America; permission conveyed by Copyright Clearance Center, Inc.*

https://www.nobelprize.org/prizes/medicine/1946/summary/

its DNA ancestors. You should never think of a mutation as DNA damage! Mutations can occur without any DNA damage whatsoever, as when DNA polymerase inserts a nucleotide residue that is not dictated by Watson–Crick base pairing (and both polymerase proofreading and mismatch repair fail to correct the misincorporation). Likewise, some forms of DNA damage do not lead to mutation, and many of the important forms of DNA damage lead to mutation only when the relevant repair mechanism(s) fails.

The relationship between damage and mutation is even more interesting. For example, damage induced by *ultraviolet* light (UV) in *E. coli* causes mutation at some frequency when the two major repair mechanisms are insufficient to correct the damage. The interesting point is that the mutations induced by UV are not passive, but are actually induced by the cellular response to UV and created by a specific DNA polymerase that helps the cell survive the UV damage (see Chapter 12). *E. coli* mutants that are deficient in this cellular response or in the specific DNA polymerase show no UV-induced mutations! Indeed, when both major repair pathways and this mutagenic response pathway are defective, a single UV-induced lesion in the genome kills *E. coli* (but does not induce a mutation). This is not some unusual situation seen only in a bacterial model system, but rather a common paradigm in cellular responses to DNA damage.

8.3 Spontaneous DNA damage

The N-glycosidic bond that links the deoxyribose sugar to the base is susceptible to breakage in a hydrolytic reaction that is temperature dependent (Figure 8.1A). This hydrolysis results in an abasic site, which is also called an *ap*urinic/apyrimidinic (AP) site. In fact, purine residues are about a 100-fold more susceptible than pyrimidines, so AP sites are mostly apurinic. AP sites formed by this spontaneous hydrolytic pathway are among the most common forms of spontaneous DNA damage, and it is estimated that the average human cell suffers many thousands of such AP sites per day. AP sites are also generated as a transient intermediate in one of the

(A)

(B)

(C)

(D)

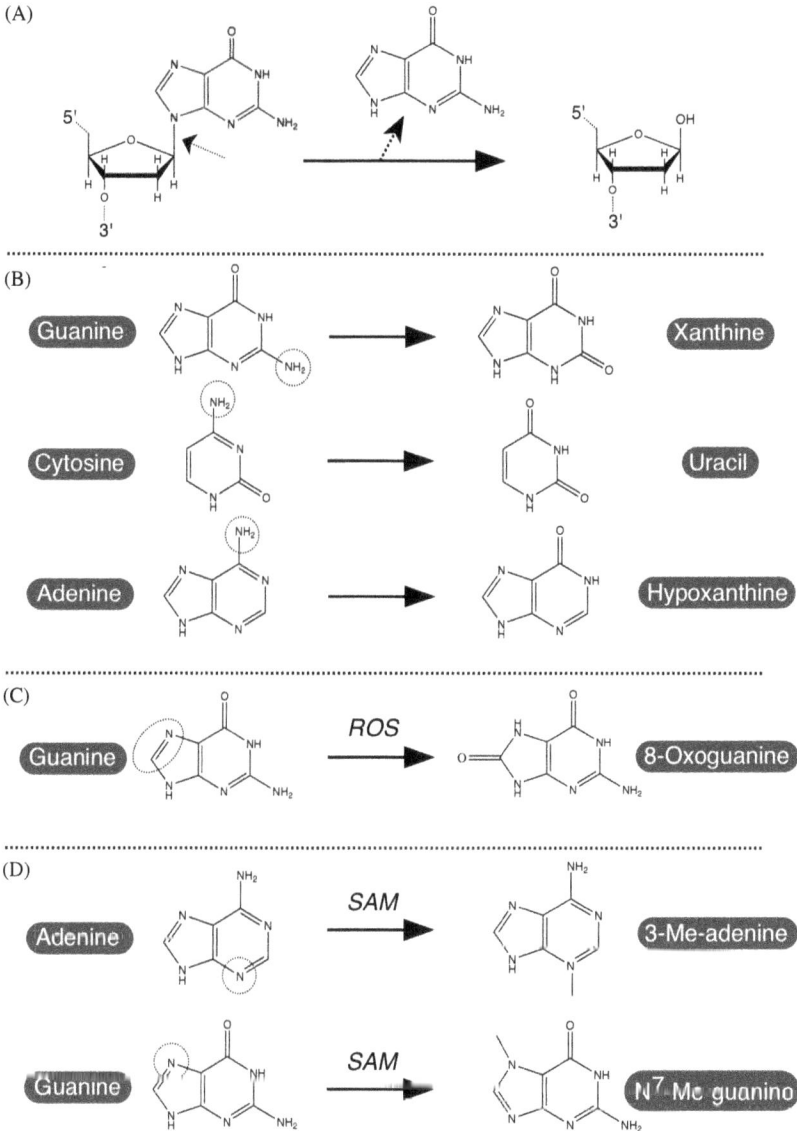

Figure 8.1. Common forms of DNA damage. Important causes of DNA damage are highlighted and described in more detail in the text. The sites of damage on the starting nucleoside (panel A) or base (panels B through D) are indicated by arrow or dotted circles. ROS, reactive oxygen species; SAM, S-adenosyl methionine.

branches of excision repair (see Chapter 10). A secondary form of DNA damage can also occur at a small fraction of AP sites — the aldehyde form of the sugar at the AP site can undergo a β-elimination reaction that leads to a single-strand break.

Several DNA bases with exocyclic amino groups undergo a spontaneous deamination reaction, which is both pH and temperature dependent. Guanine residues deaminate to xanthine, cytosine residues to uracil, and adenine residues deaminate to hypoxanthine (Figure 8.1B). It has been estimated that human cells suffer a few hundred spontaneous cytosine deaminations per day, although direct measurements of this frequency are lacking. The deamination of cytosine to uracil is particularly interesting from an evolutionary standpoint. Recall that RNA molecules incorporate uracil residues while DNA uses the closely related thymine base. Scientists believe that thymine was adopted in DNA during evolution precisely because deaminated cytosine residues would be indistinguishable from uracil residues if uracil were normally used in DNA. Instead, any G:U base pair in duplex DNA is clearly the result of deamination of the C, rather than misincorporation of G at what should be an A:T (A:U) base pair.

In spite of this rather comforting turn of evolution, mammalian cells have also evolved to occasionally methylate cytosine in certain sequences. Many promoter regions contain repeated copies of the sequence 5′-CG-3′ (also called CpG islands), where gene transcription begins and where methylation of the C alters the frequency of transcription. The problem is that when methylated C deaminates, it becomes none other than thymine! The resulting G:T base pair cannot be repaired by the pathway that repairs G:U base pairs in the rest of the genome. The difficulty that cells have dealing with this lesion is evident from the fact that CpG islands suffer the expected C to T mutations at a relatively high frequency, and these mutations are commonly seen in human cancers.

Another very important class of spontaneous DNA damage is caused by *reactive oxygen species* (ROS). As mentioned earlier, ROS can cause about 100 different forms of DNA damage. Several of these are generated frequently, on the order of 1000 times per day in human cells, including ring-saturated pyrimidines (e.g., thymine glycol), 7,8-dihydro-8-oxoguanine (generally called 8-oxoG; Figure 8.1C), and products of lipid peroxidation such as etheno-A and etheno-C. ROS themselves are generated as by-products of aerobic metabolism, such as leakage from the electron-transport chain in mitochondrial respiration. Certain mammalian cells also form ROS intentionally, as occurs during attack of bacterial invaders by phagocytic leukocyte cells.

The detailed biology of ROS generation and defense is quite complex and beyond the scope of this chapter. However, it is worth mentioning that cells have multiple layers of defenses against ROS, including ROS-degrading enzymes such as *super*oxide *d*ismutase (SOD), catalase, and peroxidases. It is also interesting to note that ROS damage many different macromolecules in addition to DNA, and that oxidative stress is involved in multiple human diseases (e.g., cancer, neurodegenerative disease, and atherosclerosis) and contributes to the aging process.

We will see shortly that many exogenous chemicals alkylate DNA by adding methyl groups or larger moieties to particular positions on the DNA molecule. Alkylation also occurs spontaneously in the form of methylation induced by the coenzyme *S*-*a*denosyl *m*ethionine (SAM). SAM is the methyl donor in various important enzymatic reactions in the cell, but is also a weak nonenzymatic methyl donor. It has been estimated that SAM methylates the three position of adenine and the seven position of guanine at least 1000 times per day in human cells, again representing a common form of spontaneous DNA damage (Figure 8.1D).

The common generation of multiple forms of spontaneous DNA damage is thus unavoidable in living cells. Later in this book, we will discuss a variety of pathways that cells have evolved to deal with this damage, including DNA repair, tolerance, and regulatory pathways. When considering exposure to environmental chemicals and radiation, one should always keep in mind that the resulting damage is above and beyond the spontaneous damage, and that low doses of some of these exogenous agents may not even rise to the damaging level of our own endogenous DNA-damaging agents.

8.4 Damage induced by exogenous chemicals

Over 200 years ago, the English doctor Percivall Pott first linked human chemical exposure to the induction of cancer. Chimney sweeps of that era often suffered from an occupation-specific scrotal cancer, which the doctor inferred was caused by exposure to the by-products of coal burning that accumulated in the soot of the chimney. He recommended that the chimney sweeps adopt daily bathing to reduce exposure, and this indeed reduced the cancer incidence. The coal by-products were eventually shown to contain polycyclic aromatic hydrocarbons, including benzo[*a*]pyrene, which are potent DNA-damaging agents (Figure 8.2A).

Exposure to benzo[*a*]pyrene is still a concern in modern society, because it is generated by incomplete combustion of organic compounds during burning of fossil fuels, cigarettes, and meats on the barbecue. A metabolic derivative of benzo[*a*]pyrene is actually the culprit in damaging DNA. This derivative becomes covalently linked to guanine bases in the DNA, causing a dramatic distortion in DNA structure and inducing mutations at a high frequency. It should be noted that smoke from fossil fuel and cigarettes contains many different noxious chemicals, so benzo[*a*]pyrene and related polycyclic hydrocarbons are not the only DNA-damaging (and carcinogenic) compounds present in smoke. One study identified 60 different carcinogenic compounds in cigarette smoke! Benzo[*a*]pyrene is an example of a class of alkylating agents that transfer a large or "bulky" adduct (often the entire molecule) to the DNA molecule. While some repair pathways can act on bulky adducts, these tend to be very difficult lesions to repair, and mutations are often induced when translesion DNA polymerases replicate the damaged DNA (see Chapter 12).

Considered broadly, alkylating agents react with many different positions on the DNA bases, while particular agents have their own characteristic specificity. In Section 8.3 on endogenous damage, we discussed SAM, a simple (and very weak) alkylating agent that transfers only a single methyl group to particular sites on DNA. A number of much more potent methylating (and ethylating) agents

(A)

benzo[*a*]pyrene

(B)

methyl
methanesulfonate

N-methyl-*N'*-nitro-*N*-
nitrosoguanidine

ethyl
methanesulfonate

(C)

bis(2-chloroethyl)
sulfide

chlormethine

cyclophosphamide

(D)

cisplatin

psoralen

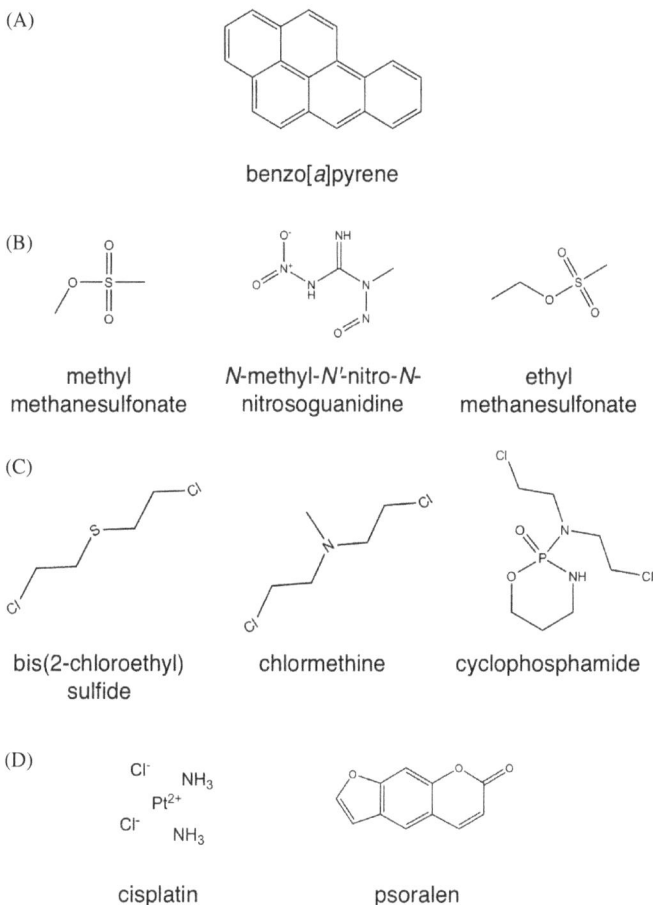

Figure 8.2. Examples of DNA-damaging chemicals. See text for the descriptions of these damaging agents and their consequences to the DNA molecule.

exist and are often used in the laboratory to induce mutation and study DNA repair. This includes methylating agents such as *m*ethyl *m*ethane*s*ulfonate (MMS) and *N-m*ethyl-*N'*-*n*itro-*N*-*n*itrosoguanidine (MNNG), and ethylating agents such as *e*thyl *m*ethane*s*ulfonate (EMS) (Figure 8.2B).

As seen in the discussion of benzo[*a*]pyrene, alkylating agents can transfer more complex moieties to DNA. In general, alkylating agents can be divided into two broad classes: monofunctional agents attach to only a single position on the DNA while bifunctional agents attach at two locations, artificially bridging the DNA. The bifunctional alkylating agents can thereby link two different locations on the same strand, in a structure called an intrastrand crosslink, or two different locations on the opposite strands, called an interstrand crosslink. Interstrand crosslinks are particularly toxic, in that they block biological processes that need to separate the DNA strands, including DNA replication and transcription.

The history of alkylating agents that cause crosslinking (also called crosslinking agents) is of interest. Alkylating agents were first used extensively as an agent of war, namely mustard gas in World War I, which killed thousands and injured countless others. The agent was named based on its mustard-like smell, and, strictly speaking, it acts as a cloud of liquid droplets rather than a true gas. Unfortunately, in spite of international conventions, sporadic use of mustard gas has continued into the modern era in Middle East conflicts.

The original mustard gas compound in World War I was bis (2-chloro-ethyl) sulfide (Figure 8.2C), but numerous related compounds have been synthesized. The general category of mustard gas compounds encompasses a family of DNA alkylating agents that cause DNA crosslinks. Some of these were later adopted as chemotherapeutic agents in the clinic. The first alkylating agent to receive regulatory approval for cancer treatment was chlormethine (also called mustine), and the most widely used alkylating agent in current chemotherapy is cyclophosphamide (Figure 8.2C). The powerful toxicity of mustard gas compounds relates to their ability to cause DNA crosslinks, that is, these are bifunctional alkylating agents.

A number of other compounds, not in the mustard gas category, also cause DNA crosslinks, including cisplatin and psoralen (Figure 8.2D). Like the abovementioned mustard gas compounds, cisplatin is useful in chemotherapy. Psoralen, which needs to be photoactivated, is used to treat skin conditions such as psoriasis under UV-light activation.

8.5 Radiation-induced DNA damage

Ionizing radiation and nonionizing UV[1] are important sources of DNA damage and are generated in abundance by the sun. Fortunately for us, most of the damaging solar rays of both types are eliminated by oxygen and ozone in our atmosphere. Indeed, many scientists believe that terrestrial life could not evolve until aquatic life forms generated enough oxygen to provide this protective barrier. Ionizing radiation is very damaging to DNA and occurs in the form of galactic cosmic rays and energetic particles from the sun. The earth's atmosphere, along with the powerful magnetic fields of the earth, shields us from nearly all of this ionizing radiation (although exposure increases at higher altitudes). Nonionizing UV from the sun, which penetrates dependent on atmospheric conditions, can be very damaging to organisms (see discussion of skin-cancer syndromes in Chapter 10). Both ionizing and nonionizing radiation are very real threats to human health that must be counteracted during space flight, particularly when planning prolonged missions to destinations like Mars.

The most common forms of DNA damage from UV involve the generation of covalent dimers of adjacent pyrimidine residues. The most common dimer involves a cyclobutane ring with two linkages between the adjacent pyrimidines, and this lesion is called a *cyclob*utane *p*yrimidine *d*imer (CPD) (Figure 8.3A). A somewhat less common lesion involves a single linkage between the adjacent pyrimidines, which is called the pyrimidine–pyrimidone (6-4) adduct or (6-4) photoproduct (Figure 8.3B). Interestingly, this less abundant UV photoproduct is responsible for most of the UV-induced

---

[1] Ionizing forms of radiation, including X-rays and gamma rays, have wavelengths of about 100 nm and less. Nonionizing UV radiation includes wavelengths from about 100 to 400 nm as follows: UV-C, 100–295 nm; UV-B, 295–320 nm; and UV-A, 320–400 nm. UV-C, used in germicidal lamps, is very damaging to DNA but solar UV-C is eliminated by the atmosphere. UV at wavelengths below about 340 nm damages DNA directly, whereas indirect pathways of damage also occur for longer wavelength UV light. UV-A is generated by blacklights, which allow visualization of fluorescence from appropriate materials.

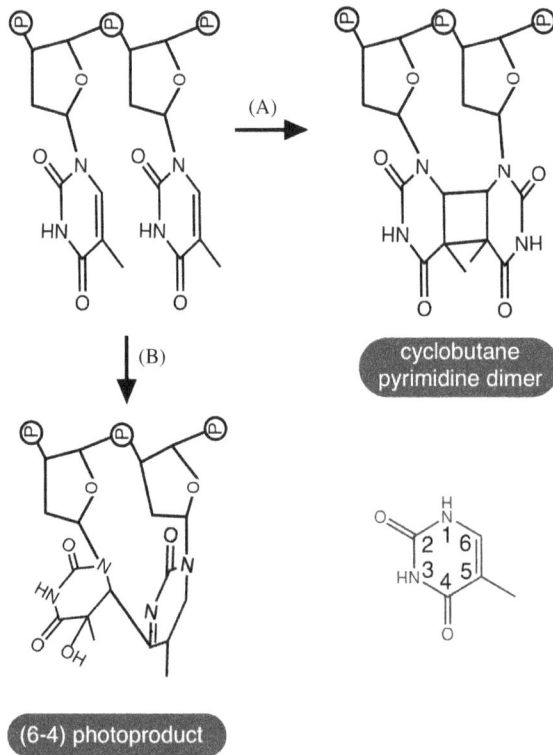

Figure 8.3. Ultraviolet (UV)-induced pyrimidine dimers. UV light induces two major pyrimidine dimer forms. The cyclobutane pyrimidine dimer involves dual linkages between the C5 and C6 carbons of the two pyrimidines (A), while the pyrimidine–pyrimidone (6-4) adduct (also called (6-4) photoproduct) involves a single C6-C4 linkage (B). The numbering system of the pyrimidine ring is shown at the bottom right for reference.

mutations that occur in *E. coli*, as we will discuss in subsequent chapters.

While these two pyrimidine dimers are the most common lesions induced by UV, they are not the only ones. Some other lesions are generated directly by adsorption of photons, while others are generated when cellular chemicals are activated by reactions with UV. Single-strand breaks can also be detected after treatment with certain wavelengths of UV. UV has been used extensively in the study of DNA damage and repair pathways, in part because it is easy and inexpensive to apply in a carefully dosed manner to model organisms in the lab.

Ionizing radiation, including X-rays and gamma rays, causes a large variety of DNA lesions and extensive damage to other molecules in the cell. As mentioned above, cosmic rays provide one natural source of ionizing radiation, as does the decay of radioactive compounds including those found in our own bodies. One important class of lesions caused by ionizing radiation involves DSBs (see "How did they test that?" at the end of this chapter). The resulting DNA fragments generally have a portion of the broken nucleotide residue or a phosphate group remaining at the 3′ end of the break site, and so these are sometimes called "dirty breaks." Ionizing radiation also causes a variety of alterations in the structures of the DNA bases, some by direct reactions of the ionizing radiation on DNA and others by generation of ROS and other indirect pathways. ROS generation can be particularly potent — it has been estimated that a single photon from ionizing radiation can produce over 30,000 hydroxyl radicals!

8.6 DNA damage by incorporation of damaged nucleotides

The abovementioned pathways can also damage nucleotide precursors, even when they are not polymerized in nucleic acids. While many damaged nucleotides would not be recognizable by DNA polymerases, certain damaged deoxynucleotides do get incorporated into DNA. The best-studied example involves incorporation of 8-oxo-dGTP, a ROS-generated modified guanine nucleotide.

Due to base-pairing ambiguity, DNA polymerases insert 8-oxo-dGTP residues opposite either template C or A (Figure 8.4), with some DNA polymerases even preferring the template A. The two possible products of this incorporation, 8-oxo-G:C and 8-oxo-G:A base pairs (boxed in Figure 8.5), are usually processed by repair pathways that will be discussed in Chapter 10. However, either of these base pairs can lead to mutation if the relevant repair pathway

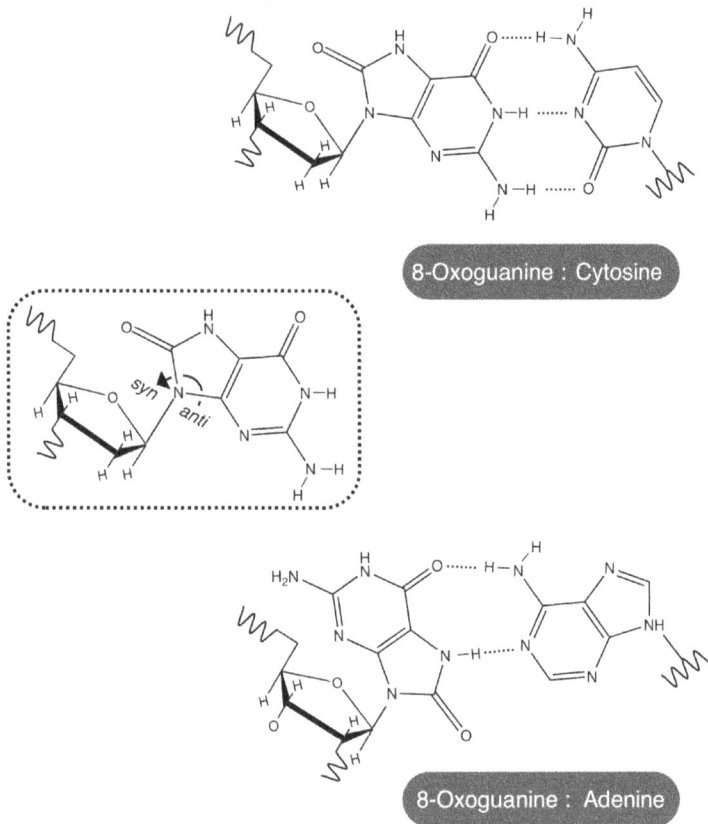

Figure 8.4. Alternate base pairing of 8-oxoguanine. The nucleotide residues within normal base pairs in DNA adopt the *"anti"* conformation, in which the deoxyribose sugar and base are in opposite orientations across the N-glycosidic bond. This is the conformation of the 8-oxoguanine:cytosine base pair at the top and the base pair depicted within the dotted box. Rotation around the N-glycosidic bond creates the *"syn"* conformation, in which the base is aligned over the sugar. The modified base in the alternate base pair 8-oxoguanine:adenine adopts this *syn* conformation to allow base pairing, as shown at the bottom. This figure was modified in part from Figure 2.13 of Friedberg *et al.* (2006).

fails. Subsequent rounds of DNA replication can generate a G:C to T:A mutation from the 8-oxo-G:C base pair or a T:A to G:C mutation from the 8-oxo-G:A base pair (Figure 8.5).

Given this mutagenic potential, it is not surprising that cells have evolved another defense to counteract the threat. Both *E. coli*

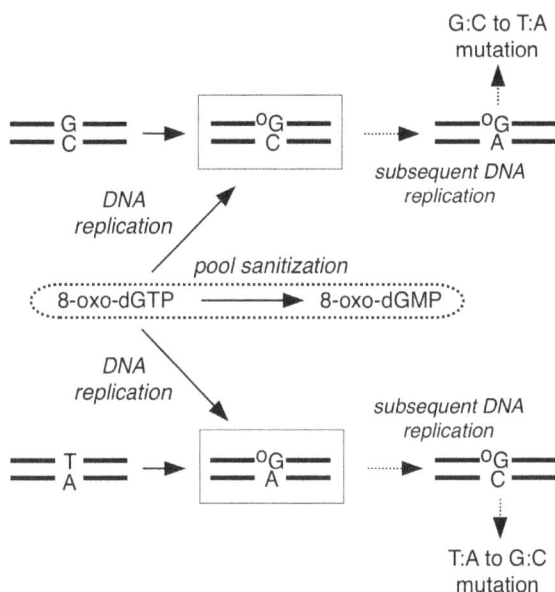

Figure 8.5. Pool sanitization prevents mutations from 8-oxo-dGTP incorporation. Reactive oxygen species (ROS)-generated 8-oxo-dGTP can be incorporated by DNA polymerase across from either a C or an A residue (see Figure 8.4 for the base-pair structures). Without intervention, base-substitution mutations can be generated via subsequent rounds of DNA replication, as shown. Two strategies have evolved to reduce this mutation burden. First, the pools of 8-oxo-dGTP precursor are greatly reduced by pool sanitization, in which the sanitizing enzyme cleaves off the two terminal phosphates to generate 8-oxo-dGMP. Second, excision repair pathways act on the 8-oxo-G-containing base pairs before the subsequent rounds of replication, as will be described in Chapter 10.

and mammalian cells have an enzyme that hydrolyzes 8-oxo-dGTP back to the monophosphate form, preventing incorporation (Figure 8.5). Amazingly, when this enzyme is inactivated by gene knockout in *E. coli*, the frequency of T:A to G:C mutations increases by several orders of magnitude. The *E. coli* enzyme can also act on two other damaged nucleotides that are derived from deoxyadenosine triphosphate (dATP). This general phenomenon is called nucleotide-pool sanitization, and is obviously very powerful in preventing DNA damage and mutations before they even start.

8.7 DNA damage from cellular enzymes

We have already discussed one form of DNA damage from cellular enzymes in Chapter 7. Recall that certain chemotherapeutic drugs can subvert DNA topoisomerases by stabilizing the covalently linked topoisomerase-DNA complex. This complex is a form of DNA–protein crosslink, major DNA damage that blocks replication and transcription. Scientists are currently working to elucidate the pathways that repair DNA–protein crosslinks in vivo. A similar covalently linked topoisomerase-DNA complex can also be formed without drug treatment, when a topoisomerase acts at the site of a particular damaged base but cannot complete its reaction. Thus, topoisomerases can contribute to a cascade of damage that initiates from spontaneous base damage (e.g., formed by ROS or alkylation).

Another class of enzymes that contributes to DNA damage and mutation is the cytidine deaminases. Recall that deamination of cytidine generates uridine, and that this reaction can occur spontaneously (see Section 8.3). The immune system of mammalian cells has adopted this pathway to generate a high frequency of variant antibody genes to recognize diverse antigens, during a process called somatic hypermutation (occurs specifically in B cells of the immune system). The enzyme cytidine deaminase (also called AID or *a*ctivation-*i*nduced *d*eaminase) deaminates cytidine residues in ssDNA, generating the expected C:G to T:A (through U:A) mutations (along with other more complex mutations). These mutations are generally targeted to the relevant segment of the antibody genes to alter the binding specificity of the region of the protein that binds antigens.

Another mammalian immune response utilizes cytidine deamination to protect against invaders such as retroviruses. This family of deaminases has the rather awkward acronym APOBEC, after the activity of its founding member as the "*apo*lipoprotein *B* mRNA *e*diting enzyme, *c*atalytic polypeptide-like" (actually, the full name is even more awkward than the acronym!). Depending on the family member, the APOBEC enzymes can deaminate C residues in RNA and/or DNA, and these enzymes have varied roles in addition to their antiviral function. Humans have 10 such enzymes, while the number varies in different mammals.

In addition to their normal roles mentioned earlier, the AID/APOBEC enzymes appear to make major contributions to cancer in human cells. Some human cancer lines show induced levels of one of these deaminase enzymes (which is normally repressed in most human cells). Furthermore, many cancer genomes show the hallmark of many mutations, often clustered, that are generated by deaminase enzymes (including the expected C:G to T:A) (also see Section 14.8).

The ultimate form of DNA damage induced by cellular enzymes occurs during the process of apoptosis, an active program of cell death that is invoked to protect the larger organism from aberrant cells. Apoptosis can be triggered by DNA damage, other cellular stresses such as hypoxia, and also by programmed cell-fate decisions during development. Apoptotic regulatory pathways are often found to be disturbed in cancer, demonstrating the importance of apoptosis in preventing cells from entering into unrestrained or aberrant growth. As you might expect, the pathways of apoptosis involve quite complex regulatory cascades that are beyond the scope of this chapter. The relevant point is that these cascades result in a carefully regulated and massive destruction of cellular DNA that ensures cell death. Chromosomal DNA is cleaved in the regions between nucleosomes, leading to a characteristic "ladder" pattern of duplex DNA fragments on an agarose gel (sizes corresponding to fragments with 1, 2, 3, 4, etc. equivalents of nucleosomal DNA).

8.8 Summary of key points

- DNA damage can involve either one or both strands.
- DNA damage and mutation are distinct — DNA damage does not always lead to mutation, and mutation is not a form of DNA damage but rather a heritable change in DNA sequence.
- Common forms of spontaneous DNA damage include breakage of the N-glycosidic bond to generate abasic sites; deamination reactions of guanine, cytosine, and adenine; ROS-mediated base

damage including thymine glycol and 8-oxoG; and alkylation of DNA by SAM.

- A wide variety of chemicals functions as alkylating agents, attaching simple (e.g., methyl or ethyl) or complex (e.g., bulky and/or crosslinking) moieties to DNA.
- Crosslinking agents have been used as lethal agents in military settings and include some useful anticancer agents.
- The most common lesions from UV are the CPD and the (6-4) photoproduct; the latter can cause mutations in *E. coli*.
- Ionizing radiation such as X and gamma rays cause a variety of DNA lesions including DSBs.
- Damaged nucleotide precursors can lead to DNA damage; for example, 8-oxo-dGTP is generated by ROS and can be sanitized (converted to 8-oxo-dGMP) by a cellular enzyme.
- DNA is also damaged by cellular enzymes including DNA topoisomerases and cytidine deaminases.

Further Reading

Alberts, B., Johnson, A., Lewis, J., Morgan, D., Raff, M., Roberts, K., & Walter, P. (2015). *Molecular Biology of the Cell* (6th ed.). New York, NY: Garland Science, Taylor and Francis Group.

Cadet, J., & Wagner, J. R. (2013). DNA base damage by reactive oxygen species, oxidizing agents, and UV radiation. *Cold Spring Harb Perspect Biol, 5*(2), 9, a012559.

Figueroa-Gonzalez, G., & Perez-Plasencia, C. (2017). Strategies for the evaluation of DNA damage and repair mechanisms in cancer. *Oncol Lett, 13*(6), 3982–3988.

Friedberg, E. C., Walker, G. C., Siede, W., Wood, R. D., Schultz, R. A., & Ellenberger, T. (2006). *DNA Repair and Mutagenesis.* (2nd ed.). Washington, DC: ASM Press.

Singh, N. P. (2016). The comet assay: Reflections on its development, evolution and applications. *Mutat Res Rev Mutat Res, 767*, 23–30.

Singh, N. P., McCoy, M. T., Tice, R. R., & Schneider, E. L. (1988). A simple technique for quantitation of low levels of DNA damage in individual cells. *Exp Cell Res, 175*, 184–191.

How did they test that?
Detecting DNA damage in cells — the comet assay

One widely used approach detects breaks in cellular DNA by visualizing the behavior of the chromosomal DNA from broken cells on a microscope slide. Singh *et al.* (1988) presented one of the earliest such studies, directly demonstrating DNA breaks after X-ray treatment of cells. Cells were immobilized within agarose on the surface of a glass slide and then treated with various X-ray doses. The embedded cells were then lysed with detergent and exposed to an alkaline condition that denatures DNA into single strands. The slides were placed for a short time in an electrophoresis chamber, where DNA migrates toward the positive anode (due to its negative charges). DNA from untreated cells remained immobile due to its overall structure (top figure, panel A), while some of the DNA from cells treated with X-rays (50 and 100 rads) migrated out of the vicinity of the cell toward the anode to form tails (top figure, panels B and C). The DNA was visualized with the dye ethidium bromide, which fluoresces only when bound to DNA. Note that the slides were returned to neutral pH after electrophoresis to promote the binding of ethidium bromide. Quantitation of DNA from many cells showed a clear dose-dependent increase in the distance traveled by the released DNA with the X-ray exposure (bottom figure). As conducted, this assay can potentially reveal single-strand breaks, double-strand breaks (DSBs), and alkali-sensitive sites. A later study introduced the name "comet assay," based on the resemblance of the DNA patterns to celestial comets, and the method is also sometimes called "single-cell gel electrophoresis." A semiquantitative estimate of breaks can be obtained from the comet assay by measuring the length of the comet tail and/or comparing the intensities of DNA within the tail versus head region. The comet assay has been modified in various ways to increase its utility, for example, to allow detection of pyrimidine dimers and oxidative DNA damage (for review, see Singh, 2016). The figures were reproduced from Singh *et al.* (1988), with permission from Elsevier; permission conveyed by Copyright Clearance Center, Inc.

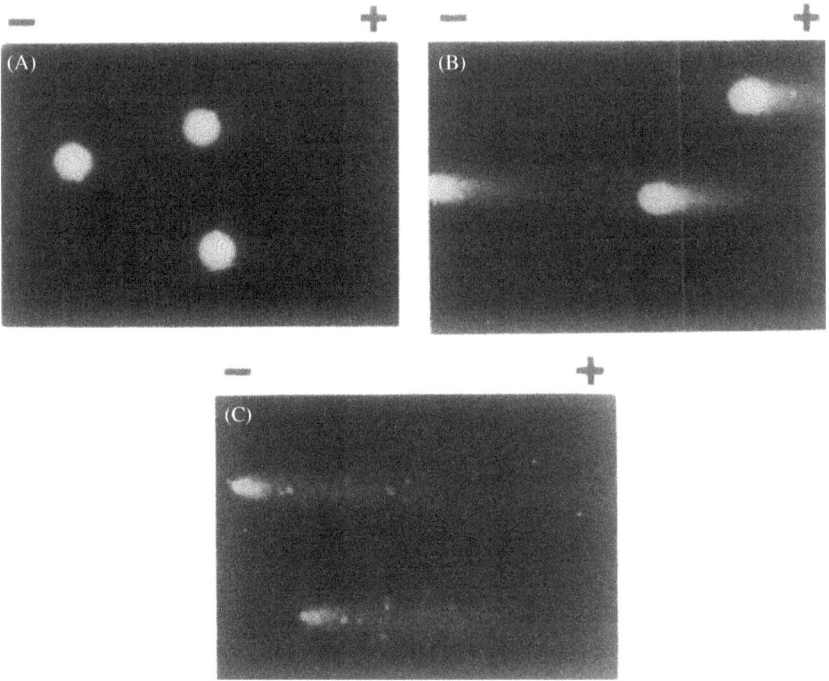

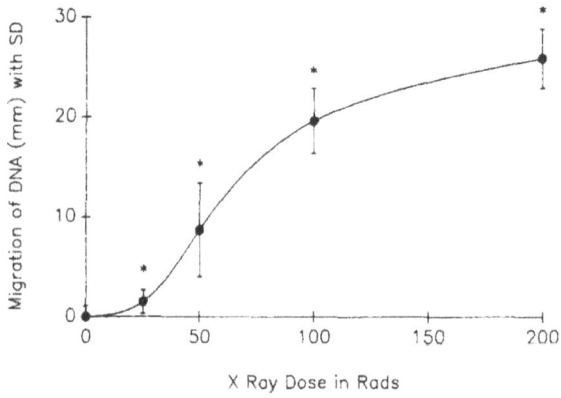

Chapter 9

Direct reversal of DNA damage

Conceptually, the simplest form of DNA repair involves chemical reversal of the damage. A number of proteins have evolved to promote a reaction that reverses the damage without any need to replace nucleotide residues (as happens in excision repair; see Chapter 10). Direct reversal is advantageous in its simplicity, generally involving only a single protein, and also in the fact that the process is error free. In this chapter, we will consider three such damage-reversal proteins that protect DNA.

9.1 Photoreactivation of UV-induced damage

Photoreactivation of *ultraviolet* light (UV)-induced DNA damage was the first pathway of DNA repair uncovered by scientists. Beginning in the 1940s, scientists discovered that the damaging effects of UV could be reversed by incubation under visible light (for many prokaryotes and eukaryotic microbes). The effect could be very dramatic — increasing survival by as much as 100,000-fold. It took some time for scientists to discover that the process was an enzyme-catalyzed process involving a family of enzymes called photolyases. These enzymes bind to the two major products of UV damage (see Chapter 8), reverse the dimerization and thereby restore the normal bases (Figure 9.1). As we will see in Chapter 10, UV photoproducts are also repaired by another pathway, nucleotide excision repair.

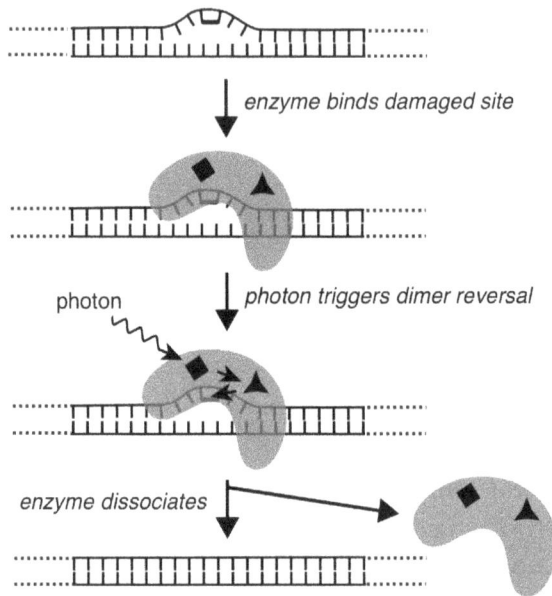

Figure 9.1. Repair of UV-induced pyrimidine dimer by enzymatic photoreactivation. Photolyase enzyme recognizes the distorted DNA containing a pyrimidine dimer. The enzyme carries two chromophores. The photoantenna (♦) captures a photon and then transfers energy to the photocatalyst (▲), which in turn transfers an electron to the damaged base dimer to trigger repair. This figure was modified from Figure 4.1 of Friedberg *et al.* (2006).

Photolyases come in two major varieties, those that act on cyclobutane pyrimidine dimer (CPD) (pyrimidine dimer photolyases) and those that act on 6-4 photoproducts ((6-4) photoproduct photolyases). While photolyases are broadly distributed in nature, some mammals, including humans, lack these enzymes. As you might predict, photolyases are very important in plants, with their intense exposure to sunlight (which both induces abundant damage and provides light for the photoreactivation reaction). Interestingly, photolyases are evolutionarily related to certain photoreceptors called cryptochromes, which are present in mammals including humans, and which play key roles in regulating circadian rhythm.

As implied earlier, the reversal reaction catalyzed by photolyases requires visible light/near UV (300–500 nm wavelengths). The enzymes carry two noncovalently bound chromophores. These chromophores are not required for the enzyme to bind specifically to the sites of DNA damage, but they play key roles in the reversal reaction (Figure 9.1). One of the chromophores, called the photoantenna, captures the photon and then transfers the excitation energy to the other chromophore, called the photocatalyst. The photocatalyst then interacts with the damaged base dimer, transferring an electron to the dimer, which triggers the final reversal reaction.

A series of crystal structures of photolyases have provided a much more detailed view of the reaction. In several cocrystal (DNA-protein) structures, the pyrimidine dimer is extruded from the DNA helix into the body of the protein (e.g., as in the Drosophila photolyase in Figure 9.2A). Structural data also show that the photoantenna chromophore is generally located some distance from the pair of

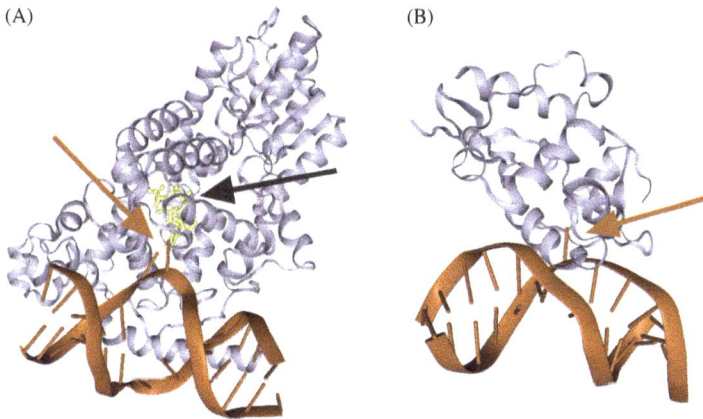

Figure 9.2.   Structures of DNA repair enzymes. The structure of Drosophila (6-4) photolyase bound to damaged DNA is shown in panel A. The pyrimidine dimer is flipped out from the DNA into the active site of the enzyme (indicated by tan arrow), and the photocatalyst (flavin adenine dinucleotide; yellow) is indicated by the black arrow. This image is from the **RCSB PDB** (www.rcsb.org) of **PDB ID 3CVU** (Maul *et al.*, 2008) (with arrows added after download). The structure of human O^6-alkylguanine DNA alkyltransferase bound to damaged DNA is shown in panel B. The alkylated nucleotide residue is flipped into the active site of the enzyme (indicated by tan arrow) to facilitate the transfer of the alkyl group to the protein. This image is from the **RCSB PDB** (www.rcsb.org) of **PDB ID 1T38** (Daniels *et al.*, 2004) (with arrow added after download).

thymines, but the photocatalyst chromophore (black arrow in Figure 9.2A) is much closer. The detailed mechanism of energy transfer in the enzyme is currently under investigation.

Multiple repair proteins (and certain other proteins) use the general strategy of flipping bases out of the helix, and we will encounter this strategy again in this chapter and in Chapter 10. Flipping bases out of the helix allows the enzyme to contact much more of the base, including parts that would be buried in the helix. This can be critical for recognition of the damage. Furthermore, base flipping brings the damaged base(s) into proximity of the active site of the enzyme, which can be located deep within the protein and yet still repair the damage.

9.2 Reversal of alkylation damage by DNA alkyl transferases

As described in Chapter 8, alkylation damage in DNA occurs both spontaneously and in response to a wide variety of environmental chemicals (including those that are naturally occurring). Given the importance of DNA alkylation damage in nature, it is not surprising that cells have multiple repair mechanisms for this damage. In this chapter, we will consider two distinct mechanisms by which proteins directly remove alkyl groups from DNA, while in Chapter 10, we will see that base excision repair also contributes to repair of alkylation damage by an entirely different strategy.

The first mechanism for removal of alkyl groups from DNA involves the transfer of the alkyl group from DNA directly to a cysteine residue in the protein (Figure 9.3). These proteins are called DNA alkyltransferases, or more formally O^6-*alkyl*guanine-DNA alkyl*transferases (AGTs) based on one of their common substrates. As with the photolyases described earlier, these proteins employ a base-flipping mechanism to access the modified bases in duplex DNA (e.g., see structure of human AGT–DNA complex in Figure 9.2B).

A surprising aspect of the alkyltransferase mechanism is that the protein is essentially "used up" in the reaction, that is, the protein is

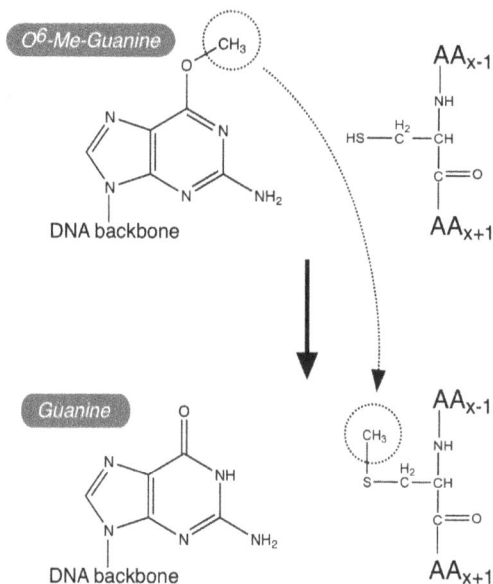

Figure 9.3. Removal of alkylation damage by DNA alkyltransferase. This diagram depicts the transfer of a methyl group from a damaged guanine to the active-site cysteine of an O^6-alkylguanine-DNA alkyltransferase.

not actually an enzyme that repeatedly catalyzes a reaction. One molecule of a DNA alkyltransferase can accept only one alkyl group from DNA, and at that point it is no longer active as an alkyltransferase. The exception that proves the rule is that one of the bacterial alkyltransferases has two different active-site cysteines, and can accept two alkyl groups. There is evidence from multiple systems that the alkylated form of DNA alkyltransferase is subjected to proteolysis, so that the inert form of the protein does not accumulate in cells.

As just implied, this is a very expensive way to repair DNA damage — an entire protein has to be synthesized (and then degraded) to repair just a single site of DNA damage (or two in the exception noted earlier). Contrast this pathway with the photolyases we discussed earlier — one molecule of *E. coli* photolyase can reverse about 50 CPD lesions *per minute* in a proper enzymatic reaction that repeats over and over again.

Alkyltransferases are found in all branches of life, and individual examples have quite different properties. For example, some alkyltransferases accept methyl groups from O^6-methylguanine, O^4-methylthymine, and even methylphosphotriester (which occur when an oxygen in the phosphodiester bond of DNA is methylated). Others are more limited in their substrate specificity, reacting with only one or two of these lesions. The individual alkyltransferases also vary in their efficiency with different alkyl groups, with methyl groups generally the preferred lesion.

An interesting aspect of one of the bacterial alkyltransferases involves its regulatory mechanism, which was actually instrumental in the discovery of this protein and repair pathway. *E. coli* has very low levels of this alkyltransferase (only a few molecules per cell), and so you would expect that this protein could handle only a very small number of alkylation lesions in the DNA of a given cell. Yet, this pathway, under the right circumstances, can help the cell survive high levels of DNA alkylating agents that produce many DNA lesions. The explanation is that the gene that encodes this protein is subject to an autoregulatory mechanism. The alkyltransferase itself becomes a very specific transcriptional activator, but only after accepting an alkyl group from DNA! The alkylated form of the protein binds to its own promoter (and that of a few other relevant genes) and stimulates transcription by RNA polymerase. This in turn increases the amounts of "fresh" (unreacted) alkyltransferase up to thousands of molecules per cell, which can thereby handle a greatly increased load of DNA alkylation damage.

Mammalian cells including humans and mice have a single DNA alkyltransferase that is homologous to the regulated bacterial protein described earlier. Mice with an engineered knockout of the gene that encodes this protein are much more sensitive to alkylating agents than their normal littermates, demonstrating the biological importance of the protein (see "How did they test that?" at the end of this chapter).

As mentioned in Chapter 8, alkylation agents are often used in cancer chemotherapy. One reason why these agents may be effective anticancer agents is that many cancers have reduced levels of alkyl-transferase due to transcriptional silencing. Furthermore, inhibitors of the human alkyltransferase appear to potentiate the action of alkylating agents in preclinical and clinical cancer trials, and scientists are currently searching for more potent and specific inhibitors for this purpose.

9.3 Reversal of alkylation damage by dioxygenases

The second mechanism that cells use to directly remove alkyl groups from DNA involves a subclass of the dioxygenase family of enzymes. Dioxygenases carry out diverse reactions in cells, using dioxygen (O_2) to carry out reactions such as cleavage of carbon–carbon or carbon–sulfur bonds and oxidation of thiols. In the case of the alkylation damage reversal proteins, the dioxygenase enzyme, called AlkB in *E. coli*, catalyzes the cleavage of the aberrant carbon–nitrogen bond in alkylated residues such as 3-methylcytosine and 1-methyladenine in DNA (and several other alkylated residues with lower efficiency). As with the two classes of damage-reversal proteins discussed earlier, AlkB utilizes a base-flipping mechanism when acting on alkylated residues in duplex DNA. The overall reaction requires α-ketoglutarate (in addition to O_2 and Fe^{++}) and generates succinate and formaldehyde as by-products (Figure 9.4).

Humans have nine homologs of the *E. coli* AlkB enzyme with divergent functions. Two have been shown to play important roles in repair of DNA alkylation damage, two are implicated in processing RNA via demethylase reactions, and the functions of others are still unknown. The finding of demethylase activity on alkylated RNA residues has raised the possibility that some of these enzymes play a role in RNA repair. In general, repair of damaged RNA is an under-studied area that seems likely to reveal multiple important biological pathways.

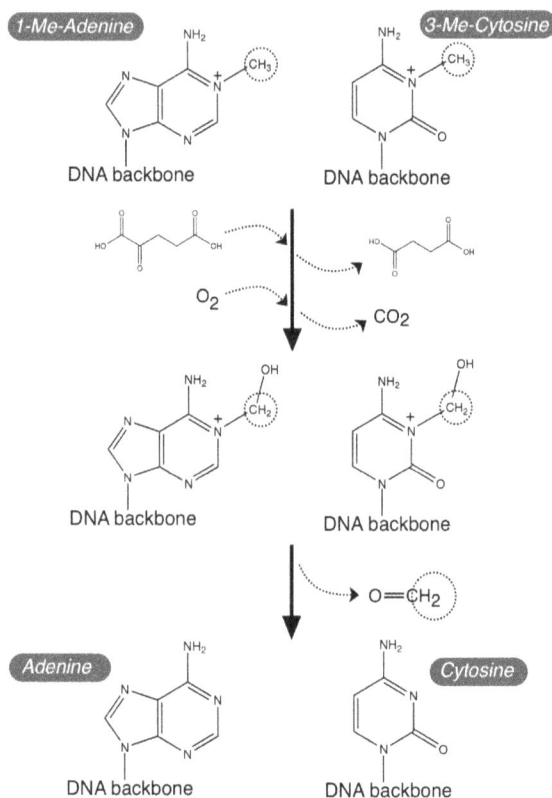

Figure 9.4. Removal of alkylation damage by oxidative demethylation. Dioxygenases such as AlkB remove the alkyl group from DNA in a two-step reaction as depicted in this figure. Removal of a methyl group from 1-methyl-adenine (left) and 3-methyl-cytosine (right) are shown (each is carried out separately). In the first step, α-ketoglutarate is converted to succinate as the methyl group is oxidized, and in the second step, the oxidized methyl group is released as formaldehyde. This figure was adapted from Begley and Samson (2003).

9.4 Summary of key points

- A few DNA lesions can be reversed directly, in processes that are relatively simple and error free.
- Photoreactivation was the first pathway of DNA repair uncovered by scientists.

- Photoreactivating enzymes capture photons using chromophores and use the resulting energy to reverse the UV-induced CPD and (6-4) photoproduct lesions.
- DNA alkyltransferases reverse DNA alkylation damage by either transferring the alkyl group to an active-site serine or by an oxidative demethylation reaction.
- Photoreactivating enzymes and DNA alkyltransferases flip the base(s) out of the helix to gain access to the damaged site.

Further Reading

Begley, T. J., & Samson, L. D. (2003). AlkB mystery solved: Oxidative demethylation of N1-methyladenine and N3-methylcytosine adducts by a direct reversal mechanism. *Trends Biochem Sci, 28*(1), 2–5.

Daniels, D. S., Woo, T. T., Luu, K. X., Noll, D. M., Clarke, N. D., Pegg, A. E., & Tainer, J. A. (2004). DNA binding and nucleotide flipping by the human DNA repair protein AGT. *Nat Struct Mol Biol, 11*(8), 714–720.

Friedberg, E. C., Walker, G. C., Siede, W., Wood, R. D., Schultz, R. A., & Ellenberger, T. (2006). *DNA Repair and Mutagenesis* (2nd ed.). Washington, DC: ASM Press.

Glassner, B. J., Weeda, G., Allan, J. M., Broekhof, J. L., Carls, N. H., Donker, I., . . . Samson, L. D. (1999). DNA repair methyltransferase (Mgmt) knockout mice are sensitive to the lethal effects of chemotherapeutic alkylating agents. *Mutagenesis, 14*(3), 339–347.

Maul, M. J., Barends, T. R., Glas, A. F., Cryle, M. J., Domratcheva, T., Schneider, S., Schlichting, I., & Carell, T. (2008). Crystal structure and mechanism of a DNA (6-4) photolyase. *Angew Chem Int Ed Engl, 47*, 10076–10080.

Pegg, A. E. (2011). Multifaceted roles of alkyltransferase and related proteins in DNA repair, DNA damage, resistance to chemotherapy, and research tools. *Chem Res Toxicol, 24*(5), 618–639.

Sancar, G. B. (2000). Enzymatic photoreactivation: 50 years and counting. *Mutat Res, 451*(1–2), 25–37.

Yi, C., & He, C. (2014). DNA repair by reversal of DNA damage. *Cold Spring Harb Perspect Biol, 5*(1), a012575.

How did they test that?
Does O^6-methylguanine DNA methyltransferase protect mice from alkylating agents?

Even though a protein may have a particular activity in the test tube, it is important to test whether this activity matches the function of the protein in cells or multicellular organisms. Glassner *et al.* (1999) tested the in vivo function of mammalian O^6-methylguanine DNA methyltransferase with regard to sensitivity to DNA alkylating agents. Using a homologous recombination approach that will be discussed in Chapter 11, they generated mice with a mutation that removes a very small region of the *Mgmt* gene that encodes the active-site cysteine. They then bred mice that were homozygous for this gene mutation (*Mgmt* −/−) and compared them to normal *Mgmt* +/+ mice. First, they found that the level of O^6-methylguanine DNA methyltransferase activity was greatly reduced in extracts of either liver or embryo fibroblasts from the double-negative mice, arguing that the *Mgmt* gene encodes the major O^6-methylguanine DNA methyltransferase in these cells (Glassner *et al.*, 1999). Next, they demonstrated that bone marrow cells isolated from the *Mgmt* −/− mice were indeed hypersensitive to a number of alkylating agents that generate modified bases that are susceptible to O^6-methylguanine DNA methyltransferase removal (BCNU: 1,3-bis(2-chloroethyl)-1-nitrosourea; MNU: *N*-methyl-*N*-nitrosurea; STZ: streptozotocin; and temozolomide; note that BCNU, STZ, and temozolomide are used clinically) (Figure, panels A, B, D, and E). As a control, these cells were not hypersensitive to mitomycin C (MMC; panel C), which generates more complex DNA lesions that are not a substrate for this protein. The hypersensitivity of bone marrow cells is notable because these cells are among the most affected by alkylating agent treatment, and also because the level of O^6-methylguanine DNA methyltransferase activity in bone marrow is relatively low. Finally, and most importantly, they showed that a measure of the lethal dose of BCNU, MNU, and STZ was 2–12 times lower in the double-negative mice than in the normal controls (again, no change seen for MMC) (Table). Therefore, this

O^6-methylguanine DNA methyltransferase protein plays the major role in protecting mice from alkylating agents. The figure was reproduced from Glassner *et al.* (1999), with permission from Oxford University Press and the United Kingdom Environmental Mutagen Society; permission conveyed by Copyright Clearance Center, Inc. The data in the table were from Glassner *et al.* (1999).

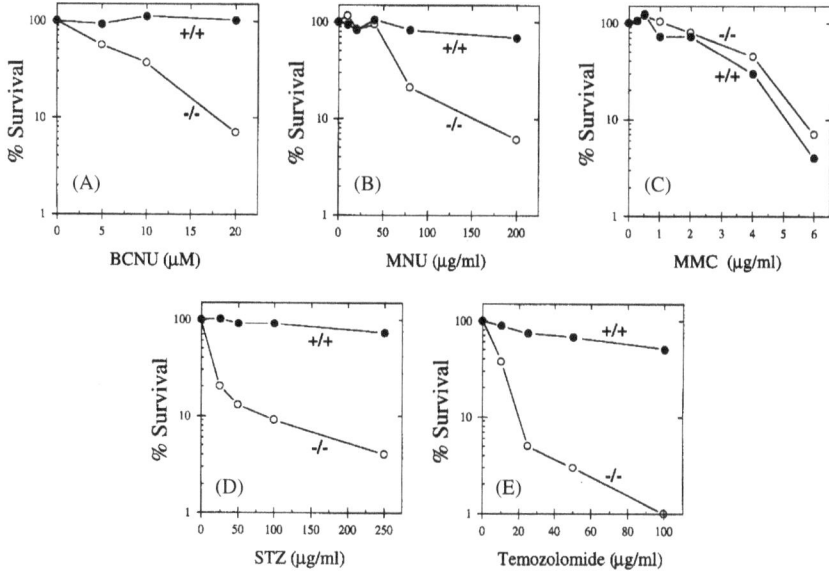

Estimated LD_{50} (mg drug/kg body weight)

MGMT Genotype	BCNU	MNU	STZ	MMC
+/+	39	107	240	14
−/−	14	9	107	14

BCNU, 1,3-bis(2-chloroethyl)-1-nitrosourea; MMC, mitomycin C; MNU, *N*-methyl-*N*-nitrosurea; STZ, streptozotocin.

Chapter 10

Excision repair — taking advantage of the complementary strand

Many DNA repair pathways in all organisms fall under the general category of excision repair. The basic strategy of excision repair is to excise the damaged region from one strand and replace it using DNA polymerase to insert the correct residue(s). This strategy takes advantage of the complementary strand to ensure that the newly synthesized residues have the correct sequence (i.e., the complement of the sequence of the complementary strand). We have already encountered one form of excision repair, namely mismatch repair, which corrects replication errors (Chapter 6).

In this chapter, the two major pathways of excision repair for damaged DNA will be explored, along with the related pathway by which abasic (AP) sites are repaired. In *base excision repair* (BER), the single damaged base is excised from the DNA prior to replacement of the residue by resynthesis. In *nucleotide excision repair* (NER), a segment of one strand that includes the damaged site (and its surrounding nucleotides) is excised and replaced by resynthesis. In general, BER repairs base lesions that only modestly distort DNA structure, such as methylated bases, while the NER pathway repairs lesions that cause more drastic structural alterations in the helix. We will see that these excision repair pathways comprise versatile pathways for correcting both endogenous and exogenous DNA damage.

10.1 Repair of AP sites

AP sites are one of the most commonly generated lesions in DNA, arising in a spontaneous fashion from hydrolysis of the N-glycosidic bond between the base and the sugar (see Chapter 8). The N-glycosidic bond of alkylated bases turn out to be more sensitive to hydrolysis than that of normal bases. This provides for a first line of defense against alkylated bases, although one that is not efficient enough to provide much protection from alkylating agents. As we will see later, AP sites are also generated frequently as intermediates in the BER pathway. In all of these situations, cells must repair the AP site, which would otherwise block DNA replication and transcription.

The pathway for repairing AP sites begins with a specialized nuclease that cleaves the phosphodiester bond on the 5′ side of the AP site (Figure 10.1A). The phosphodiester bond is cleaved on the 5′ side of the phosphate, leaving a 3′-OH end on the adjacent (normal) residue and a 5′ phosphate attached to the AP site. These enzymes are called *ap*urinic/apyrimidinic endonucleases, or AP endonucleases. Note that the AP site is still attached to one side of the DNA via its 3′ phosphodiester bond (Figure 10.1). This terminal AP site residue is often referred to as a 5′ deoxyribose phosphate, or dRP, residue.

The next step in the repair pathway removes the terminal dRP residue, generally involving an enzyme called a dRP lyase or dRPase. This enzyme cleaves the 3′ phosphodiester bond of the dRP residue, releasing the dRP residue and leaving a 5′ phosphate on the DNA (Figure 10.1B). The missing residue is usually replaced by the action of DNA polymerase inserting a single nucleotide, using the complementary strand as the template to ensure fidelity (Figure 10.1C).

The final step in repair of an AP site involves ligation of the nick to completely seal the DNA (Figure 10.1D). This is the same reaction that occurs at the end of Okazaki fragment maturation (see Chapters 2–4).

The abovementioned overview of the AP site repair pathway is an oversimplification, so let's consider some of the complexities

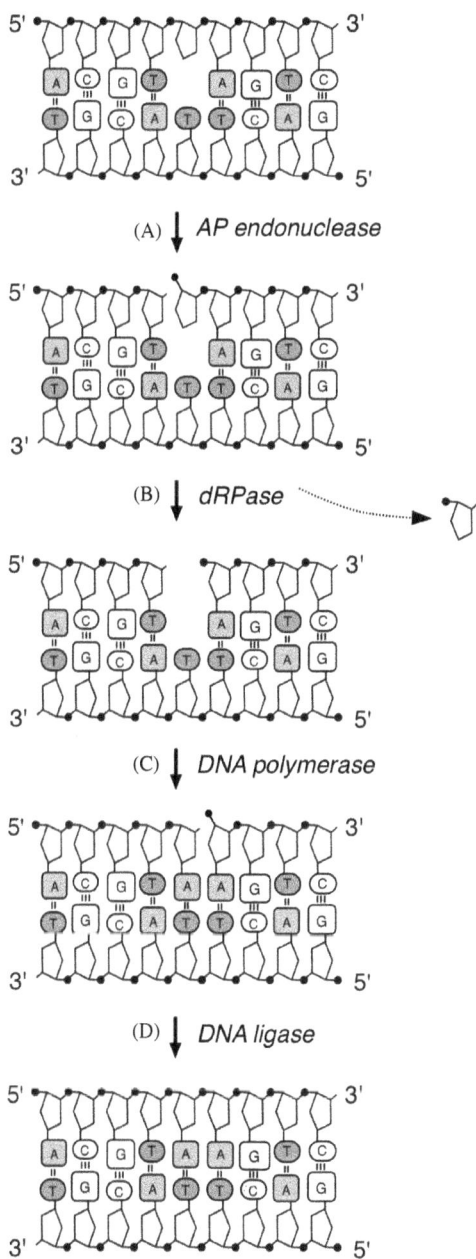

Figure 10.1.   General pathway for repairing abasic (AP) sites. The four steps in AP site repair are depicted. See text for more detailed discussion of this pathway.

and nuances of the pathway. Cells often have more than just one AP endonuclease, generally with different properties. For example, *Escherichia coli* has two AP endonucleases, one at about 10-fold higher levels than the other, but the latter is inducible by certain forms of DNA damage. Human and mouse cells have an AP endonuclease that is homologous to the major *E. coli* form, and the enzymes have similar properties. Knockouts of the mouse AP endonuclease cause embryonic lethality, reflecting the importance of repairing AP sites that occur spontaneously. Another point of interest is that AP endonucleases often have additional activities relating to DNA repair. The homologous *E. coli*/human pair mentioned earlier also has exonuclease and 3′-phosphatase activity and can remove some aberrant fragments from 3′ ends of DNA.

Regarding later steps in the pathway, many cells have multiple overlapping enzymes that remove the terminal dRP residue. Also, dRP removal and single-residue replacement steps might sometimes occur in reverse order, with DNA polymerase displacing the dRP residue as it inserts the replacement, and dRP removal occurring afterward. In fact, in mammalian cells, DNA polymerase β carries out both steps in a concerted fashion. The enzyme has dRP lyase activity in one domain and DNA polymerase activity in another domain. Interestingly, about 30% of human solid cancers have mutations in the reading frame for polymerase β presumably because increased mutation frequency caused by mutations in this enzyme contribute to cancer formation or progression.

Finally, there is an alternative pathway, called long-patch repair, in which dRPase activity is not required. Instead, DNA polymerase inserts from two to ten residues, extending the region of resynthesis and displacing a flap containing the replaced residues (Figure 10.2). Flap endonuclease then cleaves the displaced flap, and DNA ligase seals the nick, much like in Okazaki fragment maturation (see Chapter 4).

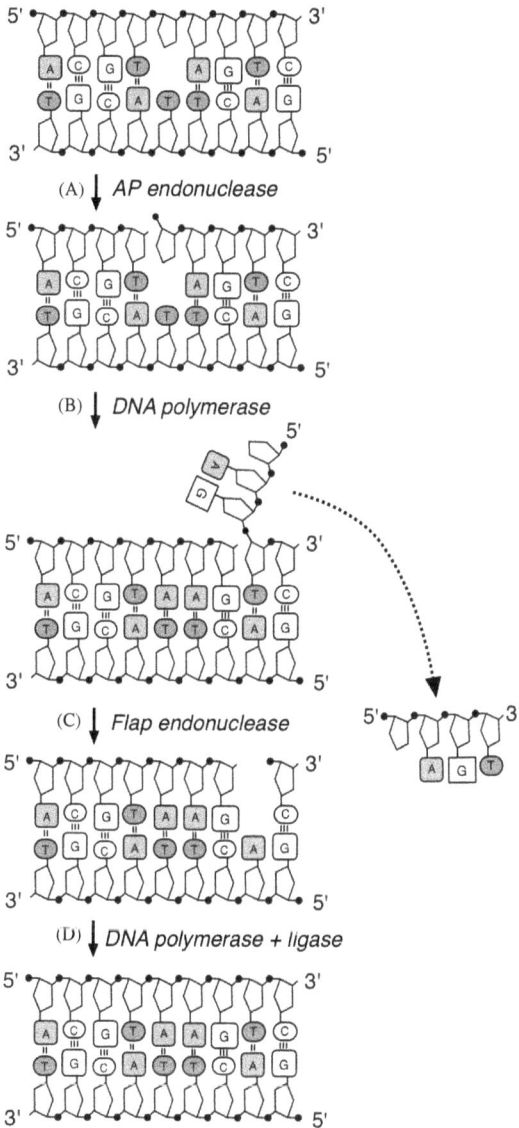

Figure 10.2. Long-patch repair of abasic (AP) sites. This pathway is initiated with phosphodiester bond cleavage by an AP endonuclease (step A). Next, DNA polymerase carries out a limited synthesis reaction from the 3'-OH group generated by the cleavage in step A — this generates a short flap of displaced nucleotide residues including the deoxyribose phosphate (dRP) residue (step B). The flap is cleaved by flap endonuclease (step C), and finally DNA polymerase, in concert with DNA ligase, fills in and seals the short gap (step D).

10.2 Repair of uracil residues in DNA — the prototype BER pathway

In 1974, BER was first uncovered in the context of repairing uracil residues in DNA (see Nobel Prize for DNA Repair on page 157). As described in Chapter 8, uracil residues arise spontaneously from the deamination of cytosine residues in DNA, and also as a result of DNA polymerase using dUTP instead of dTTP for synthesis. Cytosine deamination leads to an aberrant U:G base pair in DNA, which can result in a C to T mutation if DNA replication occurs before DNA repair.

The major discovery that revealed the BER pathway was the discovery of a bacterial enzyme that cleaves the N-glycosidic bond that connects the deoxyribose to the uracil in duplex DNA (Figure 10.3A). The bacterial enzyme was the founding member of a group of *uracil*

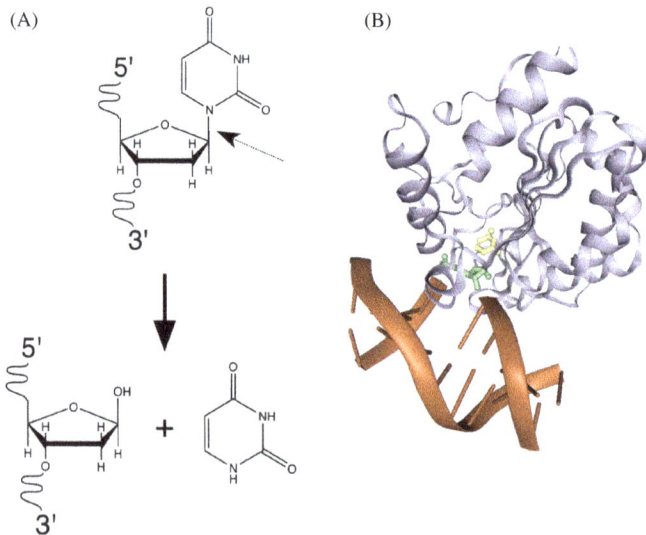

Figure 10.3. Uracil DNA glycosylase reaction. Cleavage of the N-glycosidic bond of a uracil residue in DNA is shown in panel A. The structure of human uracil DNA glycosylase bound to uracil-containing DNA is in panel B, with the uracil residue flipped out into the active site of the enzyme (uracil, yellow; DNA sugar, green). The image in panel B is from the RCSB PDB (www.rcsb.org) of PDB ID 1SSP (Parikh *et al.*, 1988).

DNA glycosylase (UDG) enzymes that are widespread in nature and that fall into several evolutionary families. UDG enzymes release the aberrant uracil from DNA and leave behind an AP site, which is repaired by the pathway described just above.

The *E. coli* UDG enzyme has been studied in great detail. The protein scans the DNA along the minor groove, searching for the aberrant U:G base pairs. At these sites, the enzyme kinks the DNA, essentially squeezing out (flipping) the uracil from the duplex into the active site of the enzyme, where the N-glycosidic bond is hydrolyzed. As you would expect, the inactivation of the *E. coli* UDG enzyme leads to a dramatic increase in C to T mutations.

Human cells have at least four different glycosylases that can remove uracil; the structure of one of these, bound to substrate, is shown in Figure 10.3B. The human UDG that is homologous to the bacterial enzyme appears to associate with the cellular replication proteins, presumably so that it can more efficiently repair uracils that are mistakenly inserted by DNA polymerase (from precursor dUTP).

10.3 Diverse DNA glycosylases expand the repertoire of BER

Both bacterial and mammalian cells have many different DNA glycosylases that recognize different kinds of damage. *E. coli* has at least eight, and at last count humans are endowed with 11 different glycosylases. In general, each glycosylase recognizes a few closely related base lesions, and there is frequently some overlap in substrate specificity between the glycosylases. Thus, in the mouse system, knockouts of a single glycosylase are often without phenotype, while double or triple knockouts can have shorter life spans and heightened cancer susceptibility.

As with UDG, the other glycosylases generally function by means of a base-flipping mechanism. Many, though not all, glycosylases have a common DNA-binding motif, consisting of an α-helix followed by a hairpin followed by another α-helix. This motif is located on top of a cleft in the enzyme that accepts the flipped-out base and promotes the hydrolysis reaction.

An interesting feature of about half of the known glycosylases is that they harbor a second activity relevant to BER; these are therefore sometimes called bifunctional DNA glycosylases. In addition to hydrolyzing the N-glycosidic bond, these enzymes can cleave the DNA chain on the 3′ side of an AP site, an activity called AP lyase (Figure 10.4B). AP lyase activity leaves the phosphate group as a 5′ end on the 3′ side of the lesion, and a β-elimination reaction converts the terminal AP site into a broken sugar residue with an aldehyde group (Figure 10.4B).

The existence of the dual enzymatic activities on the same glycosylase protein allows a concerted two-step reaction in which the lyase reaction quickly follows hydrolysis of the N-glycosidic bond. There is even a mammalian glycosylase (OGG1) that uses the detached, modified base from the glycosylase reaction (8-oxo-guanine) as a cofactor for the lyase reaction, a remarkable example of coupling of enzymatic activities in a single protein.

The lyase reaction does not, however, obviate the need for an AP endonuclease. The cleaved terminus of the DNA still contains an aberrant 3′-terminal residue that cannot support extension by DNA polymerase. As with a true AP site, the AP endonuclease cleaves the phosphodiester bond on the 5′ side of the aberrant sugar residue, leaving a 3′-OH end on the adjacent (normal) residue (Figure 10.4C). This releases the aberrant sugar-phosphate from the DNA, providing a 3′-OH group as primer for DNA polymerase and a single-nucleotide gap that is filled by the polymerase (Figure 10.4D).

Many different modified bases are recognized by one or more DNA glycosylases — as mentioned earlier, mammalian cells have at least 11 such enzymes. The substrates include various alkylated bases that result from endogenous and/or exogenous alkylating agents. They also include bases that have been damaged by oxidative reactions and fragmented bases, including the prominent lesion 8-oxo-guanine (8-oxoG; see Chapter 8).

Before leaving the topic, however, the biology of the 8-oxoG lesion is worthy of further discussion (Figure 10.5). As indicated earlier, this is a prominent lesion arising from oxidative damage. All

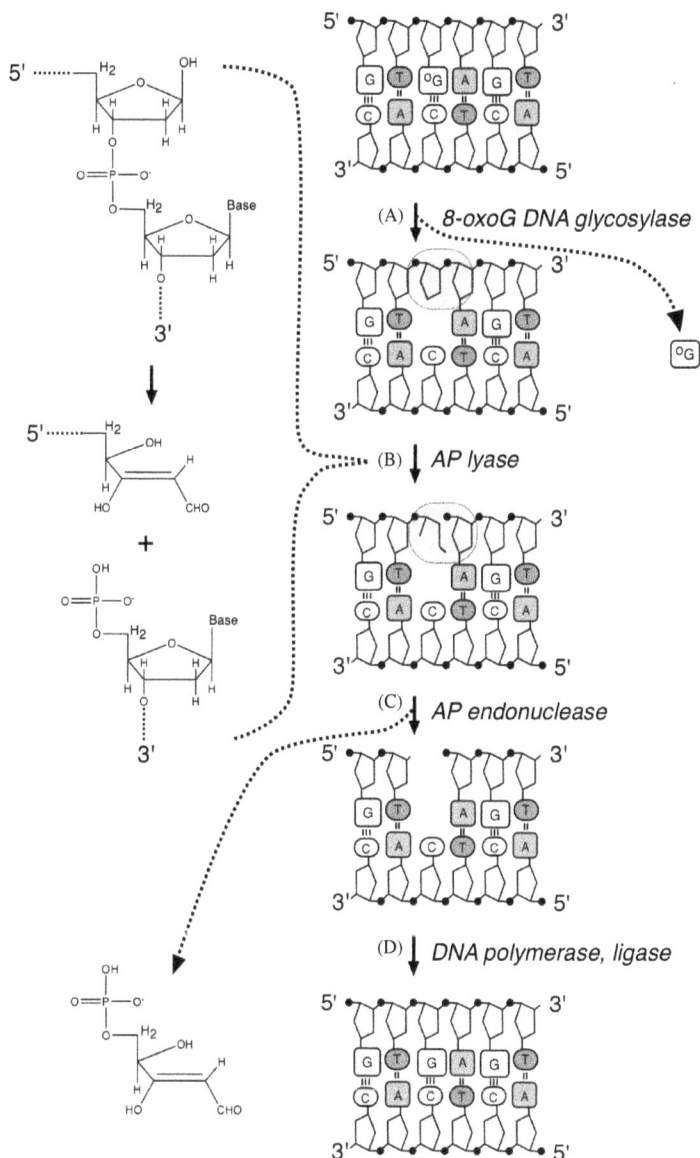

Figure 10.4. Excision repair with a bifunctional DNA glycosylase. The steps in this base excision repair pathway are depicted and described in more detail in the text. A close-up of the AP lyase reaction is shown in more detail with the relevant structures at the top left; this chemical scheme elaborates the reaction of the circled dinucleotides in the DNA on the right (top strand between steps A/B and B/C). The aberrant sugar-phosphate released by AP endonuclease in step C is shown at the bottom left.

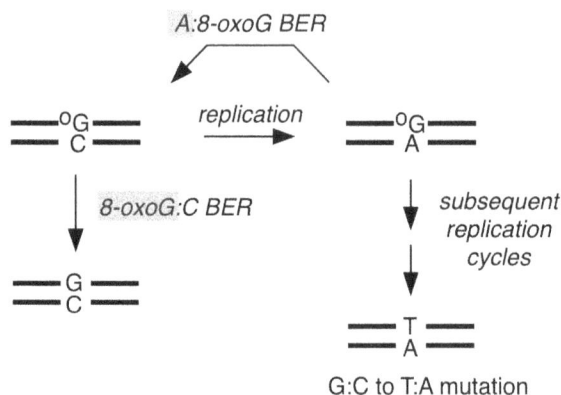

Figure 10.5. Biology of 8-oxo-guanine DNA repair and mutagenesis. One pathway of base excision repair acts on 8-oxo-guanine:cytosine base pairs, removing the 8-oxoguanine residue (left). A second pathway removes adenine residues from 8-oxo-guanine:adenine base pairs generated when DNA polymerase misincorporates adenine (top). When this latter pathway fails, subsequent replication cycles finalize a G:C to T:A mutation.

cells carry one or more glycosylases that act on the 8-oxoG:C base pair resulting from oxidative damage to the DNA. If the 8-oxoG is not repaired prior to DNA synthesis, DNA polymerase often inserts an adenine residue opposite the 8-oxo-G, resulting in an 8-oxo-G:A pair. What do cells do with this unusual base pair? It turns out they also carry another relevant glycosylase, but this one excises the A from the 8-oxoG:A pair (Figure 10.5). This provides DNA polymerase with another chance to get it right, and insert a proper C across from the 8-oxoG. BER of the A residue re-creates the 8-oxoG:C pair, which is a substrate for the first glycosylase that removes 8-oxoG residues, and thus provides a second layer of protection from oxidative damage. In *E. coli*, single mutants with a knockout of either of the two glycosylases suffer a modest increase in the expected G to T mutations, but double knockouts show a hugely synergistic effect on the rate of these mutations.

This system is relevant in human cancer. Humans have a homolog (called MYH) of the *E. coli* glycosylase that acts on 8-oxoG:A mispairs. Mutations in the gene encoding this homolog have been found in patients that constitute a subset of a particular familial cancer predisposition syndrome characterized by multiple colorectal adenomas and carcinomas. Presumably, the increased frequency of G to T mutations caused by failure to act on the 8-oxoG:A mispair contributes to colorectal cancer formation in these patients.

10.4 Bacterial NER repairs UV-induced dimers and other bulky lesions

The process of NER was first uncovered in the 1960s as a repair mechanism for UV-induced damage in bacterial systems. Two key findings from that era were that the UV-induced dimers were physically removed from the genome during this repair reaction, and that DNA synthesis occurred during repair (so-called repair synthesis). In addition, genetic experiments identified four *E. coli* genes, which, when mutated, led to a dramatic hypersensitivity to UV. The protein products of these genes were called UvrA, UvrB, UvrC, and UvrD, and these four turned out to be key proteins that catalyze NER. Later studies also showed that mutations in the gene encoding DNA polymerase I could likewise cause hypersensitivity to UV, and this polymerase was thereby implicated in bacterial NER.

The NER-defective mutants mentioned earlier also turned out to be sensitive to a wide variety of DNA-damaging agents, reflecting the ability of NER to act on many different DNA lesions. In addition to repairing CPD and 6-4 PP lesions induced by UV, NER is critical for repairing alkylation damage, particularly for larger DNA adducts (generally called bulky lesions) (see Chapter 8). NER also plays a role in repairing DNA crosslinks, and can even act on DNA-protein crosslinks involving relatively small proteins (roughly 10 kDa or less).

Biochemical studies with extracts and later with purified proteins have led to a detailed understanding of the mechanism of bacterial NER (see Nobel Prize for DNA Repair on page 157). The five proteins mentioned earlier act in a sequential manner to locate and then verify the site of damage, remove an oligonucleotide containing the damage, and then fill in the resulting gap with a short patch of DNA synthesis.

The UvrA and UvrB proteins play the critical role of identifying the site of DNA damage (Figure 10.6). A complex of the two proteins is thought to slide along duplex DNA scanning for sites of damage, driven by the ATPase activity of the UvrA protein. Structural evidence indicates that UvrA searches for possible sites of damage by testing the local flexibility of the DNA without actually contacting the lesion per se. Relatively inflexible sites are identified as candidates for damaged sites, and UvrA thereby delivers UvrB to these candidate sites. In this role, UvrA has been called a "molecular matchmaker," pairing UvrB protein to the site of possible DNA damage. The UvrB protein then attempts to verify whether the damage is real and appropriate to trigger NER. The ATPase activity of UvrB is activated to induce localized unwinding in the immediate vicinity of the candidate damage (Figure 10.6). Structural studies show that UvrB protein engages in very intimate contact with the DNA bases in the denatured region, including base flipping into the body of the protein. Several models have been proposed to explain how UvrB verifies the DNA damage, for example, one model in which a particular tyrosine in the protein stacks against the damaged base. In general, this is a particularly interesting and challenging problem, because a very wide variety of lesions can be identified by the combined action of UvrA and UvrB. Substantial evidence argues that lesions are recognized, in large part, by the DNA distortion and helix destabilization that they cause, but a full understanding of lesion identification and verification has not yet been achieved. Whatever the detailed mechanism may be, it is clear that UvrB needs to certify the damage as appropriate for the NER reaction to proceed.

Once the damage is certified, UvrB attracts the UvrC protein to the damaged site (Figure 10.6). UvrC is a DNA-damage-specific endonuclease that cleaves on both sides of the damage. Interestingly, the protein has two nuclease active sites, one for cleaving DNA in the 3′ direction from the damage and the other for the 5′ cleavage. The UvrB-damage complex dictates the sites of cleavage, which display

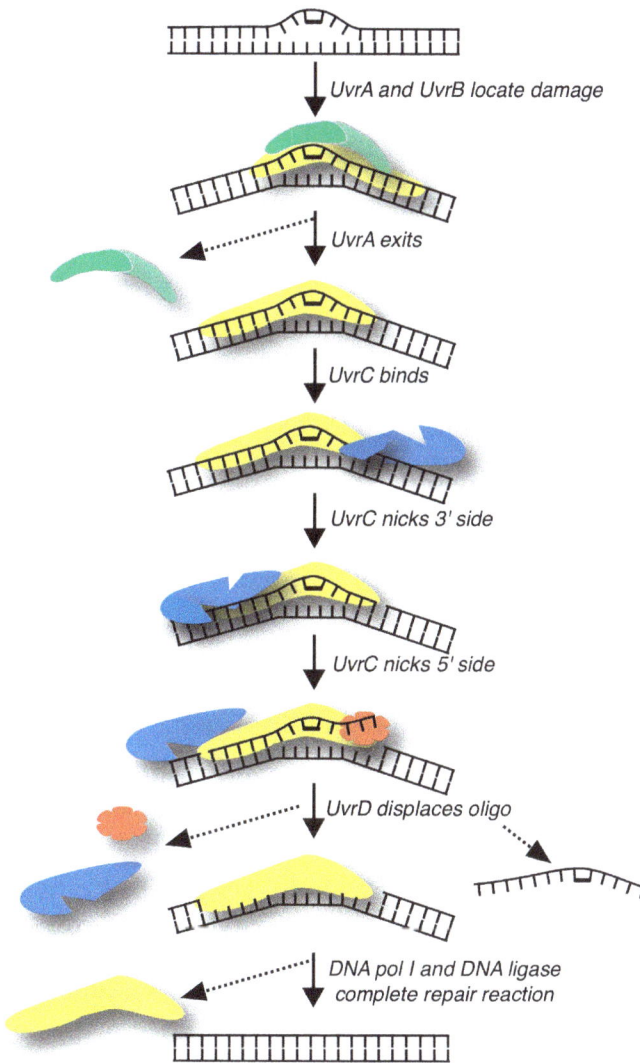

Figure 10.6. Nucleotide excision repair. The steps in bacterial nucleotide excision repair are described in detail in the text. The depicted proteins are as follows: UvrA, green; UvrB, yellow; UvrC, blue; and UvrD, orange (DNA pol I and DNA ligase are not shown). This figure was modified from Figures 7.15 and 7.20 of Friedberg *et al.* (2006).

some minor variations depending on both the DNA sequence and the nature of the DNA lesion. Incision usually occurs at the eighth phosphodiester bond in the 5′ direction from the lesion and the fourth or fifth phosphodiester bond in the 3′ direction (Figure 10.6; see "How did they test that?" at the end of this chapter).

The next key protein in the bacterial NER pathway is the UvrD protein, which is also known as DNA helicase II. UvrD is responsible for displacing the incised oligonucleotide from the DNA, and also appears to displace UvrC from the complex (Figure 10.6). DNA polymerase I cannot access the 3′ end at the gap until both the UvrC and the oligonucleotide are displaced. The polymerase then fills in the gap with repair synthesis, and this action also appears to displace UvrB from its site of binding (on the nondamaged strand). As with BER and Okazaki fragment maturation, the very last step involves ligation of the final nick by the action of DNA ligase (Figure 10.6).

While this bacterial NER system was worked out largely using the *E. coli* model system, the four Uvr proteins are very widely conserved throughout the bacterial kingdom, and thus the model is quite robust. It should be noted that this NER pathway is augmented for DNA damage that blocks transcription, as described in Section 10.6.

10.5 NER in eukaryotic systems

Cells from patients with an inherited skin-cancer syndrome played a major role in early studies of NER in mammalian systems. The disease *xeroderma pigmentosum* (XP) is characterized by extreme sensitivity to sunlight, resulting in very frequent skin cancer at remarkably early ages (see Section 14.4 for more discussion of the disease). Cells from patients with XP are hypersensitive to UV and certain other DNA-damaging agents, and show other characteristics consistent with a defect in repair of UV-induced lesions.

Scientists discovered that the cells from different XP patients had distinguishable defects in the repair pathway by means of cell-fusion studies. Human cells from two different lineages can be fused together using certain reagents, allowing what amounts to a genetic complementation test. To illustrate the principle, imagine a biochemical pathway consisting of two enzymes, call them A and B. If you fuse one cell with a defect in enzyme A with a second cell that has a defect in enzyme B, you will likely recover a normal cell because it has the functional copies of both enzymes. In this example, A and B are said to be in two different "complementation groups." If you fuse two A-deficient cells to each other, or two B-deficient cells to each other, you would generally not recover normal cells.[1] This trick was used to define the cells from various XP patients, and scientists found seven complementation groups called XP-A through XP-G with defects in NER, along with another complementation group that they called XP-V (XP *variant*), which displayed perfectly normal NER without any cell fusion. The seven NER-defective XP complementation groups defined seven proteins that participate in the excision repair reaction, while individuals in the XP-V group were eventually found to carry a mutation in the gene encoding the translesion DNA polymerase η, which plays an important role in bypass of UV-induced pyrimidine dimers (see Chapter 12). Related studies in model systems such as mouse, yeast, and fruit fly identified orthologs of the seven NER factors and greatly contributed to the development of the eukaryotic NER field.

While the seven XP proteins mentioned earlier are central in the damage-recognition steps of mammalian NER, 12 additional proteins are also involved but were not among those initially found to be missing in XP patients. Thus, in contrast to only three *E. coli* proteins needed for damage recognition in the bacterial system, 19 are involved in the corresponding steps in mammals (see Appendix Table 5 for listing of NER protein names). Furthermore, the proteins are not homologous between bacteria and mammals, indicating a separate evolutionary origin. In addition to the 19 proteins involved in damage recognition, several additional proteins are involved in the later steps of gap-filling synthesis and ligation (see below). Once you go beyond the relatively simple naming of the XP-A through XP-G proteins, the nomenclature of the various NER proteins is extremely complex, particularly since the proteins sometimes have different names in the

---

[1] There are interesting exceptions where two defective proteins of the same type can form a functional complex, usually involving homo-multimeric enzyme complexes.

different model systems mentioned earlier. We will avoid most of this complex nomenclature in the remainder of this chapter.

Surprisingly, one of the key protein complexes in eukaryotic NER has a second role in transcription. The transcription factor called TFIIH is involved in locally unwinding the DNA duplex in both transcription and NER, but the mechanisms of these unwinding reactions are distinct. TFIIH is a tight complex of 10 different subunits, which includes XPB and XPD (defined by XP complementation groups as described earlier). Ample evidence demonstrates that the roles of TFIIH in transcription and in NER are entirely independent of each other.

In a broad sense, the mechanism of NER in mammals is parallel to that in bacteria. That is, both NER systems recognize bulky lesions that distort the DNA helix, both systems excise an oligonucleotide containing the damage, and both systems replace the missing information by gap-filling synthesis and complete the reaction with ligation. However, many details of mammalian (and eukaryotic) NER are quite distinct from that in bacteria. The following is a rather cursory summary of a very complex reaction, and the interested reader is encouraged to dive more deeply into the mechanistic details and nuances (see Friedberg *et al.*, 2006 and Schärer, 2013 in Further Reading).

Eukaryotic NER begins when a two-protein complex (XPC along with a protein called RAD23B) surveys DNA and recognizes a distorted region of duplex DNA that is a possible DNA damage site. The protein complex can also bind distorted DNA such as small bubbles that contain no damage, and so this step in NER is not sufficient to verify an actual damaged site. This step is reminiscent of the role of UvrA in bacteria, providing a rapid but not entirely accurate screening for possible damage sites. Interestingly, the XPC/RAD23B protein complex is not very good at recognizing UV-induced dimers (CPD or 6-4PP), particularly when nucleosomes are present in chromatin. An auxiliary two-protein complex helps out in this regard. The complex contains the XPE protein (along with another) and directly binds to CPD and 6-4PP lesions. Patients in the XPE group with a defect in this auxiliary factor have a milder form of the XP disease, consistent with the role of this protein in augmenting damage recognition.

Once XPC/RAD23B has bound to a distorted site that might contain damage, it recruits the 10-protein TFIIH complex and several additional factors in the NER pathway. The XPC/RAD23B complex then departs, having completed its function as a matchmaker that brings TFIIH to the distorted site. As mentioned earlier, TFIIH plays a critical function in unwinding the region that contains the possible DNA damage. Two helicases within the TFIIH complex appear to play key roles in verifying DNA damage, perhaps as bases in the DNA that block translocation of the helicase along the DNA.

Two of the NER factors that are assembled along with TFIIH are specific nucleases that make the incisions necessary to release the damage-containing oligonucleotide. Both nucleases appear to recognize the single-strand/double-strand junction at the edges of the target DNA region. Recall that bacterial NER uses a single nuclease with two active sites, while eukaryotic NER uses two distinct proteins each with their own nuclease active site. The spacing of DNA cleavage is also quite different than in bacterial NER. The first incision occurs at the 22nd, 23, or 24 phosphodiester bond on the 5′ side of the lesion, while the second incision cleaves the fifth or sixth phosphodiester bond on the 3′ side of the lesion (some particular lesions lead to slight variations on these spacings) (see "How did they test that" at the end of this chapter).

After the dual incisions, the roughly 30-base long oligonucleotide containing the damage is released, along with the TFIIH that is bound to it. Replication proteins that should be familiar from Chapter 4 are involved in the gap-filling synthesis step, including DNA polymerases δ and ε, sliding clamp *proliferating cell nuclear antigen* (PCNA) and its loader *replication factor C* (RFC), and the ssDNA-binding protein RPA. There is some evidence that NER in actively replicating cells uses polymerase ε and DNA ligase I (to seal the final nick), while NER in quiescent cells employs polymerase δ and DNA ligase III (an alternative DNA ligase).

There is extensive current research on the additional complexities of completing NER within the context of eukaryotic chromatin.

It is already clear that extensive chromatin remodeling is needed to facilitate NER within chromatin, with many chromatin-modifying factors and multiple pathways implicated.

10.6 TC-NER repairs lesions that have blocked RNA polymerase

The bacterial and eukaryotic NER pathways described earlier operate throughout the genome and are sometimes called *global genome* NER (GG-NER). A distinct version of NER in both bacteria and eukaryotic cells operates on the transcribed strand of actively expressed genes. The need for so-called *transcription-coupled* NER (TC-NER) becomes obvious when you consider that the bulky lesions repaired by NER often block RNA polymerase, and blocked RNA polymerase complexes can be very stable. On the one hand, the NER machinery needs to access the blocking lesion, but the blocked RNA polymerase physically obstructs the damage site. On the other hand, transcribing RNA polymerase is an excellent sensor for this kind of DNA damage, easily identifying a single damaged site within any given transcription unit. It is therefore not surprising that nature found a way to couple the blockage of transcription with the process of NER. The importance of repairing blocking lesions in transcribed regions is emphasized by the need to successfully transcribe and translate proteins such as DNA repair and damage response factors upon DNA damage.

TC-NER was first discovered in eukaryotic systems, which then led to its discovery in bacteria. Bacterial TC-NER relies on a special helicase called the *transcription-repair coupling factor* (TRCF; also called Mfd based on a certain phenotype of mutants that lack the factor). TRCF recognizes blocked RNA polymerase and then pushes both the RNA polymerase and the nascent transcript off the DNA by ATP-dependent translocation along the template (Figure 10.7A). The factor remains at the site of damage, attracting the UvrAB complex to begin the process of NER.

Eukaryotic TC-NER is similar in outline, and the eukaryotic TRCF factor even has (rather distant) homology to the bacterial

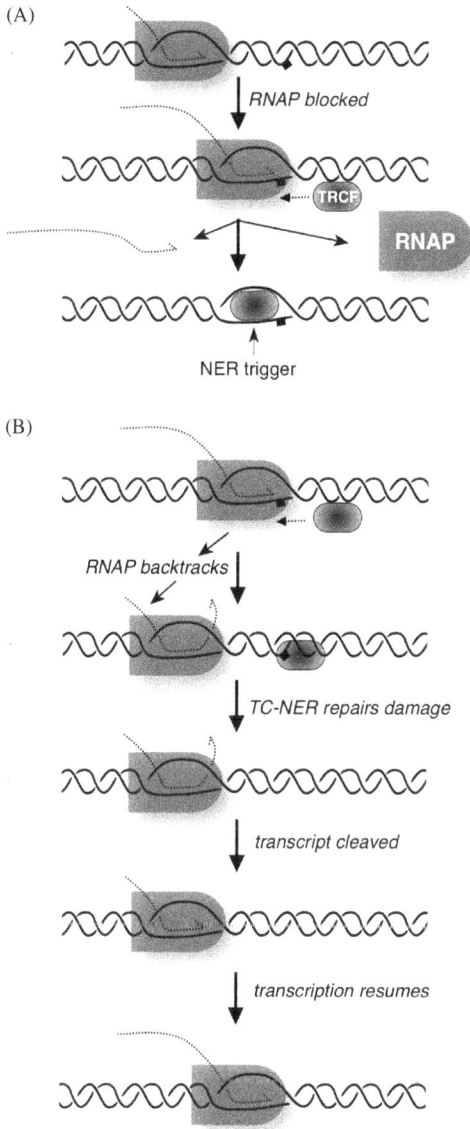

Figure 10.7. Transcription-coupled nucleotide excision repair. The bacterial transcription-repair coupling factor (TRCF; also called Mfd) displaces RNA polymerase and the transcript when RNA polymerase is blocked by a lesion (panel A). The factor then recruits the bacterial nucleotide excision repair factors to trigger repair. In eukaryotic cells, TRCF apparently causes RNA polymerase to backtrack from the site of damage, allowing transcription to resume after the nucleotide excision repair reaction (panel B).

factor. As in bacterial TC-NER, eukaryotic TRCF displaces the blocked RNA polymerase and attracts the NER machinery to the site of damage (Figure 10.7B). There is evidence, however, that the fate of the RNA polymerase might be different in eukaryotes than in prokaryotes. Recent experiments indicate that eukaryotic RNA polymerase II, the major RNA polymerase for protein-coding genes, is moved in the backward direction along the template DNA by TRCF, providing room for the NER reaction (Figure 10.7B). This would displace only the 3′ end of the transcript but keep RNA polymerase II engaged with the template strand and an upstream region of the transcript. After the damage is repaired, the excessively long (unpaired) 3′ end of the transcript would be cleaved near the RNA polymerase II, the polymerase would engage the new 3′ end and resume transcription through the repaired region. These gymnastics essentially save the transcript from being discarded, which may be particularly important in the case of long eukaryotic transcription units. Another notable difference from the bacterial system is that the initial scanning protein complex in the global pathway, in this case XPC-RAD23B, is not required for the eukaryotic TC-NER pathway. After the early damage-recognition step, the subsequent steps in eukaryotic TC-NER appear identical to those in the global eukaryotic pathway.

Mammalian TC-NER involves additional proteins whose functions are currently under intense scrutiny. These proteins clearly play regulatory roles in the overall pathway but there is still much to learn about their detailed roles. There is evidence that these additional proteins might lead to different fates for RNA polymerase depending on the situation, adding complexity to the TC-NER pathway. For example, and likely at a low frequency, RNA polymerase II might be degraded under some blocking conditions or induced to transcribe through the blocking lesion under other conditions (like a translesion DNA polymerase; see Chapter 12).

It is worth noting that another devastating human disease, Cockayne syndrome, is caused by defects in the TC-NER factors. This syndrome leads to neural disorders, premature aging, and a mean age of survival in the early teens, and cells from these patients are defective for the TC-NER pathway (also see Chapter 14). Indeed, two of the protein factors involved in TC-NER, CSA and CSB, were identified as being defective in Cockayne syndrome patients (CSB is the eukaryotic TRCF factor discussed in the paragraph above). The severe symptoms of Cockayne syndrome attest to the importance of the TC-NER pathway in mammals.

10.7 Summary of key points

- Excision repair removes damage from one strand and replaces it, using the opposite strand as template.
- BER replaces a single damaged residue, including common lesions such as AP sites, uracil residues, and 8-oxo-guanine residues.
- The specificity of BER pathways is provided by glycosylase enzymes that recognize particular damaged bases and cleave the N-glycosidic bond connecting the damaged base to the DNA sugar.
- NER acts on bulky lesions that distort DNA, such as UV-induced pyrimidine dimers, covalently attached complex chemicals, and crosslinks (DNA–DNA and DNA–protein).
- In both bacteria and eukaryotes, NER involves multiple proteins that scan for and verify DNA damage, excise a patch of >10 bases including the damage, and replace this patch by the action of DNA polymerase.
- Damage recognition in eukaryotic NER is very complex, involving at least 19 proteins including the TFIIH complex that is also involved in transcription.
- TC-NER is triggered when RNA polymerase is blocked by DNA damage, and leads to both damage repair and rescue (or removal) of the blocked RNA polymerase.
- Two deadly human diseases involve a defect in NER, XP (defect in GG-NER), and Cockayne syndrome (defect in TC-NER).

Further Reading

Friedberg, E. C., Walker, G. C., Siede, W., Wood, R. D., Schultz, R. A., & Ellenberger, T. (2006). *DNA Repair and Mutagenesis* (2nd ed.). Washington, DC: ASM Press.

Kisker, C., Kuper, J., & Van Houten, B. (2013). Prokaryotic nucleotide excision repair. *Cold Spring Harb Perspect Biol*, *5*(3), a012591.

Krokan, H. E., & Bjørås, M. (2013). Base excision repair. *Cold Spring Harb Perspect Biol*, *5*(4), a012583.

Mu, D., Hsu, D. S., & Sancar, A. (1996). Reaction mechanism of human DNA repair excision nuclease. *J Biol Chem*, *271*(14), 8285–8294.

Parikh, S. S., Mol, C. D., Slupphaug, G., Bharati, S., Krokan, H. E., & Tainer, J. A. (1998). Base excision repair initiation revealed by crystal structures and binding kinetics of human uracil-DNA glycosylase with DNA. *EMBO J*, *17*(17), 5214–5226.

Sancar, A. (2016). Mechanisms of DNA repair by photolyase and excision nuclease (Nobel Lecture). *Angew Chem Int Ed Engl*, *55*, 8502–8527.

Schärer, O. D. (2013). Nucleotide excision repair in eukaryotes. *Cold Spring Harb Perspect Biol*, *5*(10), a012609.

Vermeulen, W., & Fousteri, M. (2013). Mammalian transcription-coupled excision repair. *Cold Spring Harb Perspect Biol*, *5*(8), a012625.

How did they test that?
What are the excision products of NER?

Mu, Hsu, and Sancar (1996) compared excision products generated by purified nucleotide excision repair (NER) proteins from *E. coli* and human cells (proteins needed to complete repair, such as DNA polymerase, were not included). They constructed a substrate by ligating together six synthetic oligonucleotides (panel A). The substrate has a cholesterol adduct (panel B) near the middle, a radioactive residue (*) a few nucleotides 5′ from the adduct, and a covalently linked biotin at one of the 3′ ends. Although cholesterol is not a biologically relevant DNA adduct, cholesterol-modified DNA is a good substrate for NER and cholesterol can be conveniently incorporated into oligonucleotides. The internal radioactivity is used to visualize the DNA, and the biotin is useful for characterizing excision products (see below). In addition to the two NER reactions, a control was performed with DNA alone. One question is whether the excised oligonucleotides are released from the NER protein–DNA complex. To find out, the scientists mixed each of the three reaction products with magnetic beads that contain streptavidin, which binds tightly to the biotin (panel C). Using a magnetic particle concentrator, they separated any released material (unbound, U) from the materials that bound to the beads (B). Then, all reaction products were subjected to denaturing polyacrylamide gel electrophoresis, with size standards (M) in the first lane (panel D). Two notable conclusions emerged. First, the excised fragments from *E. coli* NER were 12–13 nucleotides long, while those from human NER were 26–28 nucleotides long (panel D). These sizes are consistent with experiments that mapped the locations of the 5′ and 3′ incisions. Second, while excised fragments from *E. coli* NER were in the bound fraction (B), those from human NER were unbound (U). Previous studies had shown that completion of repair is needed to release excised fragments in the *E. coli* system, explaining that result. It was surprising to find that excised fragments from human NER are released, and later studies showed that they are released in a protein-bound form. Panel D of the figure was reproduced from Mu *et al.*

(1996), with permission from the American Society for Biochemistry and Molecular Biology; permission conveyed by Copyright Clearance Center, Inc. Panels A and B were modified from figures of Mu *et al.* (1996).

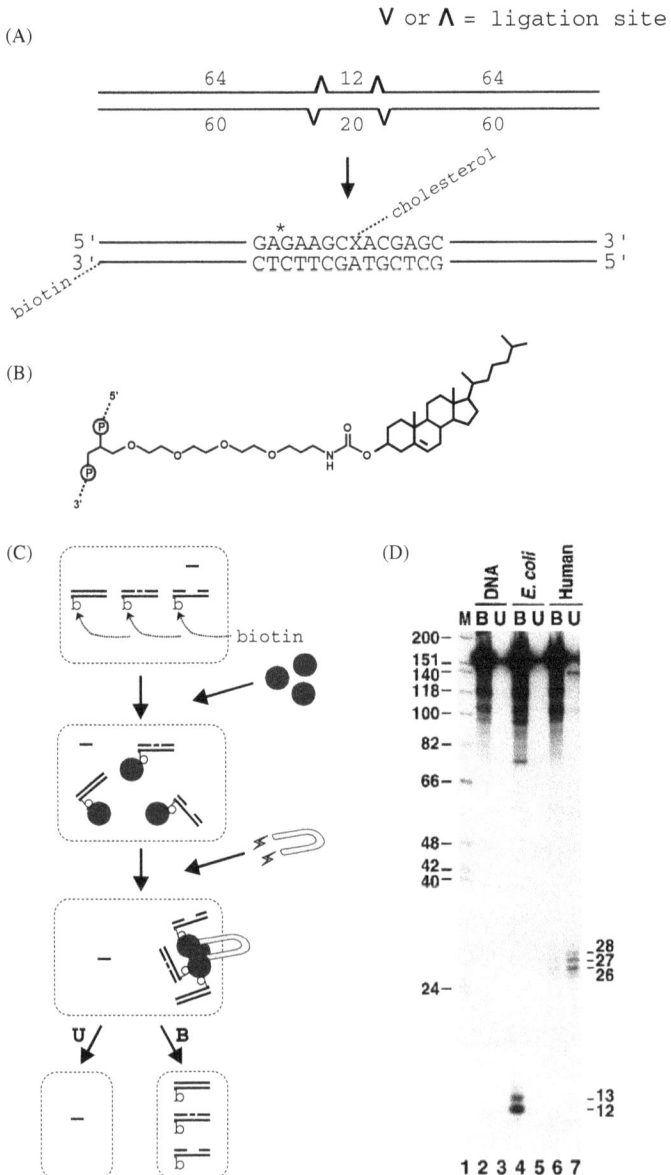

Chapter 11

Repair of double-strand breaks

Double-strand breaks (DSBs) in DNA constitute a very serious threat to the genome. A DSB in a circular bacterial chromosome converts the DNA into a linear form, which creates multiple problems including exposure to exonucleases that degrade linear DNA. Such exonucleases normally provide a defense against certain incoming viral DNA molecules in bacteria. In eukaryotic cells, a DSB splits the affected chromosome into two and can thereby lead to the loss of one of the chromosome arms. Furthermore, in all cells, a DSB can interrupt an important gene and thereby lead to disturbances in cell growth or even cell death.

As described in Chapter 8, one important cause of DSBs is ionizing radiation. In addition, various cellular processes are known to create DSBs, such as the errant action of endonucleases or the replication of DNA containing a single-strand break. Surprisingly, cells intentionally create DSBs during meiosis (see below) and during other programmed recombination events.

Given the potentially dire consequences of DSBs, two major processes have evolved to repair these breaks. Homologous recombination utilizes a homologous unbroken DNA as a template to accurately repair DSBs, and should more realistically be viewed as a collection of pathways with distinct features (see below). The second process, nonhomologous end joining (NHEJ), simply ligates two broken ends together without using a homolog for guidance. Sometimes, the two

249

broken ends are from different DNA molecules, leading to a gross rearrangement such as a chromosomal translocation. The results are more efficacious when the two broken ends were generated from the same DSB, but even in that case, NHEJ can result in the gain or loss of a few base pairs at the site of ligation. Thus, NHEJ is an inherently mutagenic process, while homologous recombination provides a largely error-free pathway for repairing DSBs. As described later, the vast majority of NHEJ events use a defined set of proteins that generally avoid joining ends from different chromosomes. This pathway is called *canonical* NHEJ (c-NHEJ), to distinguish it from another pathway introduced in the next paragraph.

In addition to these two major processes, two nonconservative pathways can repair DSBs that occur in between DNA repeats. These are called *single-strand annealing* (SSA) and *alternative*-NHEJ (alt-NHEJ), and both result in loss of the intervening DNA between the repeats (see Section 11.10). It should be noted that this chapter provides only a brief overview of the expansive field of recombination and DNA-break repair (see Haber, 2013, for a comprehensive review of the field).

11.1 The machinery at the heart of homologous recombination

The essential central step in homologous recombination reactions is the search by one DNA molecule for a homologous partner. Even in a human cell, this search process is capable of scanning the roughly 6.6-billion base pairs of nuclear DNA to find a correct match and thereby trigger the process of homologous recombination. Let us take a closer look at this amazingly effective search for homology.

For the search process to work, one of the participating DNA molecules must have a single-stranded region, which is generally at its 3′ end, and we will discuss the generation of this single-stranded end later. During the search itself, this single-stranded 3′ end probes vast amounts of duplex DNA for a region where it can correctly base-pair with its complementary sequence (Figure 11.1A). Such base-pairing displaces the strand of the duplex that is identical (or homologous) to the invading single strand, and the resulting

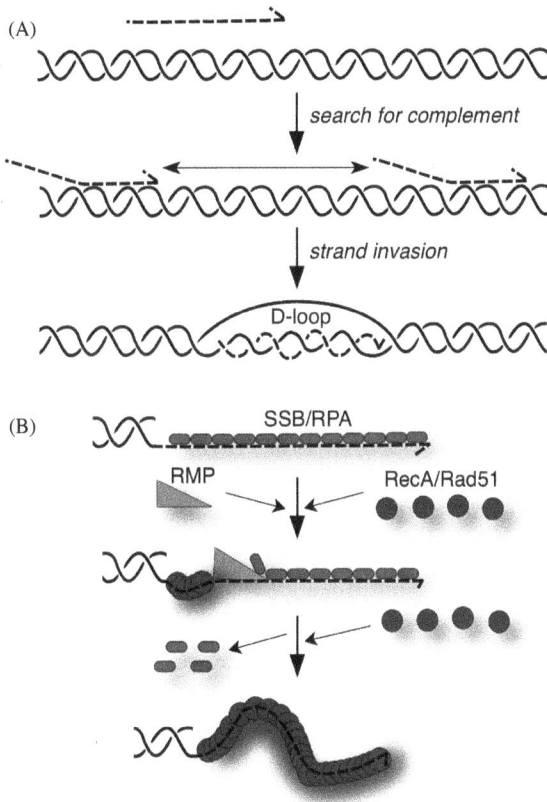

Figure 11.1. Strand invasion. During strand invasion, a single-stranded segment (dotted line) is used in an alignment search to find a complementary region in duplex DNA (panel A). Proper alignment results in a D-loop in which the strand with the identical sequence in the duplex is displaced. The proteins relevant for strand invasion include the RecA (bacterial) or Rad51 (eukaryotic) strand-invasion protein, the recombination mediator protein (RMP) that mediates loading of the strand-invasion protein, and the SSB (bacterial) or RPA (eukaryotic) protein that binds ssDNA (panel B).

intermediate is therefore called a *d*isplacement loop or D-loop. The overall reaction that creates the D-loop is called, not surprisingly, strand invasion.

In the test tube, two ssDNA molecules that are complementary to each other can, under the right conditions, align their complementary sequences and essentially zip up into a duplex, a process called annealing. However, ssDNA by itself is not capable of invading duplex

DNA in the strand-invasion reaction described earlier. Instead, all cells have a remarkable enzyme that promotes the strand-invasion reaction. The enzyme was first discovered in bacteria, where mutations that inactivate the gene for this enzyme were found to reduce the frequency of homologous *re*combination by five to ten thousand-fold. This RecA protein was found to bind as a filament to the single-stranded 3' end and promote the strand-invasion reaction, fueled by adenosine triphosphate (ATP) hydrolysis (Figure 11.1B). A structural model for the RecA filament (without DNA) is shown in Figure 11.2A. A homologous protein was later found in eukaryotic cells, and for historical reasons, the eukaryotic homolog was named Rad51 protein (knockout mutants are *rad*iation sensitive; a second eukaryotic homolog is specific for meiotic recombination; see below). The RecA and Rad51 proteins are structurally very similar and likely catalyze strand invasion by a highly conserved mechanism.

The mechanisms of homology search and strand invasion have been under intense scrutiny for several decades. Recent studies have shown that the ssDNA–RecA complex sample segments of duplex DNA for possible base-pairing, and then jumps to a different duplex DNA segment through three-dimensional space when base-pairing is not achieved. Once a segment of complementary base-pairing is achieved, the RecA filament is able to "zipper" longer stretches of the ssDNA into the duplex DNA to form even longer D-loops. Structural studies using X-ray crystallography and other methods have led to detailed models for how exactly the bases are aligned and exchanged during the test for local base-pairing, but much remains to be learned about this fascinating reaction.

Figure 11.2. Key proteins in homologous recombination. The structure of the *E. coli* RecA protein, in a model for the active filament, is shown in panel A. The grayscale diagram at the left depicts the order of the seven RecA subunits in the structure just to the right, to assist in visualizing the dimensionality of the filament. A space-filling representation is on the left, while the rightward images show a backbone representation of the protein with space-filling versions of the bound adenosine diphosphate (ADP). Both filament models are rotated 180° as indicated by the arrows to show the "backside" of the structure. In these images, the RecA protein

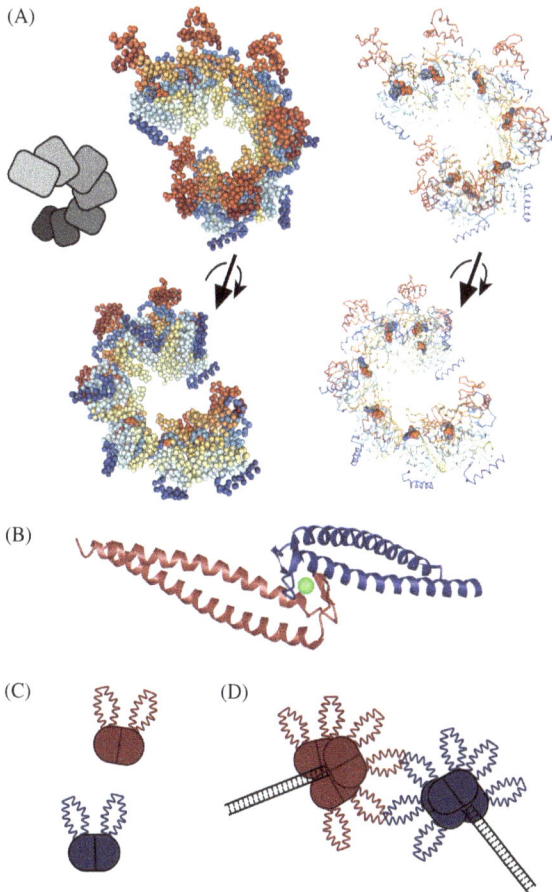

Figure 11.2. (*continued*) chains are each color coded with the rainbow scheme, with the N-terminus in blue and the C-terminus in red (and intervening regions shaded between blue and red). The structural images in panel A are from the RCSB PDB (www.rcsb.org) of PDB 1N03 (VanLoock *et al.*, 2003). The structure of the zinc-hook region of a Rad50 dimer is shown in panel B. In these images, the protein is shown in cartoon form (one subunit red and the other blue), while the bound ligand is indicated by the sphere (green). The ligand in this structure is mercury, which takes the place of the key zinc that coordinates the dimer. The structure in panel B is from the RCSB PDB (www.rcsb.org) of PDB 1L8D (Hopfner *et al.*, 2002). The structural image in panel B was generated using the web-based visualization suite iCn3D ("I see in 3D," version 2.7.15), developed by Wang *et al.* (2019). Panel C shows a cartoon of individual Mre11 complexes, while panel D presents a cartoon depicting how the Rad50 zinc hooks of Mre11 complexes can tether two DNA ends together.

In previous chapters, we learned that ssDNA in the cell is effectively coated with ssDNA-binding proteins, particularly the SSB protein in prokaryotes and the RPA protein in eukaryotes. It therefore may be somewhat surprising that these proteins completely block the binding of the RecA or Rad51 recombination protein. We now understand that homologous recombination must be regulated so that it does not occur too frequently or in the wrong places, and this initial binding step is one of the points of regulation. To illustrate, consider what would happen if RecA or Rad51 frequently intercepted the ssDNA at the replication fork and initiated unnecessary recombination reactions. Instead, additional proteins are required to load RecA or Rad51 onto ssDNA, effectively regulating the initiation of recombination. These so-called *r*ecombination *m*ediator *p*roteins or RMPs displace the ssDNA-binding protein and replace it with the RecA or Rad51 protein (Figure 11.1B). The SSB and RPA proteins do, however, play positive roles in the strand-invasion reaction. Most notably, they bind to the displaced single-strand in the D-loop intermediate and thereby stabilize and protect this structure.

Human cells have multiple proteins with RMP activity, but the most important one is the BRCA2 protein, which plays a major role in the repair of DSBs in mitotic cells. BRCA2 (and another repair-related protein called BRCA1) are proteins found to be defective in patients with an inherited hypersusceptibility to *br*east *ca*ncer. The implication of these findings is that successfully repairing DSBs protect against the initiation and progression of cancer, and this implication is now supported by extensive evidence (see Chapter 14). An evolutionarily distinct RMP protein, called Rad52, plays the role of loading Rad51 in lower eukaryotes, and bacteria also utilize a mediator protein to load RecA protein.

The proteins described so far in this section are critical in the strand-invasion reaction, which is the core reaction in homologous recombination. As we encounter the more complete sequence of reactions below, additional proteins are involved. For example, a number of different DNA nucleases and helicases play both negative and positive roles, and specialized endonucleases are often required to finalize the recombination products at the completion of the recombination reaction. In many cases, redundant activities are involved in these steps and the requirements can differ depending on the precise nature of the break, the cell type, and the growth status of the cells.

11.2 Homologous recombination in genetic exchange and meiosis

It now seems apparent that homologous recombination evolved to repair DSBs and broken replication forks (see below), but the process has clearly been co-opted for numerous other roles in diverse organisms. In bacterial cells, homologous recombination allows the exchange of genetic information, even between species and between the genomes of bacterial viruses and cells. Three major bacterial exchange pathways include conjugation (direct transfer of DNA between cells), transformation (uptake of naked DNA from the

---

1958 Nobel Prize in Physiology or Medicine (Divided)

This prize (one half) was awarded to **George Wells Beadle** and **Edward Lawrie Tatum** for their research showing that genes act by encoding enzymes which are responsible for metabolism. The second half of the prize was awarded to **Joshua Lederberg**, who along with Tatum, demonstrated that bacteria exchange genetic information through recombinational processes.

https://www.nobelprize.org/prizes/medicine/1958/summary/

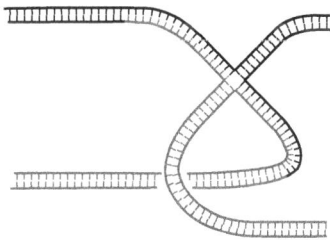

media), and transduction (transfer via a bacterial virus vector). The mechanisms of these exchange pathways are beyond the scope of this chapter.

In eukaryotic cells, a prominent role of homologous recombination is during the process of meiosis. The vast majority of eukaryotic cells are diploid, with one copy of each chromosome from the maternal and one from the paternal parent of that individual. In meiosis, those homologous chromosomes undergo homologous recombination to exchange parts, and this leads to further diversification of genomes by creating novel combinations of genetic alleles. In addition, homologous recombination during meiosis is necessary for proper segregation of chromosomes during the reductional meiotic cell division, which generates haploid gametes.

As alluded to above, meiotic recombination is initiated by self-induced DSBs. A specialized topoisomerase-like protein, called Spo11, which is specific for meiotic cells, induces protein-linked DNA breaks (Figure 11.3A). Subsequent steps result in cleavage of the covalently bound protein off the broken DNA ends, while later steps serve to repair these breaks and generate the exchange of parental DNA chromosomal segments in recombinant chromosomes.

The mechanism of meiotic recombination has been studied in great detail, particularly in model eukaryotic organisms such as yeast species, and illustrates a major pathway to repair DSBs. Following the creation of the DSB during meiosis, the broken ends must be prepared for strand invasion. A protein complex called MRN (mammalian cells) or MRX (yeast cells) is important for cleaving off the covalently attached Spo11 protein and resecting the end in preparation for the strand-invasion reaction (Figure 11.3B). The names of these protein complexes reflect the constituent protein subunits, *M*re11, *R*ad50, and either *N*bs1 (mammalian) or *X*rs2 (yeast).

The MRN (MRX) protein also plays an important role in tethering the two broken ends together, acting rather like a molecular "Velcro" (Figure 11.2B–D) (see "How did they test that?" at the end of this chapter). The Mre11 subunits bind to DNA ends, and a coiled-coil region of the Rad50 subunits extend outwards some 500 angstroms away from the break. At the tip of the extension, the chain

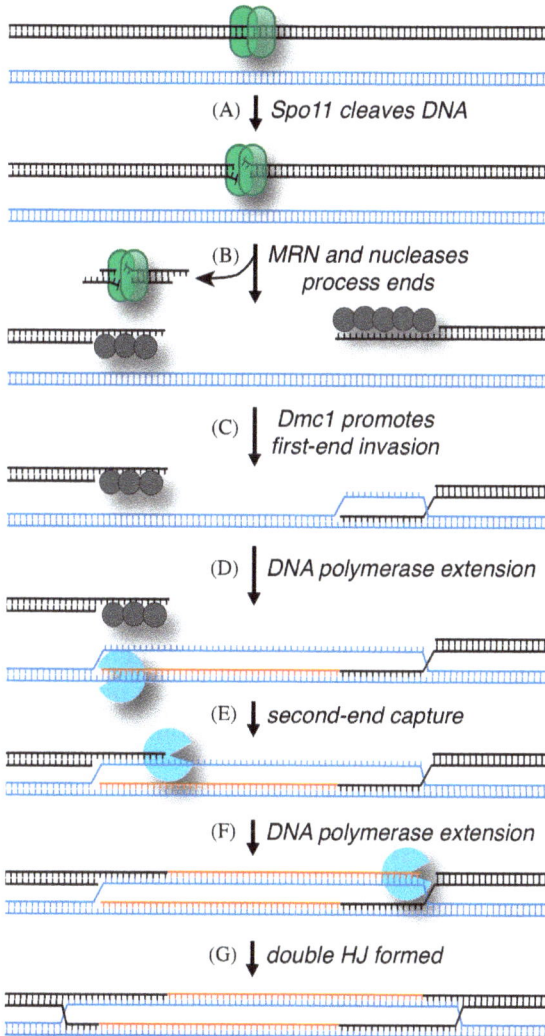

Figure 11.3. Pathway of homologous recombination in meiosis. See text for detailed description. Green ovals, Spo11 protein; gray balls, Dmc1; blue "pac-man," DNA polymerase; two homologous chromosomes color coded in black and blue; newly synthesized DNA in red.

folds back on itself to induce the coiled coil. Within this tip, a short amino acid sequence (two cysteines separated by two other residues) can interact with another Rad50 extension by mutual binding of zinc, potentially tethering two broken ends at a distance of 1000 angstroms (nearly 50 times the DNA diameter) (Figure 11.2B–D). The tethering action may also help to secure the two broken ends to a homolog and thereby coordinate the repair reaction. Genetic studies show that the tethering activity of MRN (MRX) is important for the success of the recombination reaction, and the tethering property has been conserved in evolution all the way from certain bacterial viruses (which carry an MR homolog) through lower eukaryotes such as yeast and right up through mammals including humans.

We will see later that the MRN (MRX) complex also participates in c-NHEJ and in DNA damage signaling. MRN (MRX) itself does not appear to have a 5′ to 3′ exonuclease activity to carry out the end resection and create 3′ single-stranded ends. It does, however, have an endonuclease activity that clips off a segment of the 5′ end at a break,[1] and it interacts with other proteins to prepare the 3′ single-stranded ends, including overlapping exonucleases with a 5′ to 3′ directionality.

Following end processing in meiotic recombination, the meiosis-specific strand-invasion protein, Dmc1 (Rad51 homolog), induces the strand-invasion reaction to create the D-loop intermediate by the process described earlier (Figure 11.3C). The D-loop is next extended in length by a limited DNA polymerase reaction (Figure 11.3D). This D-loop extension exposes a longer stretch of the displaced complementary strand, which allows an annealing reaction that "captures" the second broken end (Figure 11.3E). Again, DNA polymerase steps in, this time extending the captured 3′ end using the originally displaced complementary strand as template (Figure 11.3F).

The subsequent steps involve the formation, migration, and resolution of a special intermediate structure called the Holliday

---

[1] This is the endonuclease activity that clips SPO11, along with a short segment of DNA, from DNA ends during meiosis.

junction, named after the scientist Robin Holliday who first proposed the intermediate. In order to fully appreciate these steps, we need to take a "time-out" on meiotic recombination and focus on the Holliday junction itself.

11.3 Holliday junctions and pathways that process them

In most recombination models, including the one we have been discussing, the Holliday junction is represented as two strands that cross between two helices that are exactly parallel (coaxial) to each other (e.g., the two junctions at the bottom of Figure 11.3). Does this depiction provide an accurate structural view of a Holliday junction?

Scientists have made great progress in probing the actual three-dimensional structures of Holliday junctions, for example, by X-ray crystallography and other physical methods. One important conclusion is that Holliday junctions adopt different configurations/conformations, depending in part on the ionic conditions. In the (relative) absence of positively charged ions, particularly divalent cations, junctions adopt an open or extended conformation with the four duplex arms splayed apart from each other, as you might expect from repulsion between the highly (negatively) charged backbones (Figure 11.4A, left). However, when the charged phosphodiester backbones are neutralized by cations such as Mg^{++}, junctions shift to a side-by-side configuration, in which pairs of arms stack upon each other to form two nearly continuous double helices that are coaxial with each other (as in Figure 11.4A, right).

You might be confused by the fact that the crossed strands (red and blue) in this side-by-side structure change directions at the junction, while they continue in the same directions in the Holliday junctions of the recombination model of Figure 11.3. Indeed, this unexpected directionality of the crossed strands, although supported by structural data, met with resistance from some scientists when it

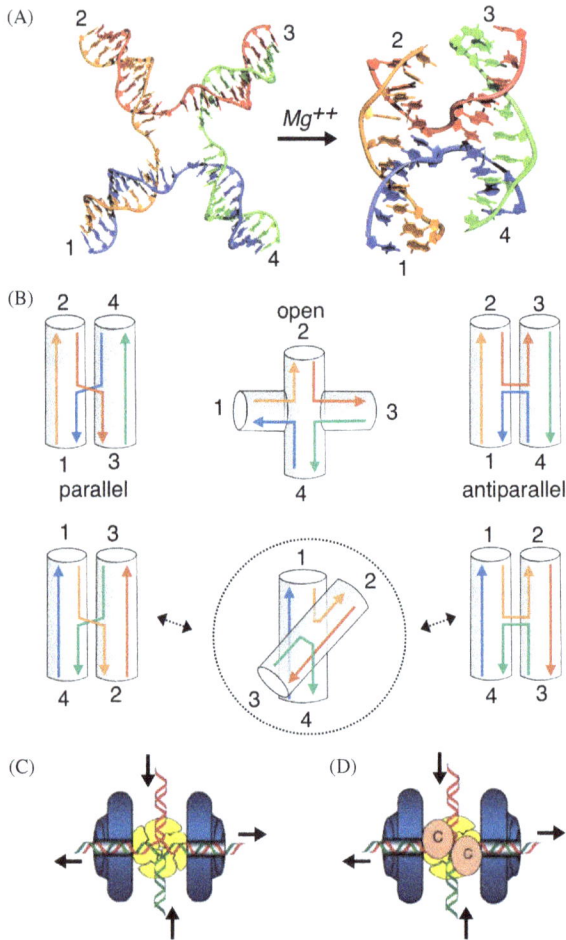

Figure 11.4. The Holliday junction. In panel A, the magnesium-induced transition between the open (left) and a side-by-side configuration (right) is depicted. This panel is reproduced from Müller *et al.* (2010), except that numbers 1 through 4 were added to identify the arms (permission from Royal Society of Chemistry, Great Britian, conveyed by Copyright Clearance Center, Inc.). Panel B shows schematic drawings of the different possible conformations of the Holliday junction. The strand colors match those in panel A for comparison, as do the arm numbers. See text for discussion of the conformations. This panel was modified from Lilley (2000). The bacterial Holliday junction migration protein RuvAB is depicted in panel C, and the junction-cleaving protein RuvC is added in panel D. The directions of migration of the four DNA arms are indicated with arrows. Panels C and D are reproduced from van Gool *et al.* (1998) (with minor modification to enhance the arrows; permission from John Wiley and Sons, conveyed by Copyright Clearance Center, Inc.; as stated by van Gool *et al.* (1998), the coordinates to generate these cartoons were from the structures in PDB entries 1C7Y and 3IGT).

was introduced.[2] It is helpful in this regard to consider the theoretical possibilities for the behavior of the crossed strands once the open or extended structure is collapsed by addition of cations. Each of the four helical arms (labeled 1 through 4) must stack with one of its immediate neighbors, for example, helix 1 can stack with either helix 2 or helix 4 (Figure 11.4B). For each of the two stacking arrangements (1/2 and 1/4), there are two possible pathways for the crossed strands. In one, the crossed strands continue in the same direction (as in the junctions in Figure 11.3); this is called the parallel conformation because the overall directions of the two helices are parallel to each other (Figure 11.4B, left). In the other orientation, the crossed strands change direction; this is called the antiparallel conformation (Figure 11.4B, right). In reality, the parallel and antiparallel conformations are two extremes, and you can imagine intermediate states between the extremes if you visualize one of the stacked helices rotating out of the plane of the paper (like the junction within the dotted circle in Figure 11.4B).

The available structural data argues that (charge-neutralized) Holliday junctions generally adopt a "mostly" antiparallel side-by-side conformation, but with some rotation toward an intermediate state (rather like the one in the dotted circle). There is evidence that the exact conditions, and even the DNA sequence near the junction, can perturb the favored conformation and also that any given junction has some flexibility in conformation. Furthermore, proteins that bind to Holliday junctions can impose a particular conformation, as we will see later.

On first blush, it might be difficult to imagine how recombination models such as the one in Figure 11.3 can work with antiparallel side-by-side conformations of Holliday junctions. However, the difficulty is mitigated by remembering that duplex DNA is quite flexible, bending in space, and forming loops and other complex

---

[2] In his excellent review article about DNA junctions, David Lilley (2000), with tongue at least partly in cheek, stated: "The chemical structure of DNA has been in existence almost as long as life itself, and the ability of DNA to fold into a structure that suggests a satisfying mechanism of recombination was probably not a selective pressure when DNA was first adopted as the genetic material."

paths (e.g., see Chapter 7). Presentations of recombination models generally show nicely parallel and orderly DNA helices like the ones in Figure 11.3 because they are much easier to draw. The two sets of Holliday junction drawings in Figure 11.5A show how recombination models might look with a parallel stacked (top) versus an open configuration (bottom).

An important feature of two homologous DNAs connected by a Holliday junction is that the branch point can migrate along the DNAs in one direction or the other (Figure 11.5A shows rightwards migration). Because the two DNA molecules are homologous to each other, this "branch-migration" reaction does not change the overall number of base pairs in either of the linked DNA molecules. For every base pair lost on one side of the migrating junction, another is formed on the other side.

Branch migration can occur in a spontaneous fashion. Remarkably, this process is about 1000-fold faster under the low ionic conditions that favor the open conformation compared to conditions that favor stacked conformations. Thus, the drawing of branch migration with the open conformation of the Holliday junctions (Figure 11.5A, bottom) is the more realistic representation of the event. Even under conditions that favor a stacked conformation, evidence suggests that the junction must transiently shift to an open conformation to undergo a branch-migration step.

While branch migration can occur spontaneously, there is substantial evidence that specialized helicase-related proteins are needed to catalyze branch migration within cells. The structure of a bacterial Holliday junction branch-migration protein in complex with a Holliday junction is particularly informative (schematic of the structure is shown in Figure 11.4C). Notice that the Holliday junction in this protein–DNA complex is in the open conformation, the one that favors branch migration — this conformation is dictated by the protein. The helicase-like protein drives two of the helices outward, essentially sucking the other two helices inward toward the branch point (arrows indicate directions of DNA helices).[3]

---

[3] Lovely animations of this reaction can be found on line, for example, at https://www.sheffield.ac.uk/mbb/ruva-2016/ruva-010.

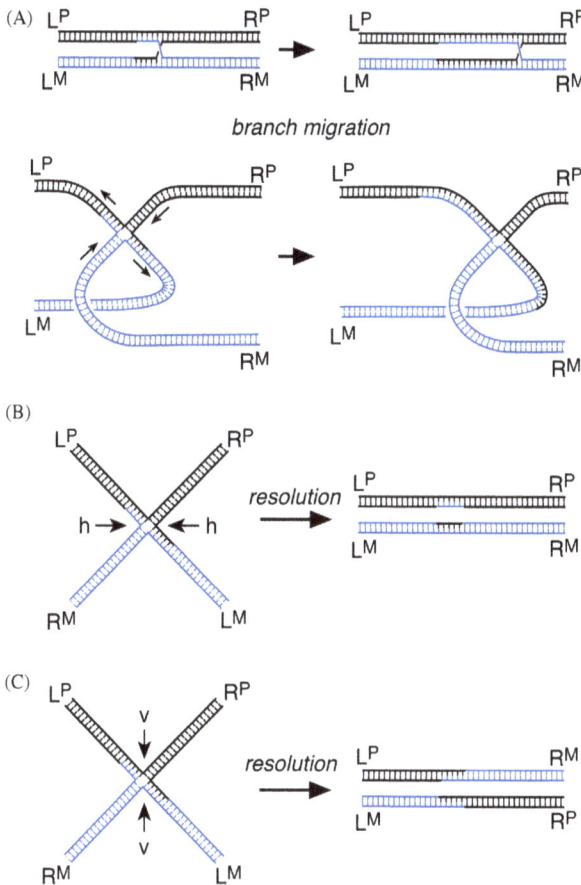

Figure 11.5. Holliday junction transformations. DNA duplexes are labeled with letters to indicate the left (L) and right (R) sides of a particular chromosomal DNA, and also with superscripts to indicate maternal (M; blue) or paternal (P; black) origin. Rightward branch migration in panel A moves the branch point along the DNA and thus increases the amount of heteroduplex DNA adjacent to the Holliday junction. The process is shown with the parallel conformation often shown in recombination models at the top of the panel, and with the more structurally realistic open conformation at the bottom of the panel. The directions of the DNA arms with respect to the junction are indicated with arrows in the drawing of branch migration with the open conformation. Resolution by cross-strand cleavage is depicted in panels B and C. Cleavage always occurs between two oppositely oriented strands, which can be either in a horizontal (h; panel B) or vertical (v; panel C) disposition with respect to each other. Note that in either case, all resulting chromosomes have one left chromosome side and one right side. Only the vertical cleavage results in a crossover of maternal and paternal chromosomes (panel C).

11.4 The final stages of meiotic recombination

We can now return to a proper consideration of the mechanism of meiotic recombination. After the second-end extension, the rightward branch of the structure is essentially a Holliday junction with a single nick at the top (Figure 11.3F). Branch migration of the junction in one or the other direction moves the nick into simple duplex DNA, where it can be sealed by DNA ligase. A slightly different branch-migration event can also convert the leftward branch point into a simple Holliday junction (Figure 11.3G). This generates an intermediate with two Holliday junctions, which has been verified to exist during meiotic recombination. This particular pathway of DSB repair is often called the *d*ouble *H*olliday *j*unction (dHJ) pathway, based on this complex intermediate.

The Holliday junctions in recombination intermediates must be resolved for the reaction to be complete and for chromosomes to segregate from each other. One important pathway for resolving such intermediates involves cleavage of two of the four strands in the crossed-strand structure by specialized endonucleases (Figure 11.5B–C). Following cleavage, DNA ligase seals the resulting nicks to complete the reaction. Recall that meiotic recombination serves to exchange parts of the two parental chromosomes into new recombinant configurations. The nature of the resolutions of the dual Holliday junctions can explain these recombinant outcomes. The cross-strand cleavages can occur in two different orientations, let's call them horizontal or vertical ("h" and "v" in Figure 11.5B–C). If the two junctions of a meiotic recombination intermediate are resolved in opposite orientations, one horizontal and one vertical, then the resulting chromosomes will each have a "crossover," with one end being from the maternal side and the other end from the paternal side (Figure 11.6A). If the two junctions are resolved in the same orientation as each other, no crossover is formed (with respect to the distal regions of the chromosomes), and only a small region between the junctions has been altered in any way (Figure 11.6B).

Recent studies have shown that another pathway of Holliday junction resolution, called dissolution, is actually the predominant pathway during meiotic recombination (Figure 11.6C). During dissolution, a DNA helicase and a type I DNA topoisomerase collaborate

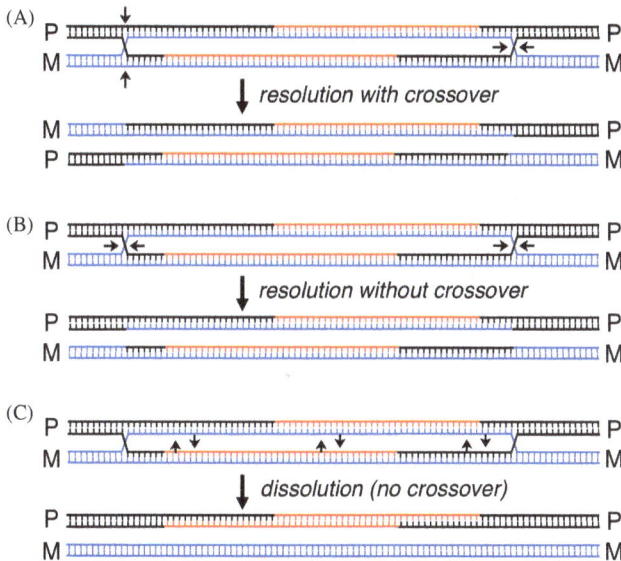

Figure 11.6. Resolution of the double Holliday junction intermediate in meiotic recombination. The double Holliday junction intermediate is identical to the last product shown in Figure 11.3 (original paternal DNA in black, maternal DNA in blue, and newly synthesized DNA in red). Cleavage of the two cross-strand junctions in opposite orientations (one horizontal and one vertical) results in a crossover (panel A), while cleavage in identical orientations results in resolution without a crossover (panel B; cleavage of both junctions in the vertical orientation also yields no crossover). The process of dissolution, involving a topoisomerase and a helicase, separates the chromosomal DNAs without a crossover (panel C). Chromosome arms are labeled as paternal (P) or maternal (M).

to resolve the doubly linked DNA molecules. The helicase is one of those that can catalyze branch migration (see above), and it moves the junctions toward each other. Meanwhile, the topoisomerase passes one strand of one of the helices through a break in the other strand, and thereby relieves the topological pressure that would otherwise accumulate; this eventually allows complete release of the single-strand linkages between the two molecules. Notably, this pathway does not lead to a crossover, much like the resolution pathway in which the junctions are cleaved in the same direction (Figure 11.6B). We will see later that this pathway of dissolution is important for preventing cancer in humans (see Section 14.6).

11.5 Repair of DSBs by homologous recombination in mitotic cells

Studies in model systems such as the yeast *Saccharomyces cerevisiae* first highlighted the potency of DSB repair via homologous recombination in mitotic cells. Exogenous DNA containing a selectable marker and a segment of DNA homologous to one of the yeast chromosomes can be introduced into yeast cells under the appropriate conditions. When the introduced DNA is circular, homologous recombination can result in an integration event that incorporates the circle into the chromosome at the position of homology, but this occurs at a low frequency (Figure 11.7A). When the same DNA is introduced with a DSB near the middle of the homologous segment, the frequency of this integration event increases by up to 1000-fold (Figure 11.7B)! Clearly, the recombination machinery is greatly activated by the DSB, and the DSB is repaired during the recombination event. Studies in mammalian (mouse) cells show an even larger stimulation of recombination events with exogenous DNA when a DSB is located in the homologous segment on the chromosome (Figures 11.7C and 11.7D). These latter studies were instrumental in effective gene targeting in mammalian cells, the basis for the Nobel Prize in Physiology or Medicine in 2007.

2007 Nobel Prize in Physiology or Medicine

This prize was awarded jointly to **Mario R. Capecchi, Sir Martin J. Evans and Oliver Smithies** for their advances on manipulating embryonic stem cells and DNA recombination, which led to the development of gene targeting in mice.

https://www.nobelprize.org/prizes/medicine/2007/summary/

While on the topic of gene targeting, it is worth describing a system that is revolutionizing gene manipulation in modern biology and greatly advancing the prospects for successful gene therapy in the clinic. Scientists discovered that certain bacterial cells carry an immune system that protects them from invading DNA such as that of

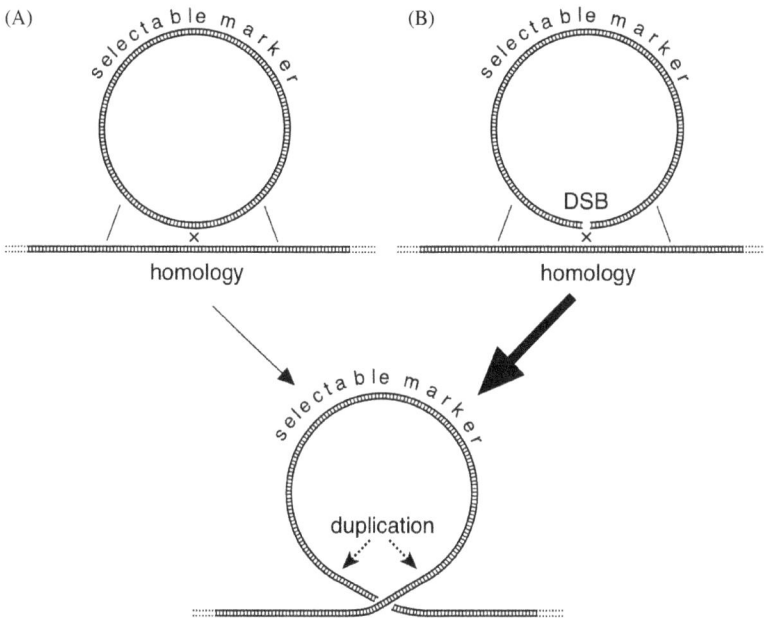

Figure 11.7. Integration of DNA containing selectable markers by homologous recombination. Circular DNA containing a segment homologous to a chromosomal region can recombine with that chromosomal region; with a crossover, the result is integration of the circle (panel A). The frequency of this event during transformation in yeast is very low, but progeny that have undergone the event can be obtained by selecting for a selectable marker on the circle (e.g., a drug-resistance gene, with the progeny selected on drug-containing media). In many systems, including yeast, a double-strand break (DSB) within the region of homology on the plasmid greatly increases the frequency of the recombination event (panel B). Transformation of mammalian cells generally utilizes linear fragments of DNA, which can integrate and replace the homologous region by means of recombination events at both ends of the fragment (panel C). Again, a selectable marker within the transforming DNA allows progeny that have undergone the gene replacement to be selected. Linear DNA fragments like this can also be incorporated into the genome at nonhomologous locations by the process of nonhomologous end joining (NHEJ), which is discussed later in this chapter. Induction of a DSB within the homologous region on the chromosome greatly increases the frequency of the homologous recombination event (panel D). In this case, repair of the DSB may be required for survival of the cells with broken chromosomes, and this recombination event provides one possible pathway for repairing the break (again, the process of NHEJ provides another pathway for repairing the broken chromosome and will be discussed later in this chapter).

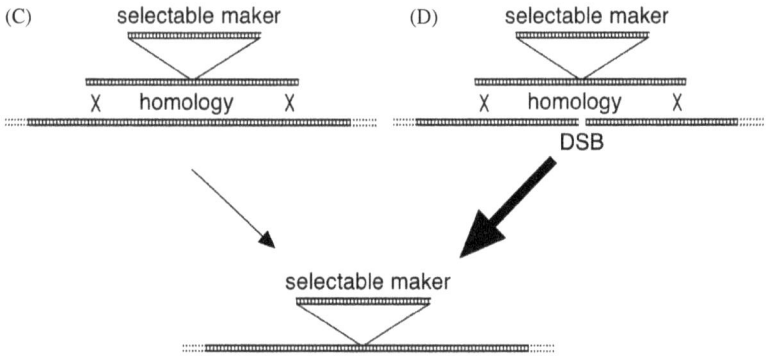

Figure 11.7. (*Continued*)

bacterial viruses. The heart of this immune system is a programmable nuclease that creates DSBs at particular sites in the invading DNA. The remarkable feature of this so-called CRISPR/Cas nuclease is that a simple RNA molecule, which is part of the nuclease, directs the DNA cleavage to a site that is complementary to a segment of the RNA. The system can therefore be reprogrammed to virtually any desired DNA sequence simply by changing the sequence of the RNA molecule, a very easy and inexpensive task. Indeed, bacterial cells have a way of changing the sequence of the RNA to that of invading DNA, which is how the system can function as an immune system in bacteria.

Regarding gene replacement and therapy, CRISPR/Cas nucleases can be readily designed to cleave near or within the site of a deleterious mutation, such as a mutation for an inherited disorder in humans. DSB repair using a normal (wild-type) version of the sequence can then correct the deleterious mutation. The wild-type version of the sequence might be provided from exogenous DNA or even from the homologous chromosome when the individual is heterozygous. The most promising diseases for quickly applying this technology relate to those affecting circulating cells, such as certain blood disorders, where the target cells can be removed from an individual, treated in the test tube with CRISPR/Cas, and reintroduced into the individual after the gene mutation has been corrected.

Let us return to the general aspects of DSB repair in mitotic cells. A number of studies in mammalian systems have revealed the importance of DSB repair to cell growth and to the viability of the organism. As already mentioned, inherited predisposition to cancer in humans can result from mutations in the BRCA2 RMP protein, and inherited mutations in other recombination proteins can cause hematological diseases and premature-aging syndromes (see Chapter 14). The most striking results were obtained in the mouse model system. Complete knockout mutations that abolish the Rad51 strand-invasion protein were found to cause embryonic lethality, and the Rad51-deficient cells accumulated chromosomal breaks prior to cell death. These results imply that DSBs occur frequently in the course of the normal cell cycle, and that homologous recombination is necessary to repair the breaks and prevent cell death. We will see below that some of the spontaneous DNA breaks that occur in cells result from broken replication forks during S phase, but other endogenous pathways also generate these dangerous lesions. As already mentioned, exogenous sources such as ionizing radiation also add to the overall burden of broken DNA.

The predominant mechanism of DSB repair in mitotic cells is somewhat distinct from that described earlier for meiotic cells. The pathway is called *s*ynthesis-*d*ependent *s*trand *a*nnealing, or SDSA, which succinctly describes how the second broken end is captured (Figure 11.8). The early steps in this pathway are, in outline, similar to the early steps in meiotic DSB repair. MRN (MRX) and 5′-specific exonucleases prepare the single-stranded 3′ ends, and then one of the resulting single-stranded 3′ ends enters the strand-invasion pathway described earlier (involving RPA, an RMP protein such as BRCA2, and Rad51; Figure 11.8A). Once Rad51 is loaded, the homology search commences. Following successful D-loop formation (Figure 11.8B), DNA polymerase extends the invaded 3′ end using the homologous DNA as template (Figure 11.8C). This is where the mitotic pathway diverges from meiotic recombination in dramatic fashion. In mitotic cells, the D-loop is a much more dynamic structure. As DNA polymerase extends the invaded 3′ end, the branch point at the back of the D-loop migrates in the same direction (Figure 11.8C). This reaction is rather like the act of transcription, except that the transiently base-paired region in the D-loop of SDSA is longer than in the transcription bubble. At some point, the unwinding of the D-loop outpaces the extension, expelling the extended 3′

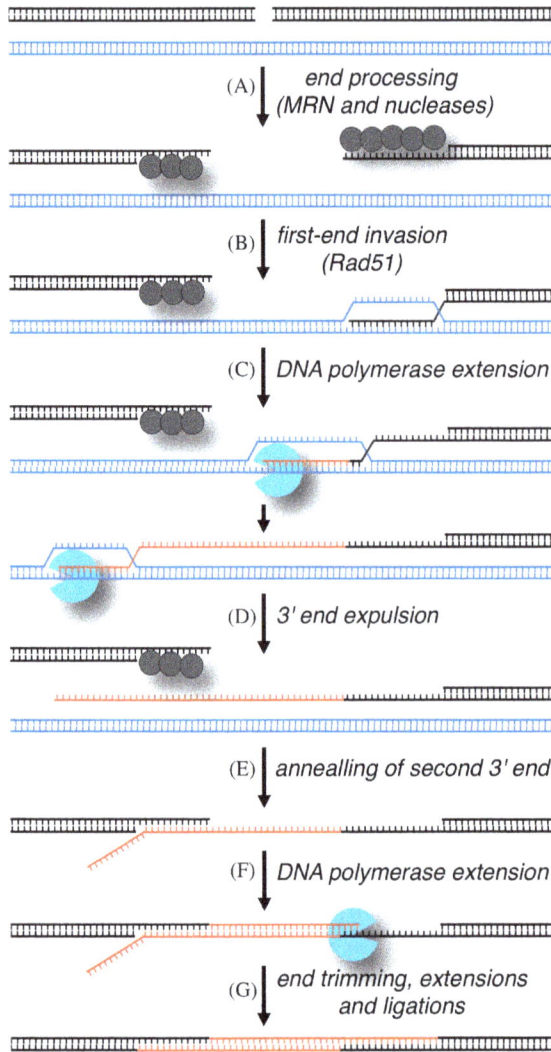

Figure 11.8. Model for synthesis-dependent strand annealing. See text for description of the steps in this model. The Rad51 protein is depicted as gray spheres and DNA polymerase as light blue pac-man; the MRN, RPA, and RMP proteins are not depicted in this diagram.

end from the invaded duplex (Figure 11.8D). With the help of one or more proteins that promote annealing, the expelled 3′ end next anneals with the second broken end (also a single-stranded 3′ end) (Figure 11.8E). Notice that these two ends must be complementary to each other, and that the DNA polymerase extension within the D-loop provides this complementarity that allows the two ends to anneal. In some systems, there is evidence that D-loop formation and extension can be an iterative process if the first round does not provide enough of a complementary sequence to bridge the original break. After the annealing reaction results in capture of the second end, the gaps are filled by DNA polymerase extension, ends are trimmed by nuclease activity, and the two resulting DNA nicks are sealed by DNA ligase (Figure 11.8F–G). Notice that the two broken ends have been reattached to each other in the same chromosome, and that the homologous chromosome is completely unchanged by the sequence of events in SDSA even though it served as a template.

An important feature of the SDSA pathway is that the repaired chromosome has not undergone a crossover, as is generated during a subset of meiotic recombination events (see above). This is likely one of the reasons that SDSA is favored in mitotic cells. Crossovers between the two homologs during mitotic growth can result in the loss of heterozygosity (LOH) in large stretches of the chromosome, as depicted in Figure 11.9. LOH can result in cell lineages that are homozygous for deleterious alleles, and indeed it has been shown that mitotic LOH contributes to cancer formation in this manner. In this sense, LOH counteracts one of the main advantages of being a diploid organism.

A number of proteins involved in DSB repair by homologous recombination (including MRN/MRX, RMP proteins, and Rad51) aggregate within small regions, called nuclear foci, after DNA damage. These foci have been visualized both in yeast and mammalian cells using recombinant proteins engineered to carry fluorescent tags, which can be viewed under a fluorescence microscope. These foci are induced when cells are treated with X-rays and other agents that induce DSBs, and each focus contains numerous copies of each of the involved proteins. Furthermore, fluorescence microscopy has shown that the foci are located right at the site of DNA breaks, and therefore presumably reflect (at least in part) the accumulation of

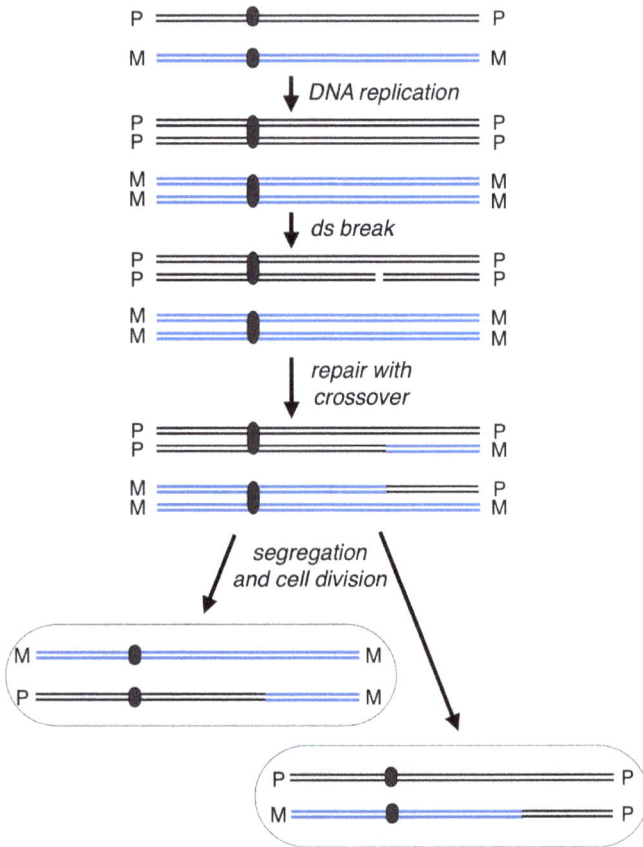

Figure 11.9. Homologous recombination can result in loss of heterozygosity (LOH). Two homologous chromosomes of a diploid are depicted, with M (blue lines) and P (black lines) indicating maternal or paternal origin of each chromosome. Following DNA replication (late in S or in G2 phase), recombinational repair of a DSB can result in a crossover between two chromatids of the partner chromosomes. Notice that this results in each of the involved chromatids having one M and one P segment. During the subsequent division process, segregation of the chromatids can occur, stochastically, in two different ways. The segregation depicted results in each daughter cell receiving one recombinant and one nonrecombinant chromosome. One of these daughter cells is homozygous for its paternal DNA in the distal segment while the other daughter cell is homozygous for its maternal DNA in that segment. If this region contains a deleterious allele in one of the parental segments, one of the daughter cells will have suffered LOH for that allele. The second possible pattern of segregation results in one daughter cell with both nonrecombinant chromosomes and the other with both recombinant chromosomes (not shown; neither daughter cell would suffer LOH for the distal region).

repair proteins on the broken DNA. Scientists believe that these are discrete repair factories, and that each factory has the potential to repair multiple DSBs. We still do not know why so many copies of each protein are needed in the foci, and there is also no direct evidence yet demonstrating that repair indeed occurs within the foci (although it seems highly likely given the above data). In addition to DSB-repair proteins, these foci contain many proteins involved in DNA-damage signaling (see Chapter 13).

11.6 Connections between homologous recombination and DNA replication

Scientists have uncovered multiple important connections between the processes of homologous recombination and DNA replication. In a broad sense, homologous recombination serves to rescue the cell from problems that occur during DNA replication, and we are still learning about the scope and importance of these pathways. In one sense, the SDSA pathway described earlier represents one connection between recombination and replication, although in this case a replication reaction that involves only leading-strand synthesis.

Nearly 50 years ago, Anna Marie Skalka, who was experimenting with a simple bacterial virus, proposed that a special pathway of homologous recombination could serve to restart broken replication forks, regenerating both leading- and lagging-strand synthesis. She recognized that DNA template nicks were dangerous, potentially causing a replication fork to break apart when the nick is encountered (Figure 11.10A; also see Section 3.7). Subsequent research has shown that certain DNA-junction-specific endonucleases can also cleave the replication fork due to its branched structure. Replication forks appear to be much more susceptible to junction-specific nucleases when the fork is somehow blocked or stalled, probably due to the loss of replication proteins from the fork and/or some disturbance in the overall structure of the protein-replication fork complex.

The pathway for restarting broken replication forks begins in much the same manner as the homologous recombination pathways mentioned earlier. The end of the broken arm is converted to a single-stranded 3′ end, and then the relevant strand-invasion protein

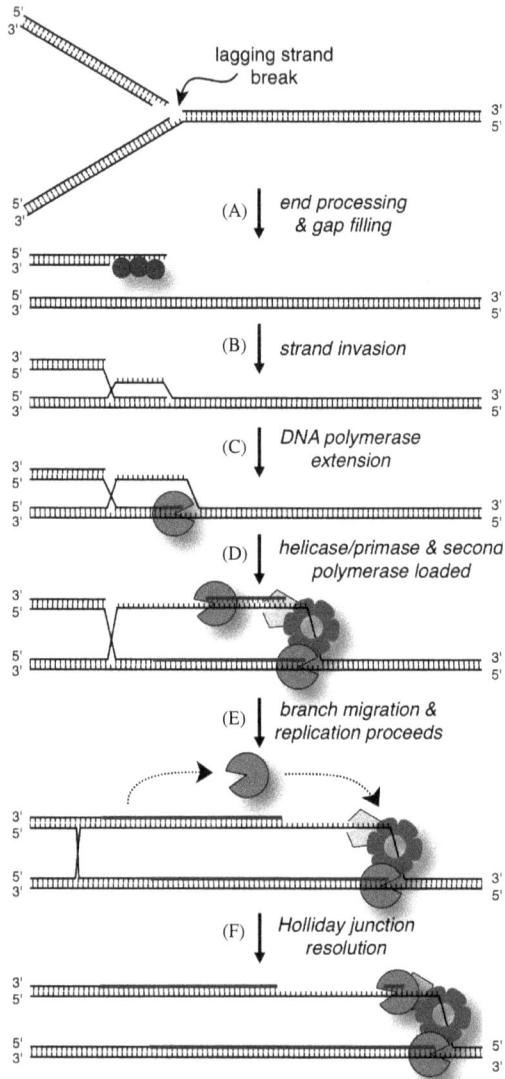

Figure 11.10.   Model for repair of broken replication fork by homologous recombination. See text for description of the steps in the pathway. Notice that the lagging-strand polymerase is shown traveling forward to the fork immediately after step E; this is simply the recycling of the lagging-strand polymerase as discussed in earlier chapters. While not depicted in this figure, the lagging strand is presumably folded around as described in the trombone model (see Section 2.5). Also note that the two strands on the broken DNA are flipped in the drawing during step B — since the DNA is a double helix rather than the two parallel lines drawn, this flipping is only a convenience for drawing and meaningless in a chemical sense.

assembles on the single-stranded region as described earlier (Figure 11.10A). The search for homology can be satisfied by strand invasion into the very same region where the end broke off, leading to D-loop formation (Figure 11.10B). Now, the pathway diverges from those described earlier. Dr. Skalka first proposed that the D-loop could become the site of assembly of a new replisome, with the invading 3′ end being used as the primer for leading-strand synthesis (Figure 11.10C). In this model, the displaced strand of the D-loop is used as a template for lagging-strand synthesis after assembly of a helicase–primase complex at the site of the leading-strand polymerase (Figure 11.10D). As the new fork progresses further down the template, Holliday junction–resolving enzymes can resolve the branch point left behind at the D-loop site, completing the process (Figure 11.10E–F). The net result is that the broken replication fork has been repaired and replication has been successfully restarted.

Many years after this initial proposal, scientists uncovered evidence that this pathway does in fact contribute to the successful completion of DNA replication. Studies in both bacteria and bacterial viruses established that D-loops can be the sites of assembly of new replication forks, and indeed bacteria produce special proteins that are needed for this assembly pathway. *E. coli* mutants that are deficient in these proteins grow very poorly, indicating that this pathway is used frequently during bacterial growth to rescue replication forks (see Section 3.7). The process is sometimes referred to as *re*combination-*de*pendent *re*plication (RDR).

Currently, scientists are working to understand the details of this process in eukaryotic cells and to understand how often the process is invoked. Because eukaryotic chromosomes are replicated bidirectionally from multiple origins, a broken or blocked replication fork does not seem as big a problem as in prokaryotes. The oppositely oriented replication fork from the adjacent origin will eventually replicate the intervening DNA as long as it is not also damaged, and so replication can potentially be completed in many situations without invoking RDR. Nonetheless, in the yeast model system, scientists were able to prove that eukaryotic cells can carry out an RDR-like process. They created a DSB in which only one of the two ends is homologous to the partner chromosome, and found that the homologous end could invade the partner and copy the remainder of the intact chromosome, as diagrammed in Figure 11.11A. This process, called *break-induced replication* (BIR), was shown to be conservative rather than semiconservative (see Section 1.2 for definition of semiconservative replication). While the leading strand is synthesized at the front of the D-loop bubble, the lagging strand is apparently synthesized after the leading-strand product is expelled at the back of the D-loop (Figure 11.11B). This mechanism is related to that of SDSA, with extension within the D-loop and expulsion of the newly synthesized strand.

Another series of experiments that supports the importance of recombination in eukaryotic DNA replication came from studies of frog oocyte extracts. These extracts are chock-full of replication proteins (and other proteins) that are present in the oocyte to allow very rapid embryonic growth after fertilization. With this high concentration of replication proteins, these oocyte extracts are able to efficiently replicate DNA that is added exogenously, providing a useful eukaryotic DNA replication model system. Scientists found that depleting the MRN protein inhibited the overall DNA replication reaction. Most importantly, the attempted replication reaction in these depleted extracts led to extensive DNA breaks, suggesting that breaks are routinely formed during DNA replication and must be repaired to allow the forks to restart and complete the process.

As mentioned earlier, discrete foci containing DSB-repair proteins are formed in yeast and mammalian cells following treatments that induce DSBs. Importantly, foci are also formed during normal S phase in cells that have not been treated with agents that cause DNA damage. This finding again suggests that the process of DNA replication results in DSBs that must be repaired by homologous recombination proteins to allow completion of the DNA replication process.

In summary, it is clear that the processes of DNA replication and DSB repair via homologous recombination are quite tightly intertwined. Indeed, many scientists argue that homologous recombination evolved initially to allow the completion of DNA replication, which was likely an even more erratic and error-prone process early in evolutionary history.

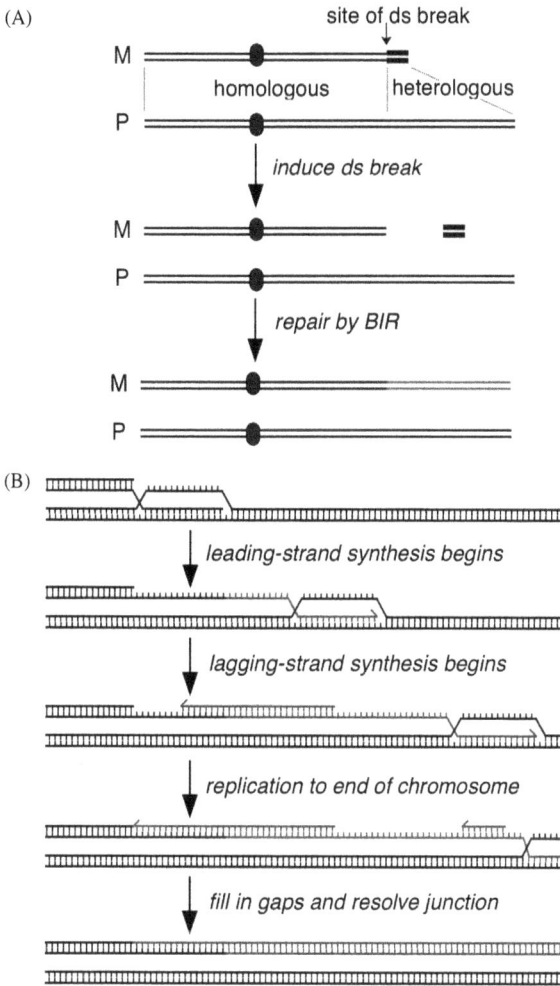

Figure 11.11. Regeneration of a missing chromosomal segment by break-induced replication (BIR). The experimental design to demonstrate this process in living yeast cells is schematized in panel A. One of two homologous chromosomes was engineered to contain a site that could be cleaved with a nuclease, and the region beyond that break site was engineered to be nonhomologous to the partner chromosome. Due to the nonhomology, conventional double-strand break (DSB) repair by homologous recombination cannot occur. Instead, at some frequency, the missing chromosome end was found to be restored after a recombination-dependent replication event. The restored DNA sequence was found to be identical to that of the partner chromosome (newly replicated DNA in gray). Panel B illustrates how some of the steps in the SDSA model can explain the regeneration of the missing chromosomal segment via break-induced replication (newly replicated DNA in gray).

11.7 The machinery of c-NHEJ

As introduced at the beginning of this chapter, a second major mechanism to repair DSBs involves end ligation during the process of c-NHEJ. Before discussing the biological roles and the regulatory balance of c-NHEJ versus homologous recombination, we will consider the protein machinery and mechanism of this pathway.

Oftentimes, the ends of a DSB are not simple ends that can be ligated as easily as complementary ends created by a restriction enzyme, and therefore end processing is an inherent aspect of c-NHEJ. For example, ionizing radiation generates many breaks without a normal 3′-OH group, with fragmented sugars near the terminus, and/or with the terminal base missing. In addition, DSBs sometimes contain flaps or loops of one strand, terminal hairpins connecting the two strands, or proteins covalently linked to the ends (similar to those in meiosis; also see Section 7.5 on DNA topoisomerase inhibitors). The process of c-NHEJ has the capacity to correct these problems, often by excising the damaged region prior to ligation (which can lead to small deletions) or by adding nucleotides to one or both 3′ ends. The net result is that joints repaired by c-NHEJ often show small deletions of a few nucleotides at the break site, and/or a few additional nucleotides that were not present in the parental DNA molecule.

The remarkable protein complex at the heart of the c-NHEJ pathway in mammals is called *DNA*-dependent *p*rotein *k*inase, or DNA–PK. The complex is named based on its kinase activity, namely the ability to phosphorylate proteins on serine and threonine residues in a reaction that depends on the presence of DNA ends. DNA–PK phosphorylates many cellular proteins in response to DNA breaks, and these phosphorylation events alter DNA repair reactions and cellular responses to damage such as cell-cycle checkpoints (see Chapter 13). DNA–PK even phosphorylates itself in response to binding to DNA ends, a modification that seems to be involved in the progression of the c-NHEJ pathway. The kinase activity of DNA–PK resides within one of the three subunits, the kinase catalytic subunit called DNA–PKcs.

The autophosphorylation reaction just mentioned modifies the DNA–PKcs subunit itself, and is greatly stimulated when the other two subunits of the DNA–PK complex engage with a DNA end. These other two proteins are named Ku70 and Ku80, names that reflect their history of discovery. A patient with an autoimmune disorder had antibodies to these two proteins, and the last name of this patient started with the letters Ku. Early characterization of the proteins showed molecular weights of approximately 70 and 80 kDa, providing the remainder of the names.

The Ku70/Ku80 heterodimer is the most abundant DNA end-binding protein in eukaryotic cells. The structure of the Ku heterodimer is a large double ring, with a cavity that slides onto duplex DNA ends (Figure 11.12). While DNA–PKcs has some ability to bind DNA ends on its own, the Ku heterodimer in the larger DNA–PK complex is the major actor in delivering the complex to DNA ends. The Ku heterodimer can accommodate simple duplex DNA ends and also ends with damaged residues and unusual structures. When one Ku heterodimer is bound to an end, about two turns of the double helix are covered. Furthermore, multiple Ku heterodimers can bind to ends, implying that the earlier bound heterodimers slide inward on the helix as new heterodimers bind.

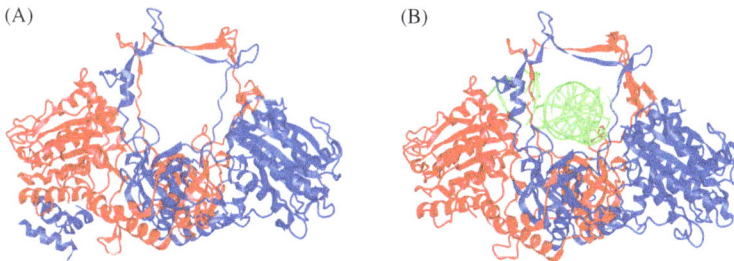

Figure 11.12. Structure of nonhomologous end joining (NHEJ) proteins, Ku70/Ku80. The structure of the Ku70/80 complex, with and without bound DNA are shown. In each panel, Ku80 is red while Ku70 is blue, and the DNA in panel B is green. The structures are from the RCSB PDB (www.rcsb.org) of PDB ID 1JEQ (panel A) and PDB ID 1JEY (panel B) (Walker *et al.*, 2001). The structural images were generated using the web-based visualization suite iCn3D (version 2.7.15), developed by Wang *et al.* (2019).

The Ku heterodimer plays multiple important roles in the c-NHEJ pathway. In addition to stimulating the kinase activity of DNA–PKcs, the Ku heterodimer protects the DNA ends from improper degradation, assists two broken ends in aligning, and serves as a platform for DNA-processing enzymes to bind to the end. The ring-shaped Ku apparently remains on the DNA ends even during the final ligation, which raises the question of how it is finally released from the repaired DNA molecule. In answer, scientists have uncovered a pathway that modifies and then destroys Ku by proteolysis.

The structures of Ku70 and Ku80 are very similar, indicating that these two proteins have a common ancestor. Indeed, Ku proteins are widely conserved in evolution, with homologs in some archaeal lineages and in a subset of bacteria. The case of bacterial species *Mycobacterium tuberculosis*, which causes tuberculosis, is informative. Rather than having a heterodimer, *M. tuberculosis* has a simple homodimer of two identical Ku subunits. The NHEJ system of this bacterial species is quite simple, and a wide variety of DNA ends can be repaired with only two proteins. In addition to Ku, the system uses a remarkable multifunctional protein that has DNA polymerase, terminal transferase,[4] exonuclease, and DNA ligase activities. This combination of activities can excise or correct the various kinds of DNA end damage and complete the end ligation to repair the breaks.

Returning to mammalian c-NHEJ, many more proteins are involved than in the bacterial system. A protein called Artemis binds to DNA–PK and has the ability to cleave unusual structures such as DNA hairpins at the end, along with exonuclease activity. Often, additional DNA bases are added to one or the other DNA end by DNA polymerase or terminal transferase activities. At least two different DNA polymerases have been implicated in this reaction in mammalian cells, and they are capable of carrying out the remarkable reaction of using a template strand from the other side of the break when necessary! Several other DNA-processing enzymes can be involved in c-NHEJ in mammalian cells, broadening the ability of

---

[4] Terminal transferase, more formally, terminal deoxynucleotidyltransferase, is a variant DNA polymerase activity that adds nucleotides to DNA 3′ ends without using a template strand.

the pathway to deal with myriad forms of end damage. Finally, a specialized DNA ligase (DNA ligase IV) with its accessory factor ligates the ends to complete c-NHEJ, and this ligase has an unusually broad specificity to ligate unusual DNA ends. The wide range of DNA-modifying activities mentioned in this paragraph should give the reader an appreciation that mammalian c-NHEJ can repair DNA breaks even when they contain diverse kinds of lesions not normally found in the DNA molecule.

An in vitro system has been reconstituted to repair relatively simple but noncomplementary DNA ends with only the DNA–PK complex, Artemis, and the two-protein DNA ligase complex. This can be considered the basal c-NHEJ pathway, and shows that the additional processing enzymes are only necessary when the ends contain particular forms of damage. The addition of appropriate DNA polymerase(s) and terminal transferase to the reaction leads to nucleotide additions at the site of the break, as might be expected.

Before leaving the description of the c-NHEJ machinery, it should be pointed out that the machinery is distinct in many lower eukaryotes including budding yeast. *S. cerevisiae* does not utilize a DNA–PKcs subunit, nor does it encode an Artemis homolog. The c-NHEJ pathway in yeast involves a Ku heterodimer, a specialized DNA ligase related to the mammalian DNA ligase IV, and a DNA polymerase that can extend across breaks as described earlier. In addition, c-NHEJ in yeast requires the MRX complex, which presumably plays some roles in end stabilization, end processing and/or coordination of the two broken ends. There is some evidence that the corresponding MRN complex in mammalian cells might contribute to c-NHEJ, but it is clearly not required for the process.

11.8 Biological roles of c-NHEJ

Many of the early studies relating to c-NHEJ in mammalian systems involved analysis of immunoglobulin gene rearrangements, particularly a pathway called V(D)J recombination. The details are complex and beyond the scope of this book, but a brief summary is appropriate. The V(D)J recombination pathway is a critical step in the generation of the vast repertoire of antibodies and antigen-binding receptors in mammals. The pathway is triggered in appropriate cells of the immune system by the induction of a protein complex encoded by the *r*ecombination *a*ctivation *g*enes, RAG1 and RAG2. The RAG1–RAG2 complex recognizes appropriate sites in the genes for antibodies and receptors and carries out a complex reaction involving DNA cleavages and ligations, including the formation of an unusual hairpin joint that links the 3′ and 5′ ends on one side of the break. Artemis cleaves this hairpin, and other components of the c-NHEJ pathway described earlier also participate in the overall reaction. Thus, the DNA-processing steps introduced earlier create variability at the joints that are generated by the final ligation. In addition, the overall pathway joins distant gene segments, bringing together particular combinations of the gene segments (called V, D, and J) from arrays of these coding segments to create mature functional antibody genes.

The link between DNA-break repair and the immune system emerged from studies of inherited human immune diseases and knockout studies in model systems such as the mouse. For example, cells from patients with human immunodeficiency syndromes were found to be hypersensitive to ionizing radiation, demonstrating overlap in the components of immunoglobulin gene rearrangement and c-NHEJ. This includes diseases associated with mutations in Artemis, DNA ligase IV, and DNA–PKcs. In another example, mouse Ku knockout lines are profoundly sensitive to ionizing radiation and are defective in repair of RAG1/RAG2-generated breaks in the immune system.

The sensitivity of c-NHEJ-deficient mammalian cells to ionizing radiation tells us that this pathway is important for cell survival after DSB formation. Additional conclusions are obtained by comparing the outcome of ionizing-radiation treatment in surviving cells that either do or do not have a functional c-NHEJ system. The most interesting finding is that c-NHEJ-deficient cells show a much higher incidence of chromosomal translocations and other rearrangements than the normal cells. This implies that, in the normal cells, the functioning c-NHEJ pathway generally repairs breaks by correctly joining the two ends generated from a particular breakage event. Thus, even if a few nucleotides were added or lost at the ligated joint, the overall chromosomal arrangement was not disturbed. In this sense, c-NHEJ is a very valuable process to maintain genome integrity, particularly since a large fraction of mammalian DNA is noncoding (and therefore the small deletions and insertions are often innocuous). In contrast, in the absence of c-NHEJ, alternative break-repair pathways are favored, and these more frequently lead to gross DNA rearrangements. One such pathway might be homologous recombination itself. Mammalian cells have many DNA repeats of various kinds scattered around the chromosomes, and DSB repair via homologous recombination can inadvertently recombine a broken end with homologous DNA in a distant repeat. We will discuss two additional pathways that can lead to DNA rearrangements in Section 11.10.

In mammalian cells, both c-NHEJ and homologous recombination play critical roles in repairing DSBs. In Section 11.9, we will consider the balance of these two pathways and why the redundancy of two pathways is critical. In contrast, in some lower eukaryotes such as the budding yeast *S. cerevisiae*, the predominant pathway of DSB repair is homologous recombination, and c-NHEJ appears to be much less important. Yeast mutants that are defective in the Ku heterodimer, for example, are not hypersensitive to ionizing radiation. However, mutants that are doubly deficient in both homologous recombination and c-NHEJ are more sensitive than single mutants deficient in homologous recombination, implying that c-NHEJ provides a backup pathway in this model system. Scientists speculate that the small genome size of organisms like yeast makes the homology search during homologous recombination much more efficient than in mammalian cells, favoring this pathway over c-NHEJ. Also, the yeast genome has relatively little repetitive DNA, minimizing the likelihood of genome rearrangements during homologous recombination.

11.9 Pathway choice in DSB repair

Once a DSB forms, how is the pathway of repair chosen? Do homologous recombination and c-NHEJ compete with each other or cooperate in some way? We have already seen some overlap in the involved proteins, namely the MRN (MRX) complex, suggesting some cooperation. Studies to date demonstrate that pathway choice varies between organisms, as already implied by the relative unimportance of c-NHEJ in yeast. In addition, a major determinant of pathway choice involves the cell cycle, and the nature of the DNA ends also likely contributes.

The influence of genome size in favoring either homologous recombination or c-NHEJ was discussed earlier. Relevant to this issue, scientists have been able to measure how long it takes to repair breaks via the two different pathways in mammalian cells, and the findings are remarkable. With ends that do not need extensive processing, repair was completed in 30 minutes by the c-NHEJ pathway but took nearly seven hours by homologous recombination! The long time required for homologous recombination presumably reflects, at least in part, the difficulty of finding the homologous DNA given the very large genome size. The more rapid repair by c-NHEJ highlights an advantage to using this pathway in mammalian cells. Nonetheless, it should be noted that this advantage is diminished when the ends require processing. Indeed, some ionizing radiation-induced breaks are repaired rapidly and some slowly, presumably reflecting the complexity of the DNA end structures. Artemis, which can cleave unusual end structures, is critical for repair of the slowly repaired lesions, consistent with the interpretation that slowly repaired ends have a complex structure. There is evidence that the MRN complex is also important for the slow phase of repair in mammalian cells, perhaps to keep the two broken ends near each other during the prolonged repair process.

The stage of the cell cycle is a major determinant of pathway choice in DSB repair. To a first approximation, c-NHEJ is dominant in the G1 phase and early in S, prior to DNA replication, while homologous recombination becomes more prevalent in S phase and

G2. This timing is very sensible. The process of DNA replication generates duplicated chromosomes connected by a centromere; the two copies are called sister chromatids. Thus, after DNA replication in S phase and throughout the G2 phase, a broken chromatid is physically connected and near to its sister chromatid, which has exactly the same sequence and is thus an ideal template for homologous recombination. Recombinational repair with a sister chromatid avoids LOH, a detrimental outcome that is possible when homologous recombination involves the homologous chromosome in a diploid organism (although the predominant break-repair pathway in mitotic cells, SDSA, avoids LOH; see Section 11.5). Another advantage of homologous recombination during S phase is that broken replication forks, essentially one-ended DSBs, can be successfully repaired to complete DNA replication (see Section 11.6). In contrast, during G1 phase, sister chromatids are unavailable, which has no negative influence on the c-NHEJ pathway. However, homologous recombination would be more challenging because the homologous chromosome is presumably more difficult to locate than the sister chromatid would be if the break occurred later in the cell cycle.

How is the cell-cycle dependency of repair pathways achieved at the molecular level? Tracing the pathway intermediates, the generation of 3′-single-stranded ends (resection) was found to differ greatly in G1 versus S and G2. Cells in S and G2 efficiently generate 3′-single-stranded ends, which are needed for homologous recombination but inhibit c-NHEJ. The inhibition of c-NHEJ can be traced to the inability of the Ku heterodimer to bind single-stranded DNA ends. These results point to the step that dictates cell cycle–dependent pathway choice, but why is resection so different across the cell cycle? A major determinant involves the *cyclin-dependent kinases* (CDKs), which drive the progression of the cell cycle. As indicated by their names, the kinase activity of CDKs is dependent on the presence of particular cyclins, proteins that are present only at certain stages of the cell cycle. CDKs that are active in S and G2 phosphorylate a broad array of proteins, including factor(s) that resect DSBs, and this phosphorylation activates the exonuclease activity for resection. The exonuclease activity is not activated in G1 phase due to the absence of the relevant CDK activity. Furthermore, the Ku heterodimer is abundant in G1, and Ku binding to DNA ends inhibits resection, providing a second layer of control. When both Ku and activated exonucleases are present (S and G2 phases), the two pathways appear to be in some competition, and the precise nature of the DNA ends may influence that competition. Again, once resection begins at a particular end, the c-NHEJ pathway is shut off, as the Ku heterodimer can no longer bind. Other pathways also appear to influence the cell-cycle-dependent regulation of pathway choice, for example, there is evidence that the protein-kinase activity of DNA–PK may exert some influence.

11.10 SSA and alt-NHEJ

Two additional pathways of DSB repair have been defined, but seem to operate at a very low frequency in normal cells. Even though their frequency is low, these pathways are still quite important because they both can lead to large deletions and rearrangements in the genome. Furthermore, these pathways can become more dominant when either of the two major pathways described earlier have been inactivated (as occurs, for example, in some cancers and inherited human diseases).

The first minor pathway is a variant of homologous recombination, with repair of the break being dependent on annealing of a rather long stretch of complementary base pairs. This SSA pathway can occur when extensive resection at a break exposes two complementary sequences, which, prior to the break, were two copies of a repetitive sequence that were distant from each other in the chromosome (Figure 11.13). The two complementary 3' ends anneal with each other, aided by a class of proteins that promote DNA annealing. This reaction is similar to the annealing that occurs in the late stages of the SDSA reaction (Section 11.5). Some of the proteins that promote DNA annealing had also been identified as RMP proteins (e.g., Rad52), which were described in Section 11.1. The net result of the SSA reaction is that the break is successfully repaired, but at the cost of a deletion of all DNA that was previously

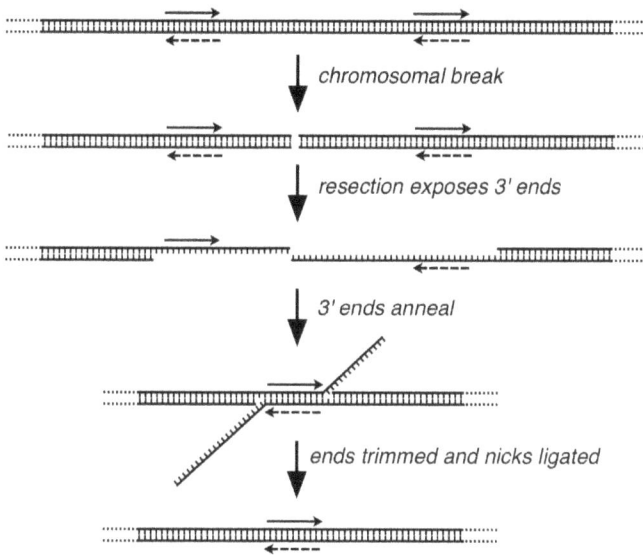

Figure 11.13. Model for break repair by single-strand annealing. See text for description of steps in the pathway. The repetitive sequences are indicated by the horizontal arrows, with the dashed arrow being the complement of the solid arrow, and with the arrowheads pointing in the 3′ direction on that strand. Trimming of the 5′ ends may occur both before and after annealing.

found in between the two copies of the repeat (Figure 11.13). SSA could also generate a translocation if the repetitive DNA segments were located on two different chromosomes that each suffered a break.

The second minor pathway of DSB repair is a variant of NHEJ, called alt-NHEJ. As with the SSA pathway, alt-NHEJ can generate both deletions and translocations. Also, as in SSA, alt-NHEJ relies on complementary base-pairing, but in this case, very short segments (<10 base pairs).[5]

The alt-NHEJ pathway operates in cell lines that are deficient for any of the factors involved in c-NHEJ, and therefore clearly employs

---

[5] These short segments of homology are often called "microhomology," and you might find literature that refers to this pathway as microhomology-mediated end joining.

a distinct mechanism(s). Scientists are currently trying to decipher the mechanism of alt-NHEJ, and indeed it is possible that multiple mechanisms are involved. One interesting finding is that the Ku heterodimer actually inhibits alt-NHEJ, highlighting a distinction with c-NHEJ. Alt-NHEJ generally requires more extensive resection than c-NHEJ to expose the complementary repeats. It is intriguing that factors involved in resection for homologous recombination also seem to contribute to alt-NHEJ.

While c-NHEJ generates some small deletion mutations during repair, alt-NHEJ tends to generate even larger deletions due to the need to expose two short complementary sequences for annealing. Furthermore, as implied in Section 11.8, alt-NHEJ has been implicated in the formation of chromosomal translocations, apparently as a result of annealing/joining events between two different broken chromosomes.

11.11 Summary of key points

- DSBs are generated by ionizing radiation and also by various cellular processes including DNA replication, repair, and meiosis.
- Homologous recombination pathways accurately repair DSBs, operating during mitosis and meiosis.
- The strand-invasion step in homologous recombination is catalyzed by strand-invasion proteins, such as bacterial RecA and eukaryotic Rad51, which are loaded with the assistance of RMPs.
- In addition to repairing DSBs, homologous recombination plays important roles in genetic exchange in many organisms and chromosome segregation in eukaryotes.
- The MRN (MRX) protein in eukaryotic cells processes broken DNA ends and also tethers broken ends together.
- The detailed pathways of homologous recombination differ between meiotic and mitotic cells.
- Holliday junctions are key intermediates in homologous recombination, and can move along DNA by branch migration.
- Holliday junctions can be resolved by cleavage of the branch junction to generate either crossover or non-crossover products,

and can also be resolved by dissolution, which utilizes a topoi-somerase and a helicase and results in non-crossover products.

- Homologous recombination is utilized in gene-targeting approaches including generation of transgenic animals and human gene therapy.
- Homologous recombination leading to crossovers causes LOH, which can contribute to human diseases, most notably cancer.
- Homologous recombination can rescue broken replication forks to allow completion of DNA replication.
- Mutations that affect a number of different human proteins involved in homologous recombination cause human diseases including predisposition to breast cancer; knockout mutations in some of these proteins cause embryonic lethality in mouse model systems.
- The c-NHEJ pathway joins broken DNA ends, frequently altering the ends prior to joining (necessary for joining unusual ends).
- The DNA–PK (protein kinase) complex that carries out c-NHEJ contains the protein kinase subunit along with the Ku70/Ku80 heterodimer, the most abundant DNA-end-binding protein in eukaryotic cells.
- Pathways of c-NHEJ are important during immunoglobulin gene rearrangement and for repairing mitotic DSBs, particularly during G1 and early in S phase.
- Alternative pathways of homologous recombination (SSA) and NHEJ (alt-NHEJ) rejoin broken ends using regions of complementary base-pairing involving copies of repetitive DNA in the genome; these pathways can generate genome rearrangements.

Further Reading

Capecchi, M. R. (2005). Gene targeting in mice: Functional analysis of the mammalian genome for the twenty-first century. *Nat Rev Genet, 6*(6), 507–512.

Chen, L., Trujillo, K., Ramos, W., Sung, P., & Tomkinson, A. E. (2001). Promotion of Dnl4-catalyzed DNA end-joining by the Rad50/Mre11/Xrs2 and Hdf1/Hdf2 complexes. *Mol Cell, 8*(5), 1105–1115.

Chiruvella, K. K., Liang, Z., & Wilson, T. E. (2013). Repair of double-strand breaks by end joining. *Cold Spring Harb Perspect Biol, 5*(5), a012757.

Friedberg, E. C., Walker, G. C., Siede, W., Wood, R. D., Schultz, R. A., & Ellenberger, T. (2006). *DNA Repair and Mutagenesis* (2nd ed.). Washington, DC: ASM Press.

Haber, J. (2013). *Genome Stability: DNA Repair and Recombination.* New York, NY: Garland.

Hays, F. A., Teegarden, A., Jones, Z. J., Harms, M., Raup, D., Watson, J., ... Ho, P. S. (2005). How sequence defines structure: A crystallographic map of DNA structure and conformation. *Proc Natl Acad Sci USA, 102*(20), 7157–7162.

Hopfner, K. P., Craig, L., Moncalian, G., Zinkel, R. A., Usui, T., Owen, B. A., ... Tainer, J. A. (2002). The Rad50 zinc-hook is a structure joining Mre11 complexes in DNA recombination and repair. *Nature, 418*(6897), 562–566.

Jasin, M., & Rothstein, R. (2013). Repair of strand breaks by homologous recombination. *Cold Spring Harb Perspect Biol, 5*(11), a012740.

Lamarche, B. J., Orazio, N. I., & Weitzman, M. D. (2010). The MRN complex in double-strand break repair and telomere maintenance. *FEBS Lett, 584*(17), 3682–3695.

Lilley, D. M. (2000). Structures of helical junctions in nucleic acids. *Q Rev Biophys, 33*(2), 109–159.

Morrical, S. W. (2015). DNA-pairing and annealing processes in homologous recombination and homology-directed repair. *Cold Spring Harb Perspect Biol, 7*(2), a016444.

Müller, J. (2010). Functional metal ions in nucleic acids. *Metallomics, 2*(5), 318–327.

Sander, J. D., & Joung, J. K. (2014). CRISPR-Cas systems for editing, regulating and targeting genomes. *Nat Biotechnol, 32*(4), 347–355.

van Gool, A. J., Shah, R., Mezard, C., & West, S. C. (1998). Functional interactions between the holliday junction resolvase and the branch migration motor of Escherichia coli. *EMBO J, 17*(6), 1838–1845.

VanLoock, M. S., Yu, X., Yang, S., Lai, A. L., Low, C., Campbell, M. J., & Egelman, E. H. (2003). ATP-mediated conformational changes in the RecA filament. *Structure, 11*(2), 187–196.

Walker, J. R., Corpina, R. A., & Goldberg, J. (2001). Structure of the Ku heterodimer bound to DNA and its implications for double-strand break repair. *Nature, 412*(6847), 607–614.

Wang, J., Youkharibache, P., Zhang, D., Lanczycki, C. J., Geer, R. C., Madej, T., ... Marchler-Bauer, A. (2019). iCn3D, a web-based 3D viewer for sharing 1D/2D/3D representations of biomolecular structures. *Bioinformatics*. doi:10.1093/bioinformatics/btz502

Yeeles, J. T., Poli, J., Marians, K. J., & Pasero, P. (2013). Rescuing stalled or damaged replication forks. *Cold Spring Harb Perspect Biol, 5*(5), a012815.

How did they test that?
Does the MRN (MRX) complex promote end tethering?

The structure of the MRN (MRX) complex suggested that it might tether together two broken DNA ends (see Figure 11.2 B–D). Chen *et al.* (2001) tested this model using biochemical approaches with *Saccharomyces cerevisiae* MRX complex and DNA ligase IV. They incubated a 400-bp DNA fragment containing short complementary 5′ ends with various proteins. With only DNA ligase IV (Dnl4; along with its partner Lif1), a substantial fraction of the DNA was ligated into monomeric circles after 90 minutes, while a 15-minute incubation produced a much smaller fraction (top panel, lanes 1 and 2). When the MRX complex or its three substituent proteins were added to the reaction, ligation became more efficient (substantial product at 15 minutes) and shifted to the form of linear dimers, trimers, and higher forms (lanes 9 and 10). All three of the MRX complex proteins were required for this effect (no product in lanes 3–8). The product identities were verified by atomic force microscopy (AFM) (not shown). Based on the DNA concentration in this experiment, the two ends of a single 400-bp DNA fragment are more likely to meet each other than to meet the ends of other DNA fragments, explaining the intramolecular nature of ligation in the absence of MRX. MRX complex must promote tethering of ends between different DNA fragments to convert ligation to the observed intermolecular mode. The scientists also used AFM of the same DNA substrate with and without MRX complex, in the absence of DNA ligase, and thereby directly visualized end tethering. DNA fragments without protein appeared as expected for 400-bp fragments (bottom panel, top left), but addition of MRX complex (small blobs) led to many instances of DNA fragments aligned end-to-end with MRX blobs in between (top right). Close-ups of molecules from the top right panel show one DNA fragment with an MRX complex at an end (bottom left) and an assembly of four DNA molecules tethered in linear fashion by three blobs (one between each pair of adjacent DNA molecules; bottom right). The figures were reproduced from Chen *et al.* (2001), with permission from Elsevier; permission conveyed by Copyright Clearance Center, Inc.

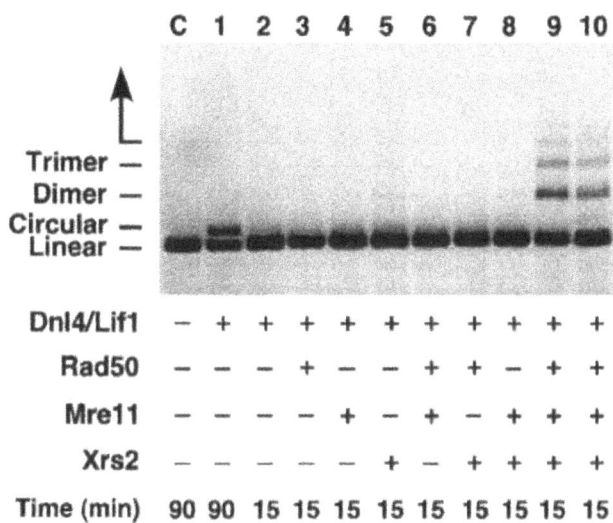

	C	1	2	3	4	5	6	7	8	9	10
Dnl4/Lif1	−	+	+	+	+	+	+	+	+	+	+
Rad50	−	−	−	+	−	−	+	+	−	+	+
Mre11	−	−	−	−	+	−	+	−	+	+	+
Xrs2	−	−	−	−	−	+	−	+	+	+	+
Time (min)	90	90	15	15	15	15	15	15	15	15	15

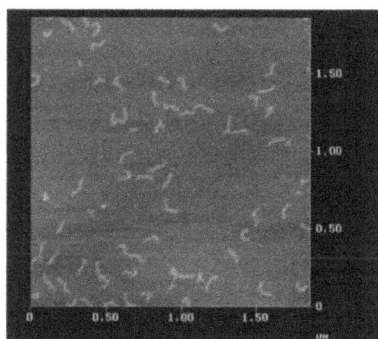

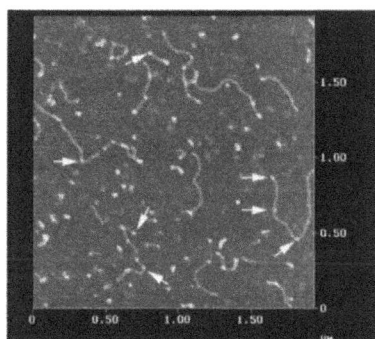

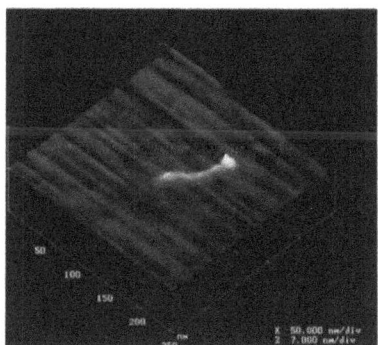

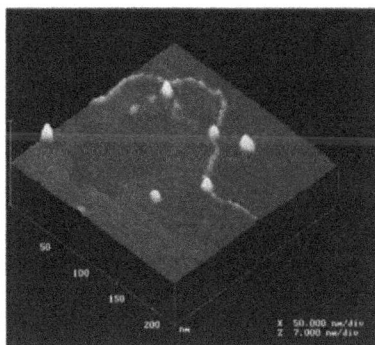

Chapter 12

DNA-damage tolerance and translesion DNA polymerases

The reversal and repair pathways discussed in the last several chapters are powerful tools that help cells and organisms survive the myriad forms of DNA damage. However, they are insufficient for vigorous growth and survival, particularly under conditions that lead to extensive DNA damage. In addition to repair pathways, all cells employ pathways that allow them to tolerate DNA damage in their genomes. In many cases, these tolerance pathways allow cell survival while providing the cell with more time and perhaps an additional cell-division cycle to properly repair the damage. Tolerance pathways are particularly important to allow the completion of DNA replication prior to cell division, leading daughter cells to inherit some level of DNA damage from their parental cell. As the name implies, a DNA-damage-tolerance pathway is one that allows the cell to survive and even propagate without actually removing or reversing the DNA damage that might otherwise lead to cell death or senescence.

Two major categories of tolerance pathways will be considered in this chapter. In the first category, DNA replication bypasses a template lesion that would otherwise block the process by temporarily switching to a homologous, undamaged template. In the second category, a class of remarkable DNA polymerases, called translesion DNA polymerases, successfully inserts residue(s) across from the template lesion, which otherwise blocks the normal replicative DNA polymerases.

Most, perhaps all, damage-tolerance pathways are carefully regulated, as will be discussed extensively in Chapter 13. In this chapter, we will make passing reference to the regulatory aspects, but it is important to keep in mind throughout this chapter that these pathways are often nonfunctional or at least moderated in normal cells with minimal levels of DNA damage. Indeed, as we will see, these pathways are inherently dangerous and can lead to genomic rearrangements and point mutations. The regulatory pathways that will be discussed in Chapter 13 can also be considered as contributing to DNA-damage tolerance. For example, we will see that DNA replication itself is often inhibited in response to DNA damage, which prevents dangerous collisions of the replication fork with blocking DNA lesions and provides additional time for traditional repair pathways to reverse the DNA damage.

12.1 Damage tolerance by template switching

Many early studies on the replication of damaged DNA utilized the bacterial model system, *Escherichia coli*, with an easily administered damaging agent, ultraviolet light (UV). These studies led to many important discoveries, including the nature of nucleotide excision repair (NER) (Chapter 10). Another important discovery was that replication of UV-damaged DNA resulted in single-stranded gaps in the daughter DNA (Figure 12.1A). In these experiments, scientists used mutant *E. coli* that were unable to perform photoreactivation and NER, which otherwise could rapidly reverse and repair the UV-induced damage. Additional experiments supported the conclusion that the single-stranded gaps were formed opposite template lesions (cyclobutane pyrimidine dimer [CPD] and (6-4) photoproduct; see Chapter 8) by the process of DNA replication, and that they formed on both the leading and lagging strands. As already discussed in Chapter 3, single-stranded gaps can be induced by template lesions when the replicative DNA polymerase is blocked by the damage but the replisome re-primes synthesis downstream of the blocking lesion.

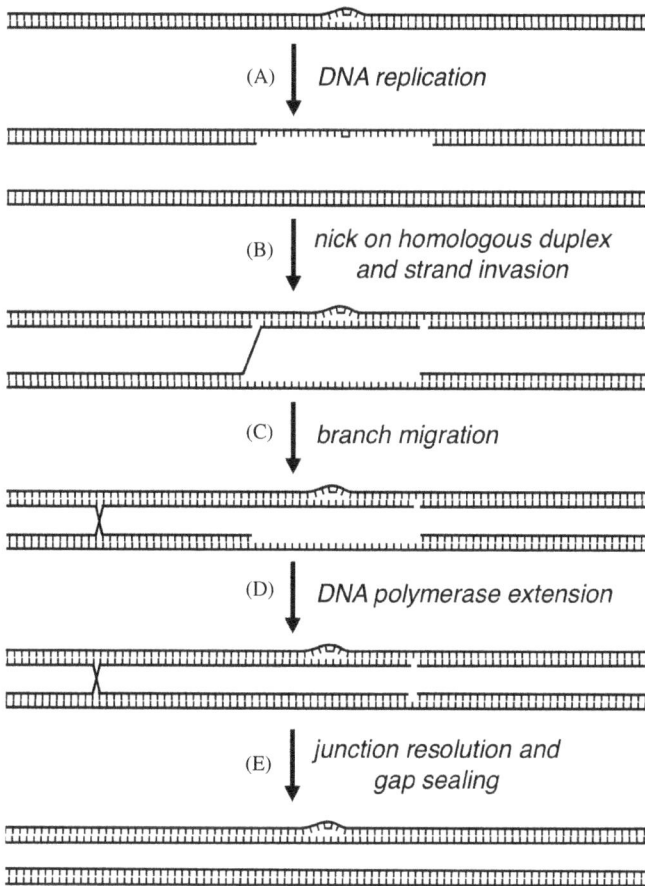

Figure 12.1. Model for daughter-strand-gap repair. See text for description of steps in the pathway. The DNA damage represented in this diagram is a pyrimidine dimer, which persists throughout the pathway and remains in the product DNA.

One of the most interesting aspects of this line of experiments is that the daughter-strand gaps disappeared upon further incubation, even though the UV-induced lesions remained. Thus, *E. coli* cells had some mechanism(s) for completing replication of the gaps even though they could not actually repair the UV-induced lesions. This process was called daughter-strand-gap repair. Because the UV-induced lesion is still present in the completed DNA molecules, the lesion itself is not repaired, only the daughter-strand gap. Thus, with respect to the UV-induced DNA damage, this process reflects a DNA-damage-tolerance pathway.

How does daughter-strand gap repair function? Additional studies showed that the process requires the key bacterial recombination protein RecA, along with RMP proteins that were subsequently shown to load RecA protein onto the (SSB-coated) single-stranded gaps. A model for daughter-strand-gap repair involves strand invasion between the corresponding locations of the two daughter chromosomes, which have identical sequences (Figure 12.1). A nick in the opposite daughter duplex is proposed to be induced prior to or during the early stages of strand invasion, allowing a 3' end from the undamaged duplex to pair within the gap of the damaged duplex. Next, branch migration forms a Holliday junction and brings the 3' end that had been blocked by the UV-induced lesion over to the sister duplex (Figure 12.1C). This is the key step for template switching, because now DNA polymerase can extend the 3' end to fill in the information missing in the original gap (Figure 12.1D). Finally, Holliday-junction resolution and ligation of nicks restores the two daughter strands to intact, unconnected duplexes (Figure 12.1E). The net result is that the gap opposite the UV-induced damage has been filled, and better yet, filled in a manner that is perfectly accurate since the identical sister duplex has the correct complementary sequence for proper templating.

A second pathway for template switching has already been discussed in Chapter 3. In this pathway, replication-fork regression places a blocked leading-strand product opposite the newly synthesized lagging-strand product, allowing extension past the blocking site (see Section 3.7). After extension, the regression is reversed to restore a normal replication fork, but one that has now extended past the damage. As discussed in Chapter 3, fork regression and the reversal of regression likely requires the active participation of DNA helicases that bind to the branch point of the replication fork, as well as the disassembly of at least some of the replication proteins from the blocked replication fork. As with daughter-strand-gap repair, the template for extension has the correct DNA sequence, and so the process is accurate and avoids the generation of mutations.

Recent estimates suggest that template switching accounts for most of the damage tolerance in *E. coli*, with the remainder due to the translesion DNA polymerases discussed later. The translesion DNA polymerases provide a significant contribution to tolerance only in cells that are undergoing an active DNA-damage response (see below). The balance between daughter-strand-gap repair and template switching by fork regression is less clear, and further studies are needed to clarify how the choice is made between these two pathways.

The contribution of template-switching pathways to DNA-damage tolerance in eukaryotes is less clear, but is currently under intensive study. The daughter-strand-gap repair pathway, as described earlier, does not seem to operate in eukaryotic cells, but there are strong hints that some kind of template switching contributes to damage tolerance. In the *Saccharomyces cerevisiae* model system, a DNA helicase called Rad5 is involved in an error-free pathway for damage tolerance, which very likely involves some kind of template switching. Rad5 protein catalyzes branch migration and fork regression, supporting the model that fork regression is employed for damage bypass. Rad5 and a partner protein are also involved in a DNA-damage-response pathway in which they modify certain proteins by covalently attaching multiple copies of a small protein called ubiquitin (poly-ubiquitination; see below and Chapter 13). One of the proteins that is covalently modified is *p*roliferating *c*ell *n*uclear *a*ntigen (PCNA) at the replication fork, and this modification of PCNA can activate the error-free tolerance pathway under discussion. Mammalian cells have homologs of the yeast Rad5 protein, and evidence suggests that a similar error-free tolerance pathway is functional in these cells.

12.2 An active process is often needed for mutagenesis

The distinction between DNA damage and mutation was discussed previously, with damage reflecting a physical aberration in the DNA molecule and mutation being a heritable change in the DNA sequence (Chapter 8). While the two are distinct, DNA damage does often lead to mutation, but how so? In the case of certain forms of base damage, the damaged base leads to a mutation in a fairly straightforward manner. These so-called miscoding lesions reflect small perturbations that do not block the replicative DNA polymerases but sometimes induce unconventional base-pairing during the polymerase reaction. For example, the oxidative lesion 8-oxo-G can form an unconventional base pair with A, leading to frequent G:C to T:A mutations (see Section 10.3).

In contrast to these miscoding lesions, many other lesions physically block the progress of replicative polymerases. These so-called noncoding lesions are either too large or unusually shaped so that they cannot fit into the active site of the replicative polymerase along with an incoming dNTP. The mechanism by which such noncoding lesions cause mutation has been a fascinating issue in the history of modern molecular biology.

The earliest studies of this phenomenon involved the mechanism of UV induced mutations in *E. coli*. A bacteriophage called λ infects *E. coli*, replicates its DNA extensively, and then produces a new crop of progeny virus particles that can go on and repeat the cycle. Genetic mutations can be studied in phage λ by analyzing the characteristics of the progeny viruses. Scientists uncovered a surprising result, which is that UV-irradiated λ particles would sustain mutations in their genome only if the host *E. coli* cells had also been irradiated prior to the λ infection! This result implied that some active, inducible process was somehow involved in the generation of mutations. Later studies identified the inducible process as the SOS DNA-damage response, which will be discussed in Chapter 13. It was also shown that the *E. coli* chromosome sustains mutations after UV treatment in a fashion that requires the SOS response, and so the above findings were not some aberration that applies only to bacterial viruses infecting *E. coli*.

How does the SOS response lead to the induction of mutations? In 1974, Miroslav Radman proposed that an error-prone DNA synthesis step inserts incorrect bases opposite template damage, a proposal that was verified years later. Two possible mechanisms were explored. The first, which turned out to be incorrect, is that some special protein(s) modify the replicative polymerases to allow them to replicate through template damage. Instead, the correct mechanism turned out to be that cells carry a special class of DNA polymerases, called translesion DNA polymerases, which inherently have the ability to insert nucleotide residues opposite blocking template lesions. When they insert incorrect nucleotide residues, mutations happen. In the *E. coli* system just discussed, the predominant translesion polymerase is carefully controlled by multiple mechanisms so that it is only active during the SOS response (see below).

In eukaryotic systems, translesion DNA polymerases are also responsible for mutations that are generated from blocking lesions. However, as we will see, the regulatory aspects are subtler, and translesion polymerases are present even in cells that have not been treated with DNA-damaging agents. We will see that the eukaryotic translesion DNA polymerases are nonetheless carefully regulated to act only in appropriate situations.

12.3 The riddles of DNA polymerases that disregard normal base-pairing rules

As discussed in the early chapters of this book, the replicative DNA polymerases have a very constrained active site that is highly restrictive for the normal Watson–Crick base pairs, and the very rare errors that are introduced are usually excised by the proofreading exonuclease activity to avoid the generation of mutations. Translesion DNA polymerases violate these characteristics. They are more lenient in allowing aberrant base pairs or even no base-pairing in the active site, and most of them lack proofreading exonuclease activity.

On first blush, you might conclude that the translesion DNA polymerases are incapable of accurate DNA synthesis and always make errors. Indeed, these enzymes are sometimes called "sloppier copiers" to reflect their negligent fidelity. Given the ubiquity and frequency of DNA damage, wouldn't such a rampant mechanism of mutation generation be incompatible with the maintenance of the many essential gene-coding segments in genomes and thereby the continued propagation of organisms?

One answer to this riddle is that the translesion DNA polymerases are not always error prone, but instead can function to insert nucleotide residues opposite certain lesions in a relatively error-free manner. Of course, this raises another riddle, which is how can error-free replication be accomplished by these enzymes in the absence of proper Watson–Crick base-pairing? As we will see, particular translesion polymerases have evolved to act on a defined subset of lesions. For example, a particular translesion DNA polymerase acts on modified dG residues and has evolved a strong preference for inserting a dC residue, but without any base-pairing between the two (see discussion of Rev1 below). In a very real sense, the enzyme is cheating our friends Watson and Crick — it already "knows" which nucleotide residue to insert because it "knows" which damage it is acting upon. This example illustrates an important concept, which is that translesion DNA polymerases have evolved to carry out very specific functions that differ between the members of the group. With this concept in mind, it is easier to appreciate the finding that human cells carry 10 different translesion DNA polymerases, many more than the number of replicative polymerases.

12.4 Introduction to translesion DNA polymerases

DNA polymerases share a common architecture that resembles a right hand, with the active site situated in the palm domain (see Chapter 2). As an incoming nucleotide pairs with the template, the fingers and thumb domains close down tightly around the base pair, and the proper configuration for catalysis can only be achieved with the compact Watson–Crick base pairs in the active site. Nearly all DNA polymerases, including the translesion polymerases, share this architecture (PrimPol is the only exception; see Section 4.9). However, DNA polymerases group into six distinct evolutionary families that differ in sequence conservation and the details of this architecture. The eukaryotic replicative polymerases belong to the B family, *E. coli* DNA polymerase I and the T7 DNA polymerase are in the A family, and *E. coli* replicative polymerase III is in the C family. Translesion DNA polymerases have been found in each of these three families, but many belong to the Y-family, which specializes in translesion DNA synthesis.

While the structures of translesion polymerases differ among themselves, the active sites are generally more open than in the replicative polymerase and the finger and thumb domains tend to be smaller. The restricted active sites of replicative DNA polymerases, necessary for ensuring accurate base-pairing, also prevent many distorted DNA lesions such as UV-induced pyrimidine dimers from entering the site or contributing to a productive base pair. In contrast, the more open active sites of translesion DNA polymerases can accommodate a larger variety of template bases and even covalently linked adjacent residues such as in the CPD and (6-4) photoproduct lesions. For many of the translesion DNA polymerases, the more open active site also translates into more promiscuous acceptance of incoming deoxynucleotides across from the template damage. With these general ideas stated, it is important to quickly point out that every translesion DNA polymerase has its own unique characteristics, some of which will be described below.

12.5 Translesion DNA polymerases in *E. coli*

The major translesion DNA polymerase in *E. coli*, responsible for most of the UV-induced mutagenesis discussed earlier, is called DNA polymerase V and is a member of the Y-family. The activity of DNA polymerase V is carefully regulated by multiple mechanisms that limit its activity. As alluded to above, this enzyme is actively synthesized only during an SOS response due to transcriptional regulation (also see Chapter 13). In addition, the active subunit of DNA polymerase V is normally bound to an inhibitor protein, which is present in excess, and the enzyme is only active when this inhibitor protein is cleaved as a further consequence of the SOS response. This helps to delay the activity of polymerase V until late in the SOS response, presumably giving error-free mechanisms such as excision repair and template switching the first shots at the DNA damage (see Chapter 13). DNA polymerase V is also controlled by complex proteolysis mechanisms, which presumably ensure that the enzyme is efficiently shut off when the SOS response is complete.

As might be expected for the major translesion DNA polymerase of a cell, *E. coli* DNA polymerase V has a broad specificity for accepting damaged template bases. The enzyme accepts both major classes of UV damage (CPD and (6-4) photoproduct), and tends to incorporate a dA residue across from the first base of a di-thymine CPD adduct but dG across from the first base of a di-thymine (6-4) photoproduct adduct. This is an important finding, because it shows that the insertion specificity of a translesion DNA polymerase can be influenced by the nature of the damaged residue in the template position of the active site. Other translesion polymerases also show modified insertion specificity based on the nature of the damaged template. DNA polymerase V also extends past a template abasic site, where it tends to insert a dA residue. While DNA polymerase V has broad specificity, there is some evidence that one or both of the minor translesion DNA polymerases of *E. coli* (DNA polymerase II and IV) contribute to damage tolerance with a small subset of template adducts. Both of these polymerases are also induced by the SOS response, so again nearly all translesion DNA polymerase activity in *E. coli* is restricted to periods of DNA-damage response.

How does DNA polymerase V or either of the minor translesion polymerases of *E. coli* access the DNA template after a lesion has blocked the replicative polymerase? Recall that the replication clamp in both *E. coli* and eukaryotic cells behaves like a toolbelt, bringing the replicative polymerase, Okazaki-fragment processing enzymes, and mismatch repair enzymes to their sites of action at or just behind the replication fork (Chapters 3, 4, and 6). This toolbelt function of the sliding clamp also applies to the translesion DNA polymerases (Figure 12.2). All three of the *E. coli* enzymes can bind to the sliding clamp, and do so at the same site where replicative DNA polymerase III binds. Recall that the *E. coli* sliding clamp is a dimer, and the C-terminal region of each dimer has one of these binding sites. The toolbelt model argues that one of the translesion

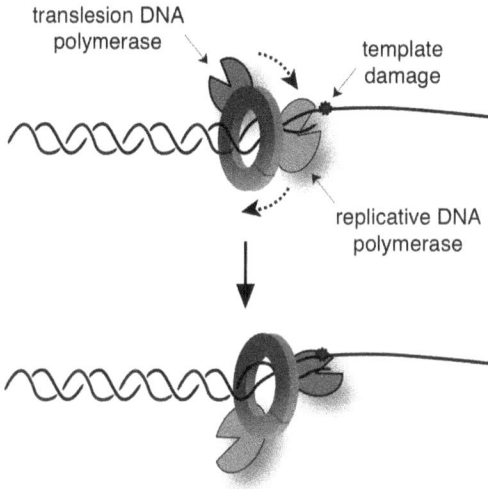

Figure 12.2. The toolbelt model for function of translesion DNA polymerases. The sliding clamp encircling DNA can bind multiple DNA polymerases, allowing switching from replicative polymerase to translesion polymerase (and back again).

DNA polymerases, usually polymerase V, can bind to one of the clamp subunits while DNA polymerase III is bound to the other clamp subunit. When polymerase III becomes stalled due to a blocking lesion, a simple polymerase switch can occur. Polymerase III is rotated away from its stalled position at the blocked 3′ end and polymerase V is rotated in to access the 3′ end for extension (Figure 12.2). After translesion synthesis by polymerase V, the switch can be reversed so that the normal replicative polymerase can re-engage downstream of the bypassed damage. The sliding clamp is thus central for this polymerase-switching model to facilitate translesion synthesis. We will see below that eukaryotic cells use a similar but even more elaborate pathway involving their sliding clamp.

12.6 Eukaryotic translesion DNA polymerases

The eukaryotic translesion DNA polymerases, like the replicative enzymes, are generally named with Greek letters; the one exception is an enzyme called REV1. Beginning with the Y-family members, the yeast *S. cerevisiae* has two homologs of *E. coli* DNA polymerase V and human cells have four. All of these enzymes are translesion DNA polymerases that lack a proofreading exonuclease. The four Y-family members in humans are REV1 and polymerases η (eta), ι (iota), and κ (kappa). Both human cells and *S. cerevisiae* have one translesion polymerase in the B family, DNA polymerase ζ (zeta). Human cells (but not yeast) have three translesion enzymes in the X family (β, beta; λ, lambda; and μ, mu) and two in the A family (θ; theta; and ν, nu). The physiological roles and biochemical properties of these various enzymes are under intensive study, but much remains to be learned. Rather than trying to summarize the latest studies on each of the 10 polymerases, we will focus on a few vignettes that illustrate some of the important principles of translesion DNA synthesis.

Human DNA polymerase η plays a well-established role in preventing skin cancer induced by exposure to UV, as will be discussed later. Correspondingly, the enzyme is able to replicate in a fairly accurate manner across from a di-thymine CPD, preferentially inserting A at both positions. Indeed, the yeast homolog is just as efficient at extending opposite a di-thymine CPD as it is at extending across from a normal di-thymine template sequence (see "How did they test that?" at the end of this chapter). Similar to *E. coli* DNA polymerase V, DNA polymerase η has an altered specificity opposite the (more distorting) di-thymine (6-4) photoproduct and preferentially inserts a dG residue across from the first (3′) thymine. This insertion specificity may contribute to the fairly frequent T:A to C:G mutations (at the 3′ position) induced by UV in vivo. Notice that once again, the nature of the damaged template base is altering the insertion specificity of the enzyme. We will return to replication of the (6-4) photoproduct in Section 12.7.

When copying a normal DNA template, DNA polymerase ι is very error prone, and has a particularly strong tendency to insert dG residues across from dT in the template (even more so than dA residues). The frequent mistakes by ι have been traced to an unusually shaped active site, which generally restricts template purine bases to the *syn* conformation (with a flipped base, rotated by 180°

around the N-glycosidic bond) and places unusual constraints on the overall shape of the allowed base pairs. The *syn* conformation of the template base promotes the formation of alternative base pairs in duplex DNA called Hoogstein pairs. As a brief introduction, Hoogstein base pairs utilize somewhat different hydrogen bond pairing schemes than Watson–Crick base pairs, but can still result in "correct" pairing interactions, as found in dA:dT and dG:dC Hoogstein pairs. Getting back to translesion synthesis, DNA polymerase ι utilizes Hoogstein pairing in its active site to allow replication past several template lesions. Most notably, the enzyme can pair a template 8-oxo-G lesion with an incoming dCTP nucleotide, leading to accurate replication opposite this very common lesion. The active sites of other DNA polymerases favor a different pairing of 8-oxo-G (again in the *syn* conformation) with incoming dATP (in the normal *anti* conformation), and this of course leads to mutations (see discussion of miscoding lesions in Section 12.2). Thus, polymerase ι has the special property of allowing error-free replication across from template 8-oxoG lesions, one of the most common DNA adducts in cells. Indeed, cell-culture experiments show that polymerase ι contributes to protection against oxidative stress and reduces the number of mutations that are generated from 8-oxo-G lesions.

Another Y-family polymerase, Rev1, has a remarkable mechanism for achieving error-free replication of certain damaged G template residues (including N^2 adducts). The enzyme inserts a C residue accurately across from these template G adducts (as well as undamaged G), but does so by actually swinging the template residue out of the active site. In its place, residue 324 (an arginine) of Rev1 is inserted in the active site and hydrogen bonds with the incoming deoxycytidine triphosphate (dCTP) nucleotide (Figure 12.3)! This allows error-free replication of this subset of damaged G residues. It is important to note, however, that Rev1 also inserts dCTP residues by the same mechanism across from abasic sites. Most abasic sites would not be generated from template G residues, and thus the special mechanism for inserting dCTP can also be mutagenic depending on the lesion being replicated.

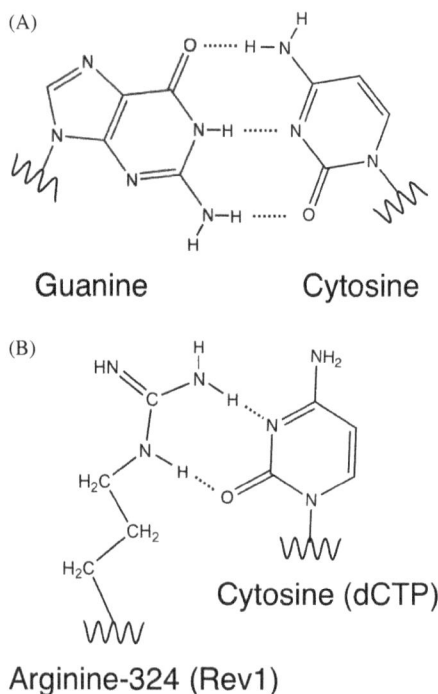

Figure 12.3. Pairing of cytosine within the active site of DNA polymerase Rev1. The structure of the normal G:C base pair is represented in panel A. The pairing of an arginine residue of Rev1 with the cytosine base of (incoming) dCTP is represented in panel B.

The abovementioned examples illustrate how translesion DNA polymerases can contribute to error-free replication of DNA containing damage, but still have the potential to cause mutations depending on the nature of the lesion and the translesion DNA polymerase that acts upon it. As mentioned earlier, human cells have 10 different translesion DNA polymerases, and we have much to learn about the detailed roles of each within the cell. However, biochemical studies show that there is overlap between the enzymes that can extend opposite most given template lesions, and many of these extension events occur in an error-prone manner that could potentially cause mutations. Later in this chapter, we will consider how translesion DNA polymerases access particular lesions and compete for their replication, and also go into more detail about their biological roles including the induction of new mutations in the genome.

12.7 One-step versus two-step pathways

To this point, we have ignored an important aspect of many translesion synthesis events. While most translesion DNA polymerases have the special capability of inserting a nucleotide residue opposite some damage or an abasic site, this insertion event often creates a poorly matched 3′ end for the subsequent extension. It turns out that polymerase active sites that are good at inserting poorly matched nucleotides are often quite poor at extending such mismatched ends. The solution nature has found for this problem is that many translesion events occur in two discrete steps. The first nucleotide is incorporated opposite the lesion by an "inserter" polymerase, while a specialized "extender" polymerase completes the reaction by incorporating one or a few additional nucleotides downstream from the damage (Figure 12.4).

A given translesion event can thus be characterized as either a one-step or two-step reaction. A prominent example of one-step translesion synthesis is the replication through a CPD by DNA polymerase η, which was described earlier. The more distorting (6-4) photoproduct, on the other hand, generally utilizes a two-step process (Figure 12.4). After polymerase η (or another polymerase) inserts a residue at the first position, an extender polymerase inserts the second residue (generally along with a few additional residues downstream). DNA polymerase ζ is the best characterized extender polymerase, participating in many two-step translesion events in eukaryotic cells. This polymerase has a reasonably high fidelity, similar to that of the replicative polymerase α, and is generally very ineffective at actual translesion synthesis (with the exception of completing replication across from the (6-4) photoproduct as described earlier).

As mentioned earlier, extender polymerases often insert a few extra nucleotide residues past the site of damage, and do so with a

DNA pol δ/ε blocked at (6-4) photoproduct

inserter polymerase (η)

extender polymerase (ζ)

DNA pol δ/ε extension resumes

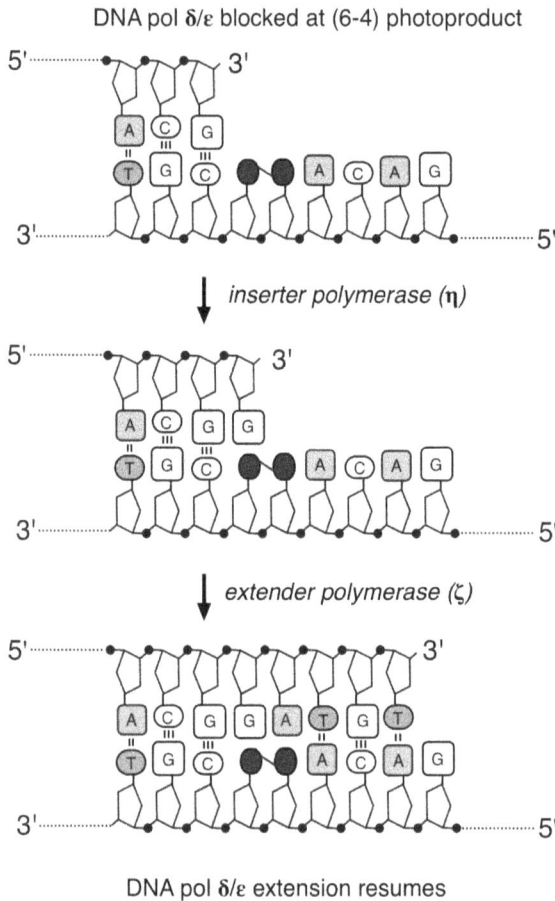

Figure 12.4. Two-step translesion synthesis. The (6-4) photoproduct in the template is indicated by the two (attached) filled circles as bases. See text for description of steps in the pathway.

relatively high accuracy. Why would it be important to extend a few bases out beyond the damage? An interesting explanation might be provided by studies of the *E. coli* replication enzymes. The replicative DNA polymerase III requires a region of about five correct base pairs at the 3′ end to extend a primer, and if this region has mismatches, the enzyme preferentially degrades the region with its proofreading exonuclease activity. If this degraded region were to include the residues inserted by translesion DNA polymerases, a futile cycle would ensue. Therefore, action of an extender polymerase could be required for the replicative polymerase to resume synthesis. A useful way to think about this is that the extender polymerase is providing a "stamp of approval" for the replicative polymerase to proceed.

12.8 Control of translesion synthesis in eukaryotic cells

In most eukaryotic cells that have been studied, translesion polymerases are synthesized even in cells that have not been exposed to DNA-damaging agents, quite different than the *E. coli* model system discussed earlier. Nonetheless, translesion synthesis is still regulated in eukaryotic cells, but the regulation usually occurs in a more local rather than global fashion.

The major focus of this local regulation is the state of the replicative sliding clamp, PCNA. The details are fairly complex, but a simplified overview follows. PCNA can be modified by covalent addition of either of two related proteins, ubiquitin and SUMO (*small ubiquitin-like modifier*), and the ubiquitin can be added in either single copy (mono-ubiquitination) or multiple copies (poly-ubiquitination). These modifications are triggered when the replicative polymerase is blocked, and scientists are studying the details of how the triggers function, how the system chooses between the three different modifications, and the proteins involved in these activation steps. There is some evidence that the details differ between various eukaryotic cells, adding to the complexity. We already learned that poly-ubiquitination activates an error-free tolerance pathway in *S. cerevisiae* (Section 12.1). In the same organism, mono-ubiquitination of PCNA is needed to recruit translesion DNA polymerases to the stalled replication complex. Poly-ubiquitination and SUMOlyation might also play roles in translesion synthesis, and there is likely extensive cross talk between the pathways.

While the regulatory details may be complex, the bottom line is that the *S. cerevisiae* translesion DNA polymerases are only allowed to access 3′ ends at replication forks when PCNA is covalently modified. PCNA, like the bacterial sliding clamp, presents multiple binding sites to DNA polymerases. PCNA is a trimer, and so potentially three DNA polymerases can compete for access to a single PCNA ring. For translesion synthesis, the toolbelt model is again likely to be relevant, with translesion polymerases exchanging onto the 3′ end as a blocked replicative polymerase swings out of the site.

The toolbelt model may also be relevant when considering the question of how a particular translesion DNA polymerase is "chosen" for a particular form of damage. There is currently no evidence suggesting a special mechanism for specifically recruiting one particular polymerase when a particular lesion blocks replication, say polymerase η for a UV-induced CPD. Instead, a favored model is that the PCNA toolbelt provides a platform for the translesion polymerases to essentially compete with each other, and polymerases that cannot extend across from the CPD would be unable to extend even if they are the first to bind. The toolbelt model also fits the two-step translesion model discussed earlier, allowing a progressive exchange of two different translesion polymerases before the replicative polymerase is exchanged back onto the extended 3′ end. There is much more to learn about how different translesion polymerases are chosen for particular lesions, and it would not be surprising to learn that the regulation between the different polymerases is even more sophisticated than we currently understand.

When cells have suffered higher levels of blocking damage or other replication inhibition, a more global response is invoked — the ATR-dependent checkpoint activation. We will discuss this checkpoint in more detail in the next chapter, but one aspect is particularly relevant for the current discussion. The global checkpoint induces the synthesis of an alternative sliding clamp, called the 9-1-1 checkpoint clamp. This clamp is recruited to RPA-coated gaps in the replication fork, and like ubiquitinated PCNA, also recruits translesion DNA polymerases.

12.9 Biological importance of translesion DNA polymerases

How important are translesion DNA polymerases in surviving DNA damage caused by exogenous and endogenous sources? What are their precise biological roles? The phenotypes of various knockout mutants provide a first approximation to answer these questions. Considering model systems such as *E. coli* and *S. cerevisiae*, mutations that eliminate DNA polymerase V or polymerase ζ, respectively, cause a moderate sensitivity to UV radiation. Notably, NER-deficient mutants are more sensitive than translesion DNA polymerase mutants, which is perhaps not surprising since NER repairs UV damage while translesion polymerases only provide tolerance without repair. As we will discuss later, inactivation of DNA polymerase V or ζ also profoundly affects mutagenesis.

Moving to mammalian systems, two findings are particularly notable. First, the inherited human disease XP-V (*x*eroderma *p*igmentosum *v*ariant) was found to result from knockout mutations in human DNA polymerase η, which is particularly adept at replication across UV-induced lesions (see above). Patients with XP-V suffer an extremely high frequency of skin cancers as a result of exposure to the UV in sunlight. Second, mouse knockout lines that are defective in DNA polymerase ζ are embryonic lethal, supporting a very important role(s) for this polymerase in mammalian cells and in development. Recall that polymerase ζ is the major "extender" polymerase, playing a role in many different two-step translesion events.

The general rationale for the biological importance of translesion DNA polymerases should be evident from the examples discussed in this chapter. Several examples highlighted the fact that extension by translesion polymerases sometimes occurs in an error-free manner, even without use of the normal Watson–Crick base-pairing. Such error-free replication allows completion of DNA replication and provides the cell another chance to repair the damage with more conventional pathways, such as NER, without suffering a mutation in the genome. Even when a mutation is induced by a translesion event, the outcome may still be less detrimental than having the process of DNA replication blocked, which can result in cell death and/or chromosomal rearrangements. We will even discuss evidence that the induction of mutations after DNA damage may have beneficial effects in some situations (see below). It is also worth pointing out that a very large fraction of mammalian genomes does not code for proteins, and so the majority of point mutations would not alter any encoded proteins. Even mutations within a protein-coding sequence quite often have little or no effect on protein function.

Translesion DNA polymerases have also been recruited into specialized roles in mammalian systems. One major example is that certain translesion polymerases are responsible for generating diversity in the immune system, by creating somatic mutations in specific regions of antibody genes. Another example is that DNA polymerase λ has been shown to function in the c-NHEJ pathway. Finally, there is evidence implicating specific translesion DNA polymerases in repair pathways such as *b*ase *e*xcision *r*epair (BER).

There is overwhelming evidence that a large fraction of the mutations induced by noncoding lesions are caused by translesion DNA polymerases, as discussed in some detail for *E. coli* in Section 12.2. In addition, these polymerases contribute to spontaneous mutagenesis, because translesion-polymerase-knockout mutations in *E. coli* and *S. cerevisiae* strongly reduce the rate of spontaneous mutations. These results argue that translesion polymerases are engaged when replication forks are blocked by endogenous lesions, and perhaps relate to the embryonic lethality of polymerase ζ knockouts in the mouse system (see above).

Finally, the mutagenic action of translesion polymerases likely contributes to the evolutionary process by providing increased diversity. The evidence is strongest in bacterial systems, where increased mutation frequencies can, for example, increase the progression of antibiotic resistance and perhaps the evolution of pathogenesis. Cancer progression within an individual can also be thought of as a form of cellular evolution. Mutations generated by translesion DNA polymerases very likely contribute to diversifying the cancer genome to promote anticancer-drug resistance, metastasis, and more rapid proliferation. This possibility is currently under intensive investigation and could potentially lead to sophisticated therapies that target, in part, the involved polymerases.

12.10 Summary of key points

- DNA-damage-tolerance pathways allow cells to survive and propagate in spite of unrepaired DNA damage and provide substantial increases in resistance to DNA-damaging agents.
- One general pathway of damage tolerance involves template switching, as occurs in daughter-strand gap repair and fork regression in prokaryotes; the pathways of template switching in eukaryotic cells are less clear but under intense investigation.
- An active process is required for generating mutations with certain DNA-damaging agents including UV; in bacteria, this is the SOS DNA-damage response.
- These damage-induced mutations are generated by translesion DNA polymerases, which insert nucleotide residues using more lenient rules of base-pairing (or even ignoring the rules altogether).
- Translesion DNA polymerases are error prone, but in some cases function in a relatively error-free manner by acting on particular DNA lesions and inserting the correct residue without using conventional base-pairing rules.
- Translesion DNA polymerases have a more open and permissive active site than conventional DNA polymerases.
- The insertion specificity of a given translesion DNA polymerase often depends on the nature of the damaged template base.
- Many translesion events occur by a two-step process involving an inserter polymerase and an extender polymerase, notably the extender DNA polymerase ζ in eukaryotic cells.
- In eukaryotic cells, translesion DNA polymerases are present even without DNA damage but are activated upon damage by complex posttranslational processes.
- In humans, mutations that inactivate translesion DNA polymerase η cause an extremely high frequency of skin cancer (from UV damage), and in mice, inactivation of translesion DNA polymerase ζ causes embryonic lethality — both results attest to prominent roles of translesion polymerases in cell survival and propagation.

- Evidence also supports roles of translesion DNA polymerases in generating diversity in the immune system, repairing breaks through c-NHEJ, and inducing mutagenesis central to evolution.

Further Reading

Atkinson, J., & McGlynn, P. (2009). Replication fork reversal and the maintenance of genome stability. *Nucleic Acids Res, 37*(11), 3475–3492.

Bell, S. D., Mechali, M., & DePamphilis, M. L. (2013). *DNA Replication.* Cold Spring Harbor, NY: Cold Spring Harbor Press.

Chatterjee, N., & Siede, W. (2013). Replicating damaged DNA in eukaryotes. *Cold Spring Harb Perspect Biol, 5*(12), a019836.

DePamphilis, M. L., & Bell, S. D. (2011). *Genome Duplication.* New York, NY: Garland Science.

Friedberg, E. C., Walker, G. C., Siede, W., Wood, R. D., Schultz, R. A., & Ellenberger, T. (2006). *DNA Repair and Mutagenesis* (2nd ed.). Washington, DC: ASM Press.

Fuchs, R. P., & Fujii, S. (2013). Translesion DNA synthesis and mutagenesis in prokaryotes. *Cold Spring Harb Perspect Biol, 5*(12), a012682.

Goodman, M. F., & Woodgate, R. (2013). Translesion DNA polymerases. *Cold Spring Harb Perspect Biol, 5*(10), a010363.

Johnson, R. E., Prakash, S., & Prakash, L. (1999). Efficient bypass of a thymine-thymine dimer by yeast DNA polymerase, Pol eta. *Science, 283*(5404), 1001–1004.

Kuzminov, A. (1999). Recombinational repair of DNA damage in Escherichia coli and bacteriophage lambda. *Microbiol Mol Biol Rev, 63*(4), 751–813.

Lehmann, A. R., & Fuchs, R. P. (2006). Gaps and forks in DNA replication: Rediscovering old models. *DNA Repair (Amst), 5*(12), 1495–1498.

McIntosh, D., & Blow, J. J. (2012). Dormant origins, the licensing checkpoint, and the response to replicative stresses. *Cold Spring Harb Perspect Biol, 4*(10), a012955.

Nicolay, N. H., Helleday, T., & Sharma, R. A. (2012). Biological relevance of DNA polymerase beta and translesion synthesis polymerases to cancer and its treatment. *Curr Mol Pharmacol, 5*, 54–67.

Sale, J. E. (2013). Translesion DNA synthesis and mutagenesis in eukaryotes. *Cold Spring Harb Perspect Biol, 5*(3), a012708.

How did they test that?
Does yeast Rad30 protein have DNA polymerase activity on damaged templates?

Prior genetic studies had identified *Saccharomyces cerevisiae* mutants that are hypersensitive to UV light, and some of these were mapped to a gene called *RAD30* (*rad*iation sensitive). Johnson, Prakash, and Prakash (1999) constructed a fusion protein between glutathione-*S*-*t*ransferase (GST) and the reading frame of the *RAD30* gene. GST fusions allow a protein to be purified efficiently, because they generally bind to chromatography columns with covalently linked glutathione and can be specifically eluted from the column by adding free glutathione. The researchers thereby purified the GST-Rad30 fusion protein away from other proteins that would interfere with their assays. They first found that the GST-Rad30 protein has DNA polymerase activity with undamaged DNA templates. In the experiment shown here, the scientists compared activity using two different 75-nucleotide long templates with the same sequence, except that one contained a centrally located pair of normal T residues while the other contained a *cis-syn* thymine dimer (position indicated by T*T* in the figure; note that only the central part of the template sequence is shown in the figure). The reactions also contained radioactively labeled 30- or 44-nucleotide long primer (P-30 or P-44, respectively); P-44 provides a 3′ end immediately before the TT sequence. Reaction products were separated on a high-resolution gel and visualized based on the radioactivity in the primer. As expected in this control, replicative polymerase Pol δ extended all the way to the end of the undamaged template but was completely blocked just before the thymine dimer of the damaged template (left panel). In contrast, the GST-Rad30 protein extended well past the TT sequence on both the damaged and undamaged template, regardless of which primer was used. The overall polymerase activity is not as robust as that of Pol δ, in that GST-Rad30 was ineffective at reaching the 5′ end of the template. In other experiments, the authors showed that GST-Rad30 prefers adding dA residues opposite the thymine dimer. Based on these results, the authors renamed the Rad30 protein Pol η, the seventh identified eukaryotic DNA polymerase at that time. The figures were reproduced from Johnson *et al.* (1999), with permission from the American Association for the Advancement of Science; permission conveyed by Copyright Clearance Center, Inc.

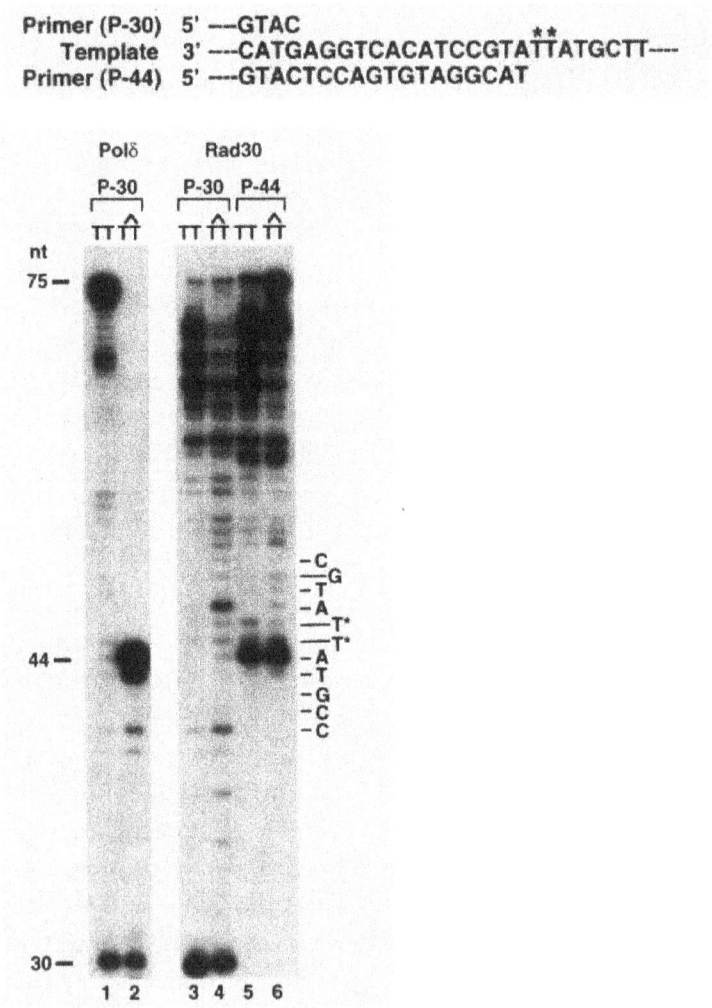

Chapter 13

DNA-damage response pathways

In earlier chapters of this book, DNA replication, repair, and tolerance pathways have been presented without much discussion of their regulation. A few examples of regulation have surfaced, including the induction of alkyltransferase synthesis in bacteria (Section 9.2) and modulation of end resection through the cell cycle in eukaryotes (Section 11.9). This chapter will provide an overview of the multifaceted regulatory pathways that modulate replication, repair, and tolerance in response to DNA damage. Hopefully, this overview will help the reader integrate an understanding of these diverse pathways and how they operate smoothly to promote cell growth and survival.

At the outset, it is important to stress that the pathways of DNA-damage response (DDR) are intricate and very complex, and that much remains to be learned. In that sense, this chapter only scratches the surface of a vibrant and active field in which thousands of informative studies are published annually. As in prior chapters, we will limit the naming and discussion of individual proteins to a handful that play particularly important roles in DDR pathways and thereby try to avoid getting "lost in the weeds." Notably, DDR pathways sometimes vary between species, implying a degree of evolutionary flexibility. In the interest of brevity, these variations will generally not be discussed. It should also be mentioned that minor and less well-studied DDRs have been identified in both prokaryotes and eukaryotes, and discussion of these are generally beyond the scope of this chapter.

13.1 The bacterial SOS response

The major bacterial DDR, the SOS response, was briefly introduced in Chapter 12 under the context of translesion DNA polymerases and the induction of mutagenesis. Here, we will consider the SOS response more broadly.

The SOS response is largely, though not entirely, a transcriptional regulatory phenomenon. A transcriptional repressor protein called LexA binds to the promoter region of a couple dozen bacterial genes, reducing or blocking their transcription under normal conditions (Figure 13.1). When cellular DNA is damaged or DNA metabolism is otherwise corrupted, the bacterial recombination protein RecA (see Section 11.1) binds to ssDNA regions that are generated as a consequence of the damage. As described in Chapter 11, RecA then plays a key role of promoting homologous recombination, for example, catalyzing *d*ouble-strand *b*reak (DSB) repair and replication fork restart. RecA, as it turns out, has another important function, namely triggering the SOS response. The trig-

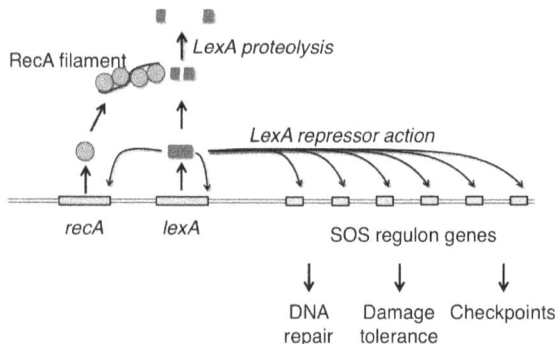

Figure 13.1. The SOS response in *E. coli*. The major DNA-damage regulatory circuit in *E. coli* is controlled by the LexA repressor and RecA protein, which acts as a "coprotease." See text for detailed description of the pathway. This figure was modified from Kreuzer, K. N. (2013).

ger is pulled when RecA, in complex with ssDNA, binds to the LexA protein and induces LexA proteolysis. The sudden loss of LexA repressor then allows expression of the genes in the circuit, thus activating the SOS response (Figure 13.1).

In the early days, the definition of the SOS regulatory circuit was greatly aided by genetic studies of specific *Escherichia coli* mutants. Knockout mutants of the RecA protein were found to be unable to induce the SOS response, and to be profoundly sensitive to DNA damage both due to the defect in SOS induction and to the direct role of RecA in repairing DNA breaks. Scientists were also able to isolate RecA mutants that constitutively express the SOS pathway at some level. This kind of mutant has amino acid substitutions that allow RecA to activate destruction of LexA even in the absence of DNA damage. The earliest LexA mutants isolated were unable to induce the SOS response, and these were eventually found to have amino acid substitutions that blocked the proteolysis reaction necessary for the SOS pathway. Initially, scientists failed to isolate simple knockout mutants that destroyed the activity of the LexA protein. This failure was eventually explained when it was discovered that one of the SOS proteins causes a profound block in cell division (see below). When this cell-division inhibitor was inactivated by mutation, LexA-deficient mutants were isolated and found to express the SOS response in a constitutive manner, as expected.[1]

Before discussing the downstream effects of the SOS response, the proteolysis of LexA deserves a closer look. Initially, the above-mentioned results suggested to scientists that the RecA protein must be a protease that destroys LexA. However, careful experimentation revealed that LexA itself was the protease, capable of destroying itself! In its native state free of RecA, the protease active site of LexA is silent. However, the binding of activated (ssDNA-bound) RecA turns the LexA protease active site on so that the protein destroys itself. Formally, RecA is therefore called a "coprotease" and LexA is essentially a latent self-protease.

---

[1] Evidently, the SOS-constitutive mutants affecting RecA (see above) express low enough levels of the SOS response to allow cell division to continue.

The coprotease function of RecA extends to two additional protein substrates that are important in the SOS story. First, we already discussed the fact that the translesion DNA polymerase activity of *E. coli* DNA polymerase V requires proteolytic cleavage of its binding partner (see Section 12.5). This cleavage event occurs in a similar fashion to that of LexA. The binding partner is again a latent self-protease, and the coprotease function of RecA activates the self-cleavage of the inhibitory binding partner, freeing active DNA polymerase V. The second additional protein that is activated to cleave itself by RecA is the repressor of the bacterial virus called λ (as well as some other bacterial viruses). The λ virus can enter a quiescent state where its DNA is inserted in the bacterial chromosome but stays essentially inert as the bacterial cell and its descendants propagate. This state is called lysogeny, and it benefits the bacterial virus because the λ DNA is propagated passively as the bacteria reproduces. However, with the induction of an SOS response, the λ repressor is activated to cleave itself, which causes the induction of transcription of the viral genes that had been heretofore repressed. The virus DNA excises from the chromosome, replicates extensively, and eventually the hapless cell bursts to release 100 or so new virus particles. Evidently, λ has evolved to use the SOS circuitry to help itself "escape from a sinking ship" when its host has been severely damaged.

Let's now go back to the SOS circuitry itself, and discuss the proteins that are transcriptionally activated after DNA damage (Figure 13.1). One is the LexA repressor itself. On first blush, that might seem confusing, since LexA protein is destroyed as part of the SOS response. However, induction of additional LexA is critical because the SOS response needs to eventually shut off, assuming the cell succeeds in repairing all the DNA damage. Once the DNA damage is repaired and RecA is no longer bound to ssDNA, intact LexA can accumulate via new transcription/translation and shut down any further SOS response. A second SOS-induced protein is RecA. An increased level of RecA during the SOS response promotes homologous recombination and allows RecA levels to keep up with the amount of ssDNA if the DNA damage is extensive.

Among the two to three dozen SOS-induced genes, many are involved in DNA repair and tolerance pathways that we have already discussed in this book. This includes genes encoding proteins involved in NER (UvrA, UvrB, UvrD, and a UvrC homolog), translesion DNA synthesis (including DNA polymerase V), and a branch-specific helicase that participates in homologous recombination. The SOS response also induces the synthesis of proteins involved in nucleotide metabolism, transcription factors, and a number of proteins that have no obvious connection to repairing or tolerating DNA damage.

The SOS response also leads to temporary inhibition of bacterial reproduction, a rudimentary form of cell-cycle checkpoint (Figure 13.1; see below for the discussion of checkpoints in eukaryotes). For many decades, it has been known that one particular SOS-induced protein directly blocks cell division, giving the cell more time to repair its DNA damage and complete DNA replication before attempting to segregate daughter chromosomes. Recent evidence suggests that additional, subtler pathways also restrain cell division during the SOS response.

While the SOS response is primarily a DNA-damage response, it has a much broader significance in bacterial physiology that even extends into modern clinical practice. The SOS pathway has been shown to promote gene transfer between bacteria via multiple pathways, and in this way contributes to bacterial evolution and the spread of plasmid-borne traits such as antibiotic resistance. In addition, the SOS pathway can be induced by exposure to a number of antibiotics, either in a RecA-dependent or RecA-independent manner. This combination of results means that treatment with antibiotics can promote the spread of antibiotic resistance via the SOS pathway, and as you undoubtedly know, increasing antibiotic resistance is a serious problem in modern medicine. There are also several examples where the SOS pathway has been shown to promote bacterial pathogenesis, including one involving the *Vibrio cholerae* pathogen that causes the devastating disease cholera.

13.2 DNA damage in eukaryotes alters cell fates: Checkpoints, apoptosis, and necrosis

Before digging into the circuitry of eukaryotic DDR pathways, let us first consider some of the major events at the cell-cycle level that are modulated by these pathways. As we will see, these events are not unique to the DDR. Some can be activated by other stresses, while others are used as intrinsic aspects of development and tissue homeostasis.

We have already encountered the phenomenon of cell-cycle checkpoints earlier in this book (Section 5.6 through 5.9). As described in those sections, the eukaryotic cell cycle is orchestrated by cyclins, proteins that vary in activity across the cell cycle, in concert with the *cyclin-dependent kinases* (CDKs), which phosphorylate target proteins in response to active cyclins. A variety of stresses, including DNA damage, can induce a temporary halt in the cell cycle, called a checkpoint, by plugging into this cyclin/CDK circuitry (Figure 13.2). In Chapter 5, we discussed the DNA replication or S-phase checkpoint, which halts the cell cycle during the S phase, and corresponding checkpoints exist at other critical junctures in the cell cycle. For example, the G1/S checkpoint controls entry into

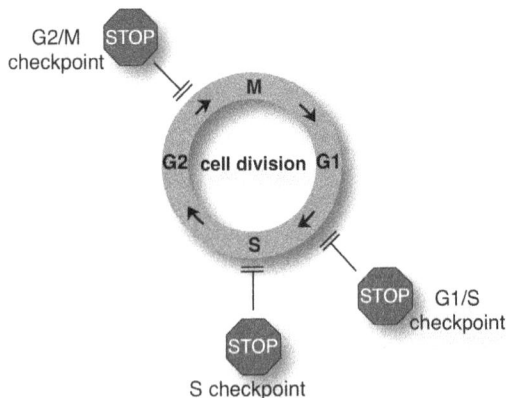

Figure 13.2. Cell-cycle checkpoints in eukaryotes. The eukaryotic cell cycle can be temporarily blocked by induced checkpoints, such as those schematized here (certain other checkpoints are not shown in this diagram).

S phase, and the G2/M checkpoint temporarily blocks entry into mitosis (Figure 13.2).

The rationale for cell-cycle checkpoints is pretty obvious when you consider the stresses that induce them. With regard to DNA damage, proceeding with the cell cycle is both more difficult and dangerous when the chromosome sustains damage. The S-phase checkpoint slows replication down to provide more time for repair pathways to mitigate the DNA damage (see below for more details). This delay reduces the number of encounters of the replication apparatus with damaged DNA, which is an inherently dangerous event. The delay in the cell cycle also provides time for the more complex repair reactions and replication-fork-reactivation pathways that become necessary when the replication fork encounters DNA damage. Ultimately, the S-phase checkpoint prevents cells from entering G2 phase and then mitosis with DNA that is only partially replicated, an event that can cause genetic instability and even cell lethality.

DDR responses also operate when cellular DNA is damaged outside of the S phase. The G1/S checkpoint halts the cell cycle just before cells start replicating their DNA. An interesting aspect of this checkpoint is that it is anticipatory. The S-phase checkpoint is activated when the process of DNA replication is judged to be blocked, a conceptually simple and direct negative regulatory event. The G1/S checkpoint, on the other hand, senses DNA damage itself and halts the cell cycle to prevent a difficult and dangerous replication process from ever beginning. Similarly, the G2/M checkpoint prevents cells from entering a mitotic division with damaged DNA. Particularly in the case of DSBs induced in G2 phase, mitotic division would lead to loss of chromosome arms in some daughter cells. Furthermore, G2 cells are better equipped than G1 cells to carry out DSB repair due to the presence of sister chromatids on each chromosome prior to mitosis.

Particularly in higher eukaryotes, apoptosis is another key cell-fate decision that can be impacted by DDRs. As you likely already know, apoptosis is an active, programmed cell-death pathway that eliminates damaged or somehow unnecessary cells from multicellular organisms. From the perspective of a multicellular organism, apoptosis can be understood as providing a great benefit to the organism as a whole even though it results in the destruction of a given cell. However, apoptosis also occurs in unicellular organisms, such as yeast, and therefore the apparent benefit of induced cell death can extend in some way to a population of cells (often genetically identical).

The induction of apoptosis by DNA damage is beneficial in that it eliminates a subset of cells with damaged genomes. Particularly since DNA damage can lead to cancer in metazoans, apoptosis is an important protective factor against tumor formation and metastasis (see Chapter 14). Many other stresses that damage cells in various ways can also induce apoptosis. Finally, apoptosis is an important element of many developmental pathways, where (undamaged) cells are eliminated from certain lineages to allow proper formation of differentiated tissue or organ systems.

The molecular pathway of apoptosis involves complex circuitry for activation and execution, which is fascinating but beyond the scope of this chapter. The downstream cellular events that lead to cell death during apoptosis include an influx of calcium, membrane blebbing, chromosome condensation, and endonucleolytic attack of the cellular DNA at the linkers between nucleosomes.

The induction of a cell-cycle checkpoint protects a cell while the induction of apoptosis destroys the cell, so how can they both be induced by DNA damage? This is again a complex area that is still being unraveled by scientists. One important observation is that the induction of apoptosis is highly variable in mammalian cells, being very dependent on the nature of the cell line, the presence or absence of activated oncogenes, and the differentiation status. Undoubtedly, the extent and the exact nature of the DNA damage also contribute to the checkpoint versus apoptosis decision.

Given high levels of DNA damage that exceed repair capabilities of the cell, another kind of cell death can occur, namely necrosis. Necrosis is generally not an active, induced, process, but rather the consequence of cells continuing to divide with unrepaired damage. In this sense, necrosis can be viewed as the consequence of failed or aborted checkpoint processes.

13.3 Activation and regulatory circuitry of the eukaryotic DDR

Three protein kinases trigger eukaryotic DDR pathways, sometimes in concert with each other (Figure 13.3). All three are large proteins (several thousand amino acids), evolutionarily related to phosphatidylinositol-3-kinase, and play a key role in sensing DNA damage or

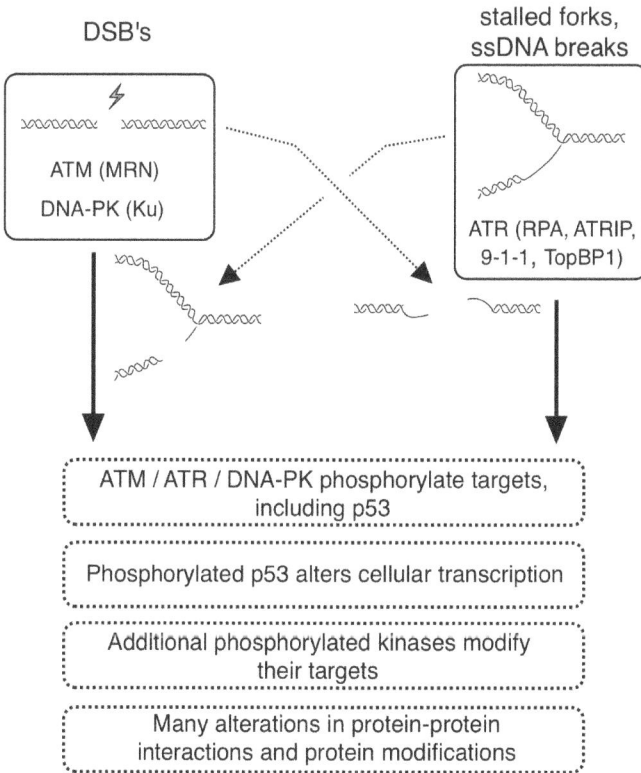

Figure 13.3. Summary of eukaryotic DNA-damage response (DDR) pathways. The kinases ataxia telangiectasia mutated (ATM) and DNA–PK trigger the primary response to double-strand breaks (DSBs), whereas ATR triggers the primary response to stalled replication forks and single-strand breaks (which lead to adjacent ssDNA regions). Some cross talk between the pathways occurs via resection from DSBs (rightward arrow) and breakage of replication forks (leftward arrow). While both major pathways result in the variety of outcomes in the dotted line boxes at the bottom, the details of these responses (e.g., complete collection of kinase targets) differ between the two DDR pathways.

aberrant DNA structures that result from the damage. We already encountered one of these proteins, DNA–PK, as a key factor that binds to DNA ends along with its partner Ku and participates in canonical *non*homologous *e*nd *j*oining (c-NHEJ) (Section 11.7). The second kinase is called ATM, a name derived from a human hereditary disease (*a*taxia *t*elangiectasia *m*utated) that causes a variety of severe phenotypes including cancer predisposition. The third, ATR, is named after its similarity to ATM and a fission yeast protein Rad3.

These DDR protein kinases recognize damaged DNA with the help of additional proteins. As already mentioned, Ku protein is the binding partner of DNA–PK, and Ku recognizes duplex DNA ends (Section 11.7). ATM also binds DNA ends, but in this case with the help of the MRN complex (specifically Nbs1 protein), which is involved in homologous recombination (Sections 11.2 and 11.5). Finally, ATR is recruited to RPA-coated ssDNA, which accumulates at blocked replication forks and other aberrant DNA structures. ATR binds to a partner protein called ATRIP (*ATR i*nteracting *p*rotein), which is required for ATR binding to RPA-coated ssDNA. In our discussion of eukaryotic translesion DNA polymerases, we encountered the alternative clamp called 9-1-1, which is induced after DNA damage (Section 12.8). This 9-1-1 clamp gets loaded by its cognate clamp loader at the junctions of double-stranded DNA and RPA-coated ssDNA (e.g., within stalled replication forks). A protein called TopBP1 in turn binds to the 9-1-1 clamp and plays a key role in ATR activation. Additional triggers have also been implicated in DDR responses, including proteins within the replisome itself such as DNA polymerases ε and α.

The earlier discussion implicates DNA–PK and ATM in the response to DSBs and ATR in the response to stalled replication forks (Figure 13.3). This is a good first approximation of the circuitry of the DDR, and as we will see, the responses to DSBs and stalled replication forks are distinct. However, there are overlaps and nuances regarding these triggers. First, extensive resection at a double-stranded end creates a long patch of RPA-coated ssDNA, allowing binding and activation of ATR. Furthermore, ATM binding to an MRN-bound double-strand end activates the very resection reaction that leads to ATR recruitment (Figure 13.3). Therefore, DSBs can lead to activation by both ATM and ATR under at least some conditions. Second, a stalled replication fork, targeted by ATR, can become broken as discussed in Section 11.6, providing a new DNA break as target for ATM (Figure 13.3). Third, there is evidence that ATM and ATR can phosphorylate each other, which may provide some cross talk between the pathways. Fourth, some of the downstream targets of ATM and ATR are shared. Finally, the relationship between DNA–PK and ATM should be considered. In Section 11.9, we learned that end resection is controlled in large measure by the cell cycle, with very limited resection in G1 favoring c-NHEJ, but extensive resection during and after DNA replication favoring homologous recombination. Correspondingly, DNA–PK and its Ku partner are active and effective in G1, but not so much later in the cell cycle when resection prevents Ku from binding. Instead, the binding of the MRN complex favors recruitment of ATM during and after S phase.

What happens once these regulators bind to their DNA targets? The immediate result is that the kinases become fully activated, triggering a cascade of further events. Each of the three regulators has its own detailed activation pathway, but these generally involve additional protein–protein interactions, autophosphorylation of the regulator, and further protein modifications (Figure 13.3). Once fully activated, the kinases induce the phosphorylation of many host cellular proteins, leading to the downstream effects discussed in Sections 13.4–13.6. In the ATM and ATR responses, hundreds of cellular proteins become phosphorylated, while DNA–PK induction has a more limited impact on targets. The DDR is a true signal-transduction pathway, encompassing a cascade of phosphorylation events by protein kinases. In particular, ATM and ATR activate several downstream protein kinases, which themselves have additional targets.

All three of the vertebrate DDR kinases, DNA–PK, ATM, and ATR, phosphorylate a key DDR regulatory protein that functions as a central hub of the DDR (Figure 13.3). Phosphorylation of the protein, called p53 (based on its apparent size in gels), increases its half-life and thus its cellular concentration, and activates the protein as a transcriptional regulator. Protein p53 regulates the expression of many proteins at the transcriptional level, and is also central to the execution of the G1/S checkpoint and apoptosis.[2] Protein p53 is famous for its role in cancer as a tumor-suppressor protein; indeed, it is often called the "guardian of the genome" due to its importance in the DDR. Patients with only one functional copy of the p53 gene present with a syndrome called Li–Fraumeni, characterized by greatly increased cancer incidence in early adulthood. Furthermore, most spontaneous human cancers have p53 defects, directly demonstrating that the normal function of p53 helps prevent cancer. Perhaps even more remarkable, elephants, which almost never get cancer, have 20 copies of the p53 gene in their genomes!

13.4 DDR activation alters both transcription and posttranslational events

As already implied in the discussion of p53, transcriptional regulation does play important roles in the eukaryotic DDR, although often in complex ways. An example of the complexity is that a subset of genes involved in the NER pathway are induced by DNA damage, but the subset that is induced varies between eukaryotic organisms and does not encompass all the NER genes. Furthermore, many genes that have no known roles in DNA repair or damage tolerance are induced by DNA damage, often for no obvious reason. A further complexity is that gene regulation is modulated in both direct and indirect ways. An example of indirect regulation is that the induction of a cell-cycle checkpoint alters expression of genes that are regulated by the cell cycle. Finally, in the yeast model system, there is little or no overall correlation between genes whose products are regulated by DNA damage and those that cause hypersensitivity to

---

[2]Yeast species do not encode a homolog of vertebrate p53; nonetheless, DNA damage in yeast induces a set of signaling events that are functionally similar to those in vertebrates.

DNA damage when inactivated. These results suggest that the baseline expression levels of many DNA repair and tolerance genes are sufficient to accomplish their functions (at least as measured in the laboratory).

All that said, there are massive transcriptional changes after DNA damage. Using modern molecular techniques, scientists observed that about a third of the yeast genes showed significant increases or decreases in transcription following a DNA-damage insult. In human cells, p53 appears to upregulate or downregulate a total of about 3% of human genes. Also keep in mind that other transcriptional regulatory proteins are modulated by the DDR, for example, by an activating phosphorylation event from one of the three above DDR kinases or their downstream kinase targets. In model systems such as yeast, and higher eukaryotic cells including humans, several specific gene-regulatory systems that respond to DNA damage have been studied in great detail, and the interested reader is encouraged to explore these systems in the literature.

In summary to this point, eukaryotic DDR responses involve a cascade of protein-kinase reactions that result in phosphorylation of hundreds of proteins. These phosphorylation events alter the activities and/or stabilities of these proteins, causing further downstream effects. A number of other posttranslational protein modifications are also executed, and some of the most important of these lead to induced proteolysis and destruction (or alteration) of specific proteins. Finally, transcriptional alterations are also induced by posttranslational modification of regulators like vertebrate p53. Note that the transcriptional alterations will generally impact the cell environment more slowly than the various posttranslational modifications, because it takes time to alter the steady-state level of specific transcripts and even more time for this to impact the overall cellular concentration of a protein gene product. As discussed in Section 13.2, the end result of a DDR can involve execution of a checkpoint response or even apoptosis, each involving their own unique cascade of translational and posttranslational events.

13.5 DNA-repair pathways are activated by the DDR

Many repair pathways are activated by the DDR, often by mechanisms involving phosphorylation (or other posttranslational modifications) of specific repair proteins. Given the large number of repair- and replication-related proteins that are targets of the DDR kinases described earlier, it seems likely that every major repair pathway is modulated in important ways by DDRs. Here are a few examples, some of which have already been mentioned in this book. First, ATM binding to an MRN-bound double-strand DNA end promotes end resection, a key step in homologous recombination (Section 13.3). Second, autophosphorylation of DNA–PK, a key factor in c-NHEJ, promotes progression of the c-NHEJ pathway (Section 11.7). Third, the 9-1-1 clamp complex is induced by the DDR and activated by phosphorylation, and the activated clamp enables reactions such as translesion synthesis (Sections 12.8 and 13.3). Fourth, RPA is phosphorylated during the DDR, likely promoting its action in various DNA-repair pathways.

The DDR also promotes DNA repair by causing an increase in deoxynucleoside triphosphate (dNTP) levels in the cell. The increased dNTP levels are created by activation of the key enzyme ribonucleotide reductase, which converts ribonucleotide diphosphates into deoxyribonucleotide diphosphates (which are then converted to the triphosphate form by a nucleotide diphosphate kinase reaction). Ribonucleotide reductase is regulated at the transcriptional level, with its specific transcriptional repressor being inactivated by phosphorylation as one of the downstream targets of the DDR-kinase cascade. However, the regulation is much more complex than this, with ATM- and ATR-directed responses stabilizing the enzyme, small protein inhibitors of the enzyme being modulated, and the location of the enzyme within the cell being redirected. With these multiple levels of regulation, the increased dNTP levels must be of great importance for successful DNA-damage responses.

There is striking evidence that the DDR is particularly important for the repair of very difficult and/or complex lesions in the DNA molecule. The best-studied examples involve repair of DNA interstrand crosslinks, a form of DNA damage that is induced by certain bifunctional crosslinking agents that are often used in cancer chemotherapy. Multiple repair pathways have been described and are currently under intensive study, and these complex pathways can each involve some combination of replication fork processing, NER, MMR, translesion synthesis, and homologous recombination, acting in a sequential and carefully orchestrated order.[3] At least one such pathway is triggered when a replication fork is blocked by an interstrand crosslink, and the repair sequence removes the damage and ultimately restarts the replication fork. The steps of these pathways are highly regulated by elements of the DDR, presumably because careful coordination of such multistep sequences is critical for success. Patients with the debilitating disease called *Fanconi* anemia, characterized by multiple developmental defects, cancer predisposition, and early death, have defects in DNA-crosslink repair. The proteins involved in the process are therefore named the FA complex, with individual protein names such as FANCA and FANCB.

13.6 The process of DNA replication is renovated by the DDR

The S-phase checkpoint has major effects on the process of DNA replication at multiple levels. An early indication of the impact of the S-phase checkpoint came from studies of the rate of DNA synthesis after radiation treatment. Normal human cells show a dramatic reduction in DNA synthesis after radiation, but cells from patients with ataxia telangiectasia, defective in ATM activity, show no such reduction and are said to exhibit radio-resistant DNA synthesis. It should be stressed that ATM-deficient cells are hypersensitive to DNA damage and display genome instability, and so this abnormal radio-resistant DNA synthesis likely contributes to the problems suffered by ATM-deficient cells.

---

[3] For recent review, see Walden and Deans (2014).

As already discussed in Sections 5.6 and 5.9, the S-phase checkpoint modulates origin usage in a complex manner. Origins close to a blocked replication fork can be activated to assist in the completion of the local replication. This activation is particularly important when the unreplicated region is located in between two blocked forks. At the same time, more distant origins and origins that tend to fire later in the S phase are temporarily repressed, presumably to prevent the generation of additional blocked replication forks while damage persists. Origin firing during S phase requires a series of phosphorylation events that are coupled to the loading of replication proteins at origins (see Section 5.8). The repression of origin activity by the S-phase checkpoint interferes with these phosphorylation events, as a downstream consequence of the DDR-kinase cascade.

In addition to these effects on origin firing, the behavior of individual replisomes is also altered with the S-phase checkpoint. The pathways and mechanisms impacting replication fork behavior in this situation are currently under intense investigation by scientists, and there is much to learn. There is some evidence that the ATR response helps to preserve the overall replisome architecture, that is, prevent the loss of key components like DNA polymerases when the fork is stalled. There is also direct evidence that aberrant replication fork structures are generated after DNA damage in the absence of a proper checkpoint response. One such structure involves extensively regressed replication forks, indicating the uncontrolled action of a branch-specific DNA helicase (see Sections 3.7 and 12.1; also see below). Presumably, the S-phase checkpoint either helps to efficiently reverse fork regression or prevent regression in the first place. Another involves broken replication forks, presumably created when branch-specific nucleases attack stalled forks. Correspondingly, the S-phase checkpoint must either help to repair broken forks when they occur or prevent fork breakage from occurring.

The S-phase checkpoint also appears to activate novel and useful pathways to repair damage and reconstitute active replication. Several important DNA helicases that can catalyze branch migration and replication-fork regression are phosphorylated during the checkpoint, activating their activity. Three such prominent helicases are the products of genes that are mutated in human inherited diseases, authenticating their important cellular function: Werner's syndrome, Fanconi anemia complementation group M, and Schimke immunoosseous dysplasia (see Chapter 14). While the details remain to be elucidated, these helicases likely activate multiple pathways for restoring a normal replication fork. We have discussed these pathways previously, namely template-switching pathways to sidestep damage (Sections 3.7, 11.6, and 12.1), fork restart by homologous recombination (Section 11.6), utilization of translesion DNA polymerases (Chapter 12), and repair of interstrand crosslinks coupled to renewal of DNA replication (Section 13.5 just above).

13.7 Chromatin structure and behavior change during the DDR

One of the most dramatic results of an S-phase checkpoint triggered by either DSBs or stalled forks is a covalent modification of nucleosomes, which occurs in a region of hundreds of kilobases surrounding the activating lesion (Figure 13.4; see "How did they

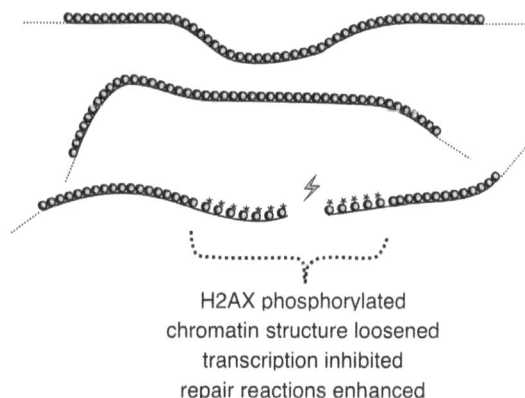

H2AX phosphorylated
chromatin structure loosened
transcription inhibited
repair reactions enhanced

Figure 13.4. Localized accumulation of phosphorylated H2AX near a double-strand break (DSB). The three lines represent duplex DNA, with the small balls representing nucleosomes and H2AX phosphorylation by asterisks.

test that?" at the end of this chapter). A minor histone variant called H2AX constitutes about 10% of the H2A histone in human cells. In the DDR, this minor variant becomes phosphorylated on residue serine-139 to form the so-called γ-H2AX, which is localized on the chromatin surrounding the break. All three of the DDR kinases discussed earlier, DNA–PK, ATM, and ATR, can phosphorylate H2AX in this way. The large local patch of γ-H2AX in the chromatin surrounding a site of damage serves to facilitate the various repair and tolerance pathways discussed earlier. This involves both a "loosening" of the chromatin structure to provide easier access as well as many specific protein–protein interactions to recruit repair machinery to the vicinity. In addition, transcription is inhibited within the modified chromatin domain, which presumably prevents complications that might occur if a transcription complex barges into a stalled replication fork, a repair complex, or a broken DNA end.

In addition to this localized effect on chromatin structure, the S-phase DDR also leads to a global change in chromosome behavior. The mechanism is currently unknown, but all chromosomes show an increased mobility around the nucleus. Increased chromosome mobility presumably facilitates multiple processes that assist in survival after DNA damage. For example, broken chromosome ends should have an easier time finding homologous sequences for recombination or partner ends for c-NHEJ. Also, DNA breaks should have an easier time migrating to the repair factories discussed in Sections 11.5 and 11.6.

2001 Nobel Prize in Physiology or Medicine

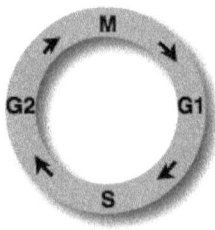

This prize was awarded to **Leland H. Hartwell, R. Timothy (Tim) Hunt** and **Paul M. Nurse** for their discoveries concerning key regulators of the eukaryotic cell cycle.

https://www.nobelprize.org/prizes/medicine/2001/summary/

13.8 Summary of key points

- The bacterial SOS pathway is induced when the transcriptional repressor called LexA undergoes self-proteolysis induced by RecA protein bound to ssDNA, leading to increased transcription of roughly 20 genes involved in DNA repair, damage tolerance, slowing cell division, and other processes.
- RecA-induced proteolysis is also involved in activating translesion DNA polymerases and inducing propagation of latent viruses such as bacteriophage λ.
- In eukaryotes, DNA damage can induce cell-cycle checkpoints, apoptosis, and necrosis.
- Three different protein kinases trigger eukaryotic DNA-damage-response pathways, DNA–PK, ATM, and ATR.
- Activation of DDR pathways involves end recognition by DNA–PK with its Ku70/80 subunits, end recognition by ATM with help from the MRN (MRX) complex, and binding of ATR to RPA-coated ssDNA.
- DNA–PK and ATM respond to DSBs and ATR responds to stalled replication forks, leading to distinct DNA-damage responses; however, there is also significant overlap between these pathways.
- ATM and ATR phosphorylate downstream effector proteins, including other kinases that extend the protein-phosphorylation cascades.
- All three of the abovementioned kinases phosphorylate p53, which functions as a central hub in the DNA-damage response and has been called the guardian of the genome.
- Activation of eukaryotic DDR results in numerous alterations in transcription and posttranscriptional events, modulating DNA repair, DNA replication, chromatin structure, and other aspects of cell physiology.

Further Reading

Ciccia, A., & Elledge, S. J. (2010). The DNA damage response: Making it safe to play with knives. *Mol Cell, 40*(2), 179–204.

Kastenhuber, E. R., & Lowe, S. W. (2017). Putting p53 in context. *Cell, 170*(6), 1062–1078.

Kreuzer, K. N. (2013). DNA damage responses in prokaryotes: Regulating gene expression, modulating growth patterns, and manipulating replication forks. *Cold Spring Harb Perspect Biol, 5*(11), a012674.

Maréchal, A., & Zou, L. (2013). DNA damage sensing by the ATM and ATR kinases. *Cold Spring Harb Perspect Biol, 5*(9), a012716.

Maréchal, A., & Zou, L. (2015). RPA-coated single-stranded DNA as a platform for post-translational modifications in the DNA damage response. *Cell Res, 25*(1), 9–23.

McIntosh, D., & Blow, J. J. (2012). Dormant origins, the licensing checkpoint, and the response to replicative stresses. *Cold Spring Harb Perspect Biol, 4*(10), a012955.

Rogakou, E. P., Boon, C., Redon, C., & Bonner, W. M. (1999). Megabase chromatin domains involved in DNA double-strand breaks in vivo. *J Cell Biol, 146*(5), 905–916.

Sirbu, B. M., & Cortez, D. (2013). DNA damage response: Three levels of DNA repair regulation. *Cold Spring Harb Perspect Biol, 5*(8), a012724.

Walden, H., & Deans, A. J. (2014). The Fanconi anemia DNA repair pathway: Structural and functional insights into a complex disorder. *Annu Rev Biophys, 43*, 257–278. doi:10.1146/annurev-biophys-051013-022737

How did they test that?
Double-strand breaks induce nearby patch
of phosphorylated H2AX

Previous studies showed that histone H2AX becomes phosphorylated after ionizing radiation exposure. Rogakou *et al.* (1999) generated an antibody that specifically binds to phosphorylated H2AX (γ-H2AX). Using western immunoblots, they showed that γ-H2AX is produced after irradiation in cells from a number of species, including humans and an Indian muntjac (with only six chromosomes, which will become important below). Next, human fibroblast cells were exposed to ionizing radiation, permitted to recover for various periods of time, and then fixed and immunostained. The subcellular localization of γ-H2AX was determined with laser-scanning confocal microscopy. The cells showed a substantial number of γ-H2AX foci after exposure to 0.6 Gy (Gray) (top left figure). The foci appeared within three minutes and many persisted through 15, 30, and 60 minutes (panels B, C, D, and E; cells in panel A are not irradiated); the number decreased by 180 minutes (panel F). The numbers of foci were substantially higher when cells were exposed to higher radiation doses (2 and 22 Gy; panels G, H). Next, the scientists created localized double-strand breaks (DSBs) by dragging a special UV laser beam over particular paths (instrument called LaserScissors^{TM}). At the highest dose (top right figure, 30% line), the line of laser exposure created an impressive collection of foci in and adjacent to the laser path. Lower doses (1% and 10%) produced a lesser number of foci. Finally, cells from the muntjac species were used to examine the localization of γ-H2AX foci when cells condense their chromosomes during mitosis; the small number of chromosomes made visualization of discrete chromosomes practical. The cells were exposed to ionizing radiation (0.6 Gy) and allowed to recover, and then cells showing mitotic chromosomes were found after immunostaining. γ-H2AX foci were evident on the condensed chromosomes (bottom figure). In a subset of cells, broken chromosomal arms (unattached to the mitotic apparatus) were found to each have a prominent γ-H2AX focus on their end (green arrow). Thus, γ-H2AX foci accumulate at the site of DSBs. Figures were reproduced from Rogakou *et al.* (1999), with permission from Rockefeller University Press; permission conveyed by Copyright Clearance Center, Inc.

X-ray-induced breaks

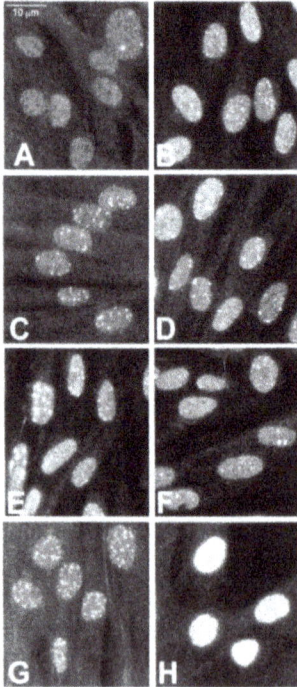

Laser-induced breaks

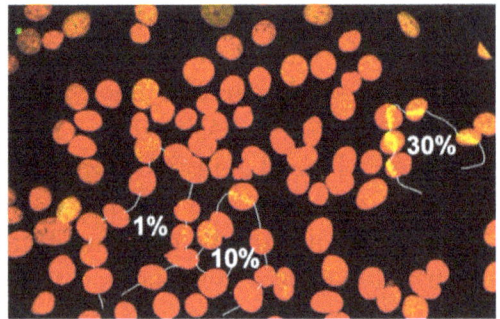

Mitotic chromosomes

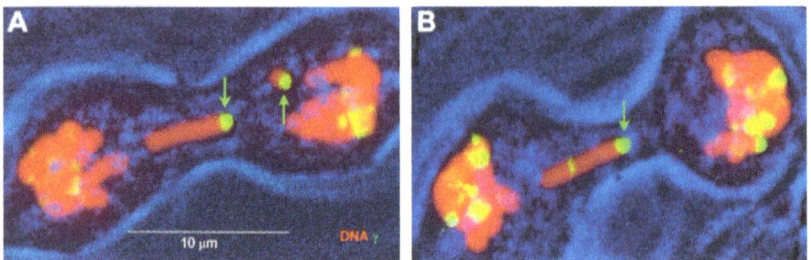

Chapter 14

DNA replication and repair in human disease

Throughout this book, we have encountered examples of inherited human diseases that are caused by mutations that affect the replication and repair machineries. In this chapter, we will pull these threads together into a more comprehensive view of the impact of these machineries on human health and disease. The clinical outcomes of inherited mutations can be both severe and widespread, with problems in development, immune response, neurological function, cancer, and aging.

Cancer predisposition is one major theme of this chapter. Cancer is nearly always dependent on the sequential accumulation of several mutations in the history of the tumor-cell lineage. In order for tumors to grow, thrive, and metastasize, multiple processes need to be inactivated or dramatically altered, including the cell response to growth controls, alteration of cell-cycle checkpoints and apoptosis, changes in nutrient uptake and metabolism, angiogenesis, and controls on cell migration to distant sites in the body. The frequency of mutation in human cells is normally low enough to make the accumulation of so many sequential mutations highly unlikely. As we will see, however, one of the early mutations in many or all cancers is one that itself increases the frequency of subsequent mutations, thus increasing the odds of the accumulation of multiple additional mutations. Cancer is thus fundamentally a genetic disease, and the lineage of each individual cancer can be considered as an evolutionary process. We will consider later how inherited mutations in replication and repair genes can, in a sense, jump-start and accelerate this evolutionary process toward cancer.

While scientists have learned a great deal about how mutations in replication, repair, and damage-response genes impact cancer and other diseases, much remains to be learned about how these defects relate to important clinical aspects. For example, most cancer-predisposition syndromes do not uniformly increase the incidence of all types of cancer, but rather show poorly understood tissue specificity. In addition, there is often uncertainty about why particular developmental defects are associated with mutations in one replication/repair gene and not another. Finally, it is worth pointing out that only a small subset of human cancers is found in patients with a known cancer-predisposition syndrome. We do not yet know how many of the remaining cancers are dependent on unrecognized mutations that modestly reduce the function of some important gene (often called polymorphisms), or a currently unrecognized predisposition gene, or perhaps on a heterozygous state for a known tumor-suppressor gene.

Later in this chapter, we will briefly review the remarkable insights that have recently begun to emerge from the use of modern DNA-sequencing methodologies on individual cancers. Sequencing results are now used to categorize tumors, sometimes providing valuable information on the etiology, prognosis, and the choice of anticancer therapy. Increasingly, knowledge about the machineries of replication, repair, and damage response is impacting and informing anticancer therapy in the clinic.

14.1 Inherited defects in replication machinery cause developmental defects

As you might expect, knockout mutations in the genes that encode proteins at the replication fork are generally lethal in model systems such as yeast and mouse. Such mutations (in homozygous form) would generally result in early embryonic lethality in humans and so would not easily be found among the spectrum of inherited human disorders. Nonetheless, certain mutations in genes in the replication process have been found in inherited human disorders.

A number of deletions that cause a syndrome due to haploinsufficiency[1] in humans remove one or another gene that encodes a replication protein (e.g., subunits of RPA and RFC clamp loader), along with other genes. However, it is not clear whether (and to what extent) the replication-protein deficit contributes to the disease or developmental characteristics of the syndrome. Recently, it was found that individuals with mutations in both copies of the gene for DNA ligase 1 suffer from immune deficiencies; these are mutations that lead to reduced (but not totally absent) DNA ligase enzymatic activity.[2]

Mutations in the prereplication complex have recently been found to cause another inherited human disease, Meier–Gorlin syndrome. This syndrome causes developmental problems including microcephaly (small head) and dwarfism in humans. Affected individuals have mutations in both copies of the gene for one of the ORC proteins or the two protein factors that help the ORC complex load the MCM complex (see Section 5.7). The severity of the disease is quite variable, likely because the individual mutations cause variable defects in the affected protein.

14.2 Dozens of human diseases relate to inherited mitochondrial defects

More than three dozen inherited human diseases are caused by improper mitochondrial function, and some of these relate to mitochondrial DNA replication and repair. As you know, the mitochondria are the powerhouses of the cell, generating cellular adenosine

---

[1] Haploinsufficiency occurs when only one copy of a functional gene in a diploid organism causes some disease or other phenotype; two functional copies are required to produce sufficient gene product for normal physiology.

[2] DNA ligase 1 plays a major role in joining Okazaki fragments and also has roles in certain DNA repair pathways.

triphosphate (ATP), and have many other important functions in biology. The diseases related to dysfunctional mitochondria vary greatly in severity and tissue pathology, due to variation in the nature and the extent of the molecular defect and by the differing functions and needs of various tissues. Tissues that require the greatest amount of energy, such as muscle, heart, brain and lungs, are generally the most severely affected. The range of symptoms and disease pathology are thus very complex, in any case not very well understood, and in many cases well beyond the scope of this chapter. Inherited mitochondrial disease is fairly common, with a prevalence of roughly one in four thousand births.

In evolution, mitochondria are derived from some early bacterial endosymbiont that invaded primordial eukaryotic cells, and mitochondria still have their own DNA genome. Like most bacterial genomes, the mitochondrial DNA is circular; like the nuclear genome, mitochondrial DNA must replicate during each cell-division cycle. The mitochondrial DNA encodes some of the mitochondrial proteins, but nuclear genes encode the majority of these proteins including those involved in mitochondrial DNA replication. Somehow, during evolution, these genes were lost from the mitochondrial DNA but acquired in the nucleus. This genetic arrangement is reflected in the inherited human mitochondrial-disease syndromes. Roughly 15% of the inherited diseases are due to mutations in the mitochondrial DNA, and these are inherited in a strictly maternal fashion, reflecting the source of mitochondria in fertilized eggs. The rest follow the classic Mendelian inheritance patterns of nuclear genes. Interestingly, most patients with an inherited mitochondrial DNA mutation have a mix of wild-type and mutant mitochondrial genomes,[3] which adds to the disease complexity for that class of patients.

Inherited mitochondrial diseases affect many different mitochondrial functions, including oxidative phosphorylation, mitochondrial protein synthesis, and replication of the mitochondrial genome.

---

[3] Human cells have multiple mitochondria and each mitochondrion contains multiple copies of mitochondrial DNA, allowing a mixture of wild-type and mutant genes even within a single cell.

Mutations affecting mitochondrial DNA replication are both clinically important and fascinating from a molecular viewpoint. The first such mutation was discovered in 2001 and affected the catalytic subunit of the mitochondrial DNA polymerase, which is called polymerase γ. The gene encoding this subunit is called *POLG*, and many different *POLG* mutations have now been found to cause a wide variety of mitochondrial diseases including devastating childhood diseases and multiple types of neurodegenerative disease. The mitochondrial DNA polymerase has an accessory subunit that promotes enzyme processivity, encoded by the gene *POLG2*, and mutations in this gene have also been found to cause mitochondrial-based diseases.

Another set of inherited mitochondrial replication disorders has been traced to mutations in the mitochondrial replicative helicase. Missense mutations in the helicase are associated with progressive external opthalmoplegia, spinocerebellar ataxia, and epileptic encephalopathy. A fascinating aspect of at least some of these helicase mutations is that they lead to a dramatic increase in mitochondrial deletion mutations, which are presumably caused by aberrant DNA replication and which contribute to the disease symptoms.

Additional mutations that cause mitochondrial dysfunction syndromes have been found in the genes that encode a translesion DNA polymerase that functions in DNA repair, an exonuclease implicated in mitochondrial DNA maintenance, several enzymes involved in generating deoxynucleoside triphosphate (dNTP) precursors for mitochondrial DNA replication, and two enzymes that function in both nuclear and mitochondrial DNA replication.

14.3 Nucleotide-repeat expansions cause neurological and developmental diseases

Depending on which types of sequences you include, repetitive DNA comprises 30%–50% of the human genome. A subset of this repetitive DNA consists of segments of short tandemly repeated sequences, for example, 30 tandem copies of the sequence 5′-CAG-3′ (5′-CTG-3′ on the opposite strand). In turn, a handful of nucleotide-repeat segments are critically important in certain human diseases, the nucleotide-repeat disorders (Figure 14.1). A recent review lists over three dozen nucleotide-repeat disorders, the most famous of which is *H*untington's *d*isease (HD) (which involves the 5'-CAG-3' repeat just mentioned). A unifying feature of these diseases is that alterations in the number of copies, usually involving expansion of the repeat length, are central to the expression of the disease.

Most of the nucleotide-repeat disorders involve trinucleotide repeats like the one in HD (Figure 14.1). However, some are known to involve different numbers of nucleotides, including tetra- and pentanucleotide repeats. The nucleotide repeats in these diseases affect the expression of a particular gene or the function of its product. Some are located within the coding region of the gene in question, some in the upstream regulatory region, and some in

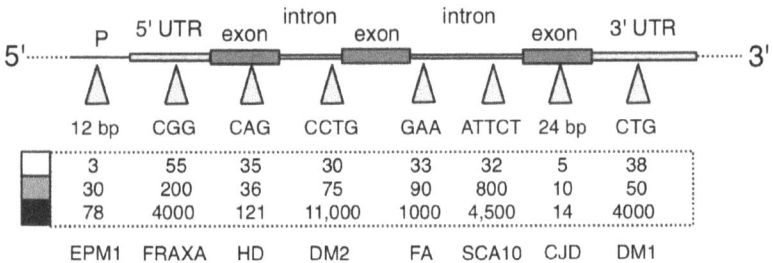

	12 bp	CGG	CAG	CCTG	GAA	ATTCT	24 bp	CTG
	3	55	35	30	33	32	5	38
	30	200	36	75	90	800	10	50
	78	4000	121	11,000	1000	4,500	14	4000
	EPM1	FRAXA	HD	DM2	FA	SCA10	CJD	DM1

Figure 14.1. Summary of some nucleotide-repeat-expansion diseases. A "generic" gene is depicted to summarize the locations of the repeats of various repeat-expansion diseases (in reality, each of these repeat loci are in distinct genes). The length or sequence of the repeating unit is shown just below the triangles. The three numbers in the dotted box represent the copy number in normal individuals (top), the lowest copy number known to be associated with disease (middle), and the highest copy number that has been observed and is associated with disease (bottom). The copy numbers for CJD is from Maizels, 2015, while the others are all from Zhao & Usdin (2015). Disease abbreviations: CJD, Creutzfeld–Jacob disease; DM1, myotonic dystrophy type 1; DM2, monotonic dystrophy type 2; EPM1, progressive myoclonic epilepsy 1; FA, Friedreich ataxia; FRAXA, fragile X syndrome; HD, Huntington disease; SCA10, spinocerebellar ataxia 10.

noncoding segments such as introns or 5′ or 3′ untranslated regions (Figure 14.1). In some cases, the disease is brought on when expansion of the repeat length leads to loss of function of the protein, while in other cases, repeat expansion leads to a toxic protein. This is the case in HD, where expansion of the 5′-CAG-3′ repeat adds to the length of a polyglutamine tract in the huntingtin protein (5′-CAG-3′ is the codon for glutamine). The normal huntingtin protein has 36 or fewer tandem copies of glutamine, but when the length increases to >36, the protein becomes toxic. By mechanisms that are still under investigation, the abnormal protein leads to neuronal death, with a progression of devastating symptoms. The symptoms of full-blown HD have been compared to those of ALS, Parkinson's and Alzhiemer's *combined.*

The biology and pathology of the various diseases related to nucleotide repeats is much too complex for this Chapter, but a few general comments are worthy of note. Most of the diseases involve neurological disorders and developmental problems, although the pathological mechanisms appear quite distinct. Comparing the different diseases, there is significant variation in the number of repeats necessary for the disease state and sometimes in the apparent mechanisms of repeat expansion (see below). As implied earlier, there is also variation in whether the repeat exerts its effect through a change in the protein, the RNA, or just the expression of the gene in question.

The inheritance pattern of nucleotide-repeat diseases has both conventional and very unusual characteristics. As with a number of other inherited diseases, the offspring of an affected individual have a 50% chance of having the disease. This pattern of inheritance is typical for autosomal-dominant mutations, where a heterozygous individual with one mutant and one normal copy of the gene would have the disease, and, on average, half of their offspring would inherit the chromosome with the mutant gene.

The unusual aspect of the inheritance relates to the generations that precede the occurrence of the disease in a particular family, and the progression of disease severity. The clinical features of this pattern were noticed before the molecular explanation became clear. The offspring in sequential generations appeared to show progressively more severe disease symptoms than their affected parent, as well as an earlier onset of the disease. This phenomenon was called "anticipation." The molecular side of anticipation involved changes in the repeat-unit copy number. Take the example of HD in Figure 14.1 (see "How did they test that?" at the end of this chapter). Most normal individuals have fewer than 28 copies of the trinucleotide repeat in each chromosome, but a subset of unaffected individuals has from 28 to 35 copies in one of their two genes encoding huntingtin. One remarkable feature is that as the number of repeats increases, the frequency of repeat-expansion events also increases. Thus, the clinically normal individuals with 28–35 copies have an increased propensity for expansion events in that gene, which increase the number of repeats in their progeny who inherit that gene. As the number increases to and somewhat beyond the threshold of 36, disease symptoms become evident but generally in a modest form and late onset. However, these individuals have an even greater propensity toward additional expansion events, and so their progeny will progressively inherit higher and higher copy numbers if they inherit the affected gene. As the number of repeats rises into the 40s and beyond, the disease expresses in a more and more severe form and earlier onset. So, at the molecular level, anticipation involves both the increase in repeat copy number and the progressively increasing propensity for further expansion.

Each of the nucleotide-repeat diseases has their own particular repeat thresholds and expansion characteristics (Figure 14.1). With another disease, called *Fra*gile *X* syndrome (FRAXA in Figure 14.1), normal individuals have up to 200 copies of the repeat, while the number of repeats in affected individuals can reach 4000. For many of the diseases, the repeat expansion occurs in steps of a few at a time, but there are also diseases where successive generations show increases of thousands of repeats compared to their parents!

Nucleotide-repeat diseases are relevant for this book because DNA-metabolic processes are responsible for repeat expansion. Strong evidence implicates DNA replication, various forms of repair, and translesion DNA polymerase action in repeat expansion (and contraction). Many different mechanisms have been proposed, largely from research in the usual model systems (*Escherichia coli,* yeast, mouse). However, scientists are still trying to test the veracity of the various mechanistic models, and determine whether different mechanisms are at play with the different forms of nucleotide repeat and (quite likely) in the different tissues. Essentially, all of the mechanistic models involve some kind of slipped-strand pairing and the formation of an unusual structure in which the repeat region pairs with itself (intrastrand). The intrastrand pairing is not a perfect Watson–Crick helix, for example, the A residues are mispaired in the 5′-CAG-3′ hairpin (Figure 14.2). Nonetheless, alternative structural configurations can accommodate this partial pairing, and some of the nucleotide-repeat sequences can form even more unusual self-pairing structures such as a four-strand structure called a "G-quadruplex" (Figure 14.2B).

The simplest models for changes in repeat length involve slipped-strand pairing during a DNA polymerase reaction. Depending on whether the template or primer strand is extruded from the helix, a repeated contraction or expansion (respectively) can result (Figure 14.2C). Note that a second round of DNA replication (or some other event) is needed to convert the extrusion-containing duplex into a heritable mutant form (last step in Figure 14.2C).

Most of the models for repeat expansion in human cells are more complex than this, and indeed repeat expansion can occur in resting cells that are not undergoing DNA replication. The processes of *b*ase *e*xcision *r*epair (BER), *n*ucleotide *e*xcision *r*epair (NER), *mis*match *r*epair (MMR), and translesion DNA synthesis have all been implicated in one or another model for nucleotide-repeat expansion. Recall that many of the diseases being discussed result in neurological problems, and also recall from Chapter 11 that oxidative DNA damage is a particularly significant problem in the brain. Thus, it has been proposed that faulty excision repair of oxidative DNA damage in neuronal tissues plays a key role in nucleotide-repeat expansion. One possible pathway involves a longer than normal excision tract, leading to the formation of secondary struc-

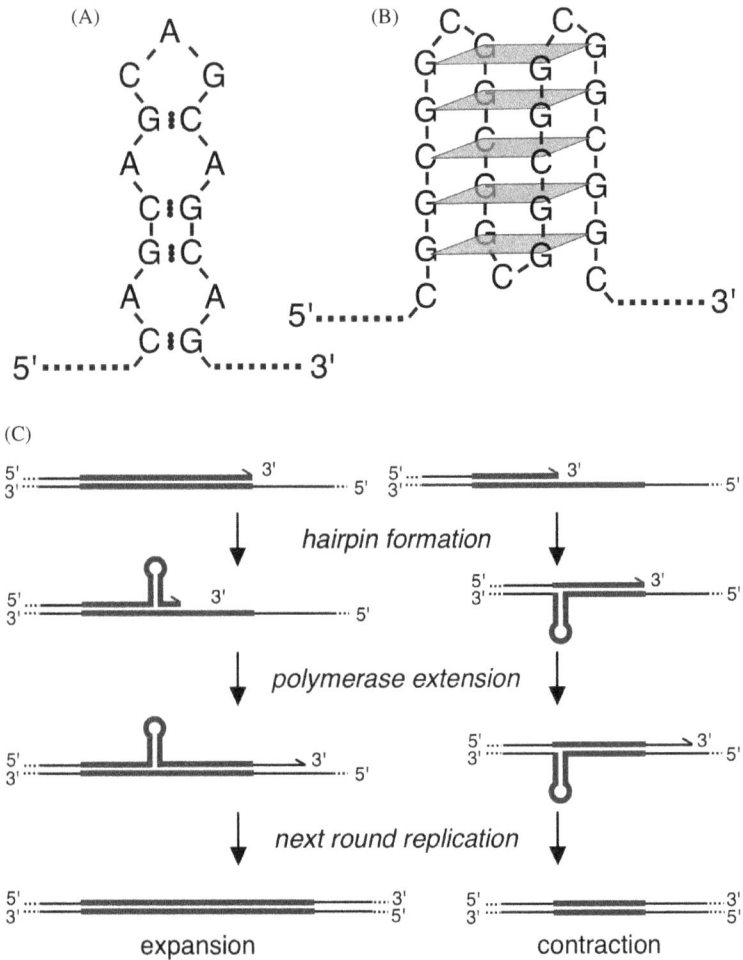

Figure 14.2. Unusual DNA structures involved in nucleotide-repeat biology. See text for discussion. This figure was modified from figures published by Mirkin, 2007.

ture on the displaced DNA strand (Figure 14.3). Normally, this displaced strand would be cleaved by nucleases like flap endonuclease FEN1, the Okazaki fragment processing nuclease (see Chapter 4). However, the unusual intrastrand pairing of these nucleotide repeats has been shown to inhibit this kind of nuclease cleavage, potentially leading to completion of the repair reaction with the

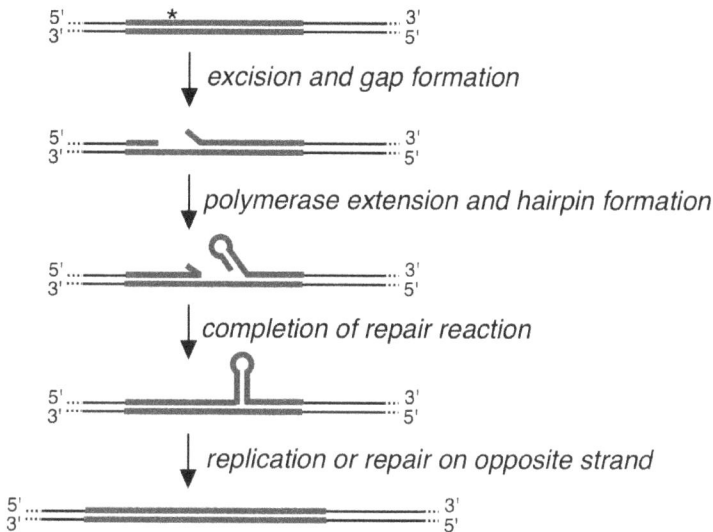

Figure 14.3. Model for repeat expansion associated with excision repair. See text for discussion. This figure was modified from a figure published by Mirkin (2007).

inclusion of the looped-out region (Figure 14.3). A subsequent cycle of DNA replication, or another repair reaction on the opposite strand, is proposed to convert this extrusion-containing duplex into a normal duplex containing an expanded number of repeats (last step in Figure 14.3). Based on the differing characteristics of nucleotide-repeat expansion in different diseases and tissues, it seems very likely that multiple mechanisms are at play in these diseases, and this discussion has only scratched the surface of an expansive and exciting field.

We focused above on repeat expansion in the germ line to explain the unusual inheritance patterns of these diseases. However, as implied earlier, repeat expansion also occurs in somatic tissues. In the case of HD, it has been shown that progressive expansion in brain cells contributes to disease progression in a patient. Thus, there is hope that some kind of "antiexpansion" therapy could be at least partially protective and alleviate some of the more severe symptoms of this horrible disease (without intervening in the germline transmission process). This hope highlights the importance of deciphering the precise mechanisms involved in nucleotide-repeat expansions.

14.4 Predisposition to sunlight-induced cancers due to NER deficiency

Xeroderma *pi*gmentosum (XP) was the first human DNA-repair disease discovered, and provided a useful paradigm for cancer formation. Patients with XP have severe sensitivity to the *u*ltra*v*iolet (UV) rays in sunlight, suffering benign, premalignant, and malignant skin cancers in exposed regions and ocular problems including neoplasms, as well as frequent neurological disturbances. The disease name relates to unusual features of the skin, xeroderma referring to a parchment-like quality and pigmentosum conveying abnormal pigmentation of the skin. The incidence of XP is roughly 1 in 100,000 births with significant variation between populations.

As described in Chapter 10, scientists interested in studying the repair of UV lesions realized that most of these clinical symptoms could be explained by defects in the repair pathway, and so began studying cells from XP patients. Sure enough, these cells were hypersensitive to killing by UV and certain other DNA-damaging agents and showed increased mutations and chromosomal rearrangements upon exposure to UV compared to cells from normal individuals. The cause of the neurological defects is still uncertain, but may relate to repair of oxidative DNA damage caused by the high oxidative activity in the brain. As described in Section 10.5, the defects in XP patient cells were traced to one of seven different NER proteins (XPA through XPG) or to the translesion DNA polymerase η, which plays an important role in bypass of UV-induced pyrimidine dimers (see Chapter 12).

The XP syndrome is inherited as an autosomal-recessive trait, meaning that affected individuals have a mutation in the relevant XP gene from each of their two parents. As is the case for other autosomal-recessive traits, consanguineous pairings result in a significant subset of XP patients in the population due to their tendency to produce homozygous offspring for any particular gene.

Regarding individuals with a heterozygous XP mutation (one mutant and one normal gene), scientists have found little or no evidence for an increased incidence of cancer. Thus, one functional copy of each of the XP genes appears to be sufficient for normal function.

What is the sequence of events that leads to cancer in XP patients? As with other cancers, a given skin cell must acquire multiple mutations to progress to cancerous forms. DNA sequencing from tumors of XP patients has revealed mutations in particular proto-oncogenes, including p53 (see Chapter 13) and others that are involved in growth regulation. Often, these mutations have occurred at pyrimidine dimer sites, which is not surprising given the defect in repairing UV-induced damage. There is some evidence that XP patients also have an increased incidence of cancer in internal organs, but the analysis is difficult because most XP patients die in the teens or earlier.

Another syndrome, called *Cockayne syndrome* (CS), is related to XP and also caused by defects in an aspect of NER. Like XP, CS is inherited as an autosomal recessive, but is much less common than XP in the population. CS patients also show heightened sensitivity to sunlight, along with severe symptoms such as developmental defects, premature aging, neurological degeneration, cataracts before the age of 3, and a mean age of death of about 12. The molecular defect in CS patients has been traced to mutations in factors involved in transcription-coupled NER (TC-NER; see Chapter 10). Without going into the complexities, there are also very rare individuals with a combined XP/CS syndrome, along with several other syndromes that are caused by very specific defects in certain NER or TC-NER proteins.

14.5 MMR defects cause predisposition to colon cancer

As described in Section 6.5, mutations in the genes that encode the MutS and MutL homologs were found to be the cause of Lynch syndrome, characterized by predisposition to colorectal cancer. Patients with Lynch syndrome are typically heterozygous for the mutation in the MMR gene, but cells in the lineage that lead to cancer in their bodies at some point become homozygous for the mutation (i.e., lose the wild-type copy of the gene). This is an example of the importance of the process called *loss of heterozygosity* (LOH) in cancer formation (see Section 11.5 and Figure 11.9).

Obviously, the loss of MMR from a particular cell lineage in humans leads to a large (100- to 1000-fold) increase in mutation frequency. As discussed at the beginning of this chapter, most cancers, including colorectal, require several mutations that cumulatively allow unbridled tumor cell growth. This is sometimes called the multihit hypothesis, and explains why a defect in MMR leads to cancer.

While Lynch syndrome is the most common inherited form of colorectal cancer, Lynch patients account for only a few percent of all colorectal cancer cases. However, many sporadic colon cancers from families with no predisposition also have characteristics similar to Lynch-syndrome cancers, and also turn out to have a defect in MMR. In these cases, the patient is homozygous for normal MMR genes, but the lineage that gave rise to the cancer evidently had some event(s) that inactivated both relevant copies of the gene. Interestingly, for many of these cancers, no mutation is found in the MMR genes even though the cells are demonstrably defective in the MMR process. This led to the discovery that about 15% of sporadic colon and endometrial cancers had undergone epigenetic silencing of the *MLH1* gene, which encodes one of the two subunits of the MutL homolog MutLα. Transcriptional silencing occurs by means of cytosine methylation at particular sequences within the promoter region of a gene, and can block most of the transcription and lead to loss of production of the protein.

The studies of MMR and cancer have led the way in understanding the relevance of genome-stability maintenance in preventing cancer. They have also been extremely useful in the clinical setting. For example, many cancers are now routinely screened for molecular defects in MMR, which provides an easier and more definitive identification of Lynch syndrome (or Lynch-like sporadic cancers) than clinical characteristics. In addition, MMR-defective cancers respond to specific chemotherapies in a different manner than MMR-proficient cancers, and therefore this information is used to tailor the most appropriate anticancer regimen for both types of cancers. Finally, the identification of a sizable fraction of cancers with epigenetic silencing of the *MLH1* gene has suggested a remarkable new approach to treatment. Certain nucleoside analogs such as decitabine inhibit cytosine methylation at promoters. Decitabine was found to reactivate expression of the silenced *MLH1* gene in those cancer cells, and these decitabine-treated cancer cells also recovered functional MMR. These findings have led to ongoing clinical trials to test whether this drug can improve the outcome for patients with this common class of cancer.

14.6 Complex developmental/cancer syndromes caused by helicase deficiencies

In Chapter 11, we discussed DNA helicases that act on branched DNA molecules, for example, catalyzing branch migration of Holliday junctions. In model systems, these enzymes have also been shown to contribute to remarkable reactions at stalled replication forks, including fork regression, lesion bypass, and replication-fork restart (see Chapters 3 and 12). Some of these helicases also play important roles in preventing recombination between two sequences that are very similar but not identical (often called "homeologous recombination"). One group of branch-migration helicases are often called RecQ family helicases, because they are evolutionarily related to a prototypical branch-specific *E. coli* helicase called RecQ.

As implied in Section 12.1, scientists are currently researching the precise roles of branch-specific helicases in assisting the process of eukaryotic DNA replication. While the data are too expansive and complex to summarize fully here, each of the RecQ family helicases has been implicated in multiple aspects of DNA metabolism. The roles often appear to be overlapping between the different helicases, and include aspects of replication initiation, elongation and termination, replication-fork restart, TLS polymerase action, repair pathways including BER, DNA-damage signaling, multiple steps in homologous recombination, and proper telomere function.

While we don't understand all the detailed and overlapping roles of these helicases in the eukaryotic cell, we know they are very important in nucleic-acid metabolism because mutational inactivation of specific helicases can cause striking cellular phenotypes. In addition, five different human syndromes are caused by mutations in three (of the five) human RecQ family helicases.

Bloom syndrome is a rare autosomal-recessive disorder characterized by cancer predisposition, dwarfism, facial distortions, reduced fertility, and hypersusceptibility to infections. Bloom-syndrome patients often develop multiple primary tumors. An interesting feature of Bloom syndrome is that the cancer predisposition is very generalized, affecting essentially all types of cancer in many tissues. Many other cancer-predisposition syndromes have a more restricted repertoire of tumor and tissue specificity.

Bloom syndrome is caused by mutations, usually knockout mutations, in the gene that encodes a RecQ family helicase called BLM. Thus, most Bloom patients have no functional BLM helicase in their cells. Given the above clinical data, BLM is apparently important in preventing cancer (i.e., its causative mutations) in most or all cells.

In the case of BLM helicase, recent studies have uncovered at least one of the key molecular functions that relate to cancer and likely some of the other clinical symptoms. For a long time, scientists knew that BLM-deficient cells show an abnormally high frequency of "sister-chromatid exchanges." These are chromosomal recombination events in which two arms of a chromosome show evidence of a crossover event that occurred after DNA replication was complete (Figure 14.4). Note that sister-chromatid exchanges reflect true homologous recombination, in that the sequences of the two arms are identical to each other. In Section 11.4, we discussed how homologous recombination can generate a double Holliday junction, and also how branch cleavage of the double Holliday junction can result in a crossover event. This is the pathway that can lead to a sister-chromatid exchange.

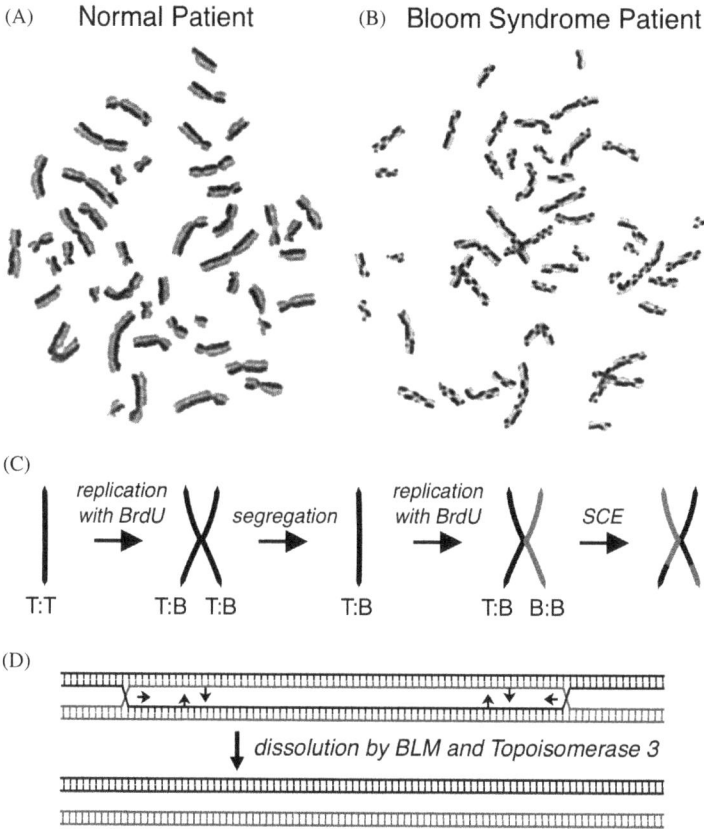

Figure 14.4. Frequent sister-chromatid exchanges in cells from Bloom syndrome patients. Sister-chromatid exchanges are evident from the color switches along the bivalent chromosomes from normal (panel A) or Bloom syndrome (panel B) cells. The micrographs in panels A and B are reproduced with permission from the Laboratory of Cytogenetics, Hospital for Sick Children, Toronto, Canada. Panel C summarizes the experimental approach used to visualize these exchanges. The cells are propagated for two cell-division cycles in the presence of the thymidine analog bromo-uridine (BrdU). The chromosomal DNA prior to these two cycles is composed of two thymidine-containing strands (T:T). After one cycle with BrdU, the DNA is substituted in one strand with BrdU (T:B). After two cycles, one daughter chromatid is T:B and the other is B:B (fully substituted with BrdU), and antibody staining can distinguish these two duplexes. Thus, a sister-chromatid exchange (crossover) in the lower arms leads to the switch in intensities of the chromosome diagramed at the right. Panel D depicts the dissolution of a double Holliday junction intermediate by the Bloom helicase and topoisomerase 3 (also see Figure 11.6C).

So, why do BLM-deficient cells show such an increased frequency of these exchanges? The recent studies alluded to above show that BLM participates in the dissolution pathway for resolving double Holliday junctions, which never leads to a crossover (Section 11.4; Figure 14.4D). Holliday-junction dissolution is catalyzed in vitro by the collaboration of BLM helicase and the enzyme topoisomerase 3 (Figure 14.4D), directly correlating with the increase in sister-chromatid exchange in vivo. Presumably, the alternative pathway of double-Holliday-junction cleavage often leads to sister-chromatid exchange and concomitant LOH, generating cells that are homozygous for mutations in cancer-relevant genes. There is also ample evidence that BLM helicase plays other important cellular roles, including in telomere biology and in replication-fork progression, and these could also play roles in inhibiting cancer formation.

Another rare autosomal-recessive disease is called *Werner* syndrome, and as in Bloom syndrome, many of the causative mutations are nulls that completely inactivate the protein (WRN helicase). Werner syndrome patients also suffer an increased susceptibility to cancer, but in this case, the cancers are specific ones that are often quite unusual in the general population.

One of the signature features of Werner syndrome is a remarkable premature aging, which is seen both at the level of the patient and in their cells. Young Werner patients suffer many symptoms common in the elderly, including cardiovascular disease, osteoporosis, cataracts, and the abovementioned cancer predisposition. With this increased disease load, most Werner patients die before the age of about 50. The premature aging appears to be related to dysfunction of telomeres, and in fact overexpression of the enzyme telomerase in cells from Werner patients can prevent the cellular aspects of senescence.

The WRN protein is a RecQ family helicase, but unlike the other RecQ helicases, also carries a domain with exonuclease activity. One proposed role for the exonuclease activity involves proofreading for certain TLS polymerases, which could increase the fidelity of TLS reactions. Like other RecQ family helicases, WRN helicase can unwind complex branched-DNA structures. Cells that lack WRN show an increased incidence of homeologous recombination (see above), leading to chromosomal deletions and translocations. Thus, WRN helicase appears to play an important role in ensuring the fidelity of homologous recombination. However, WRN-deficient cells do not show the increased incidence of sister-chromatid exchanges that are seen in BLM-deficient cells (see above). The precise roles of WRN helicase in cellular DNA metabolism are under intensive investigation, with experimental evidence supporting critical roles in telomere biology, replication-fork function, and homologous recombination.

Three other syndromes are caused by mutations in the gene encoding helicase RECQ4: *R*othmund–*T*homson *s*yndrome (RTS), Baller–Gerold syndrome, and Rapadilino syndrome. All three syndromes are characterized by developmental/skeletal abnormalities, although these abnormalities are somewhat distinct from each other in the three syndromes. One of the syndromes (RTS) also causes a predisposition to cancer (osteosarcoma) and a skin condition characterized by hypersensitivity to sunlight. The RECQ4 helicase has been implicated in multiple aspects of nucleic-acid metabolism, but one unique feature of this helicase among the RECQ family is that it localizes (in part) to mitochondria.

Two other mammalian helicases, while not in the RecQ family, also promote branch migration and play important roles in human physiology. Both appear to target the branched DNA of stalled replication forks, preventing a more catastrophic failure, and promoting the restart of replication. The first, FANCM helicase, binds specifically to a variety of branched-DNA structures including model replication forks. Accordingly, a key role of FANCM involves the recognition of replication forks that have been stalled by interstrand DNA crosslinks, the first step in DNA-crosslink repair (see Section 13.5). As the name implies, individuals with homozygous mutations in the gene encoding FANCM helicase suffer from *Fanc*oni anemia (see Section 13.5); even heterozygous *FANCM* carriers appear to suffer an increased incidence of certain (solid tumor) cancers.

The second important branch-migration helicase outside the RecQ family is called SMARCAL1. Recent studies have also uncovered a key role for this helicase in preserving genome integrity at stalled replication forks. Indeed, loss of SMARCAL1 was found to cause increased DNA-damage response (DDR) in the absence of exogenous DNA-damaging agents. Individuals with homozygous mutations affecting SMARCAL1 suffer from the syndrome called Schimke immunoosseous dysplasia, with developmental skeletal problems, kidney malfunction, and immune-system problems.

14.7 Mutations in DSB repair and DDR pathways cause complex syndromes

The proteins involved in repairing double-strand breaks (DSBs) and in triggering damage response from broken ends have significant overlaps. In particular, proteins such as the MRN complex and the Ku/DNA-PK complex play important roles in both their respective repair pathways and damage signaling. In this section, we consider inherited human syndromes in these and other proteins involved in damage repair and signaling.

We begin with the MRN complex, composed in humans of the Mre11A, Rad50 and Nbs1 proteins. MRN binds to broken DNA ends and participates in both homologous recombination and signaling through the ATM pathway (see Sections 11.2, 11.5 and 13.3). Human hereditary diseases are caused by mutations in the genes encoding two of these three proteins, and these are inherited as autosomal-recessive diseases. The very rare syndrome associated with Mre11A is called AT-like disorder (ATLD), while the Nbs1 protein is named after the associated disorder, Nijmegen breakage syndrome, which again is very rare in most populations. Both syndromes are characterized by an increased incidence of cancer, along with increased sensitivity to DNA-damaging agents, immunodeficiency and certain developmental deficiencies. Very recently, a single patient has also been found with symptoms similar to Nijmegan breakage syndrome but with a mutation affecting Rad50, and so this is likely a third case of human hereditary disease caused by MRN dysfunction. The patients with these various MRN defects likely have partially functional mutant proteins, in that complete knockout of any of the three proteins in the MRN complex causes embryonic lethality in the mouse model system.

As implied by the naming of the ATLD syndrome, the symptoms of MRN mutation in human syndromes are similar to those of ATM defects. The ATM protein is mutated in patients with the *ataxia tel-angiectasia* (AT) syndrome. As you recall, ATM is the key kinase that recognizes broken ends and triggers a major DDR pathway (see Section 13.3). AT is a fairly common syndrome (roughly 1 in 40,000 births) characterized by profound cerebellar ataxia, a dramatic increase in immune-system cancers and immune dysfunction, and progressive brain deterioration. Cells from AT patients are extremely sensitive to ionizing radiation, which led to severe clinical complications in some early AT patients who were treated with radiation as anticancer therapy.

The DDR from stalled replication forks relies on the ATR kinase (Section 13.3), mutations of which cause the human hereditary syndrome called Seckel. Like most of the diseases discussed in this chapter, Seckel is inherited as an autosomal recessive, meaning that both copies of the gene encoding ATR must be defective to trigger the syndrome. Seckel patients suffer developmental issues such as dwarfism and facial distortions, along with mental retardation. Cells from Seckel patients display increased chromosomal abnormalities, presumably due to aberrant recombination events triggered by stalled replication forks.

Increased susceptibility to *br*east and ovarian *ca*ncer in humans is caused by inherited mutations in the *BRCA1* and *BRCA2* genes. The *BRCA* gene products are unrelated proteins that both play important roles in homologous recombination and damage signaling. Patients carrying the more severe form of *BRCA1* or *BRCA2* mutations have roughly 10 times the chance of breast and ovarian cancer than normal patients. However, it is important to note that only about 5%–10% of human breast-cancer patients belong to this high-risk group.

The inheritance of *BRCA1* and *BRCA2* mutations is different than that of the syndromes described just above. *BRCA* mutations are inherited in an autosomal-dominant fashion, which means that the affected patients have one copy of the normal (functional) gene and one copy of the defective gene. As in the case of inherited susceptibility to colon cancer (Section 14.5), the increased incidence of cancer is caused because a small subset of cells from these patients loses the functional copy of the gene. For example, homologous recombination can lead to LOH (see Section 11.5), or in other cases the functional copy can be inactivated by an additional mutation or by an epigenetic regulatory event. The net result is that the precancerous cell no longer has functioning BRCA1 or BRCA2 protein, and so is unable to properly repair and/or respond to DNA breaks. There is evidence in both mouse and humans that the BRCA proteins are essential, which is to say that an embryo homozygous for a BRCA knockout does not survive.

The above examples represent some of the best-studied and best-known syndromes that relate to proteins involved in DSB repair and DDRs, but they are not the only ones. A search through the literature will reveal human hereditary syndromes associated with defects in numerous other such proteins, including p53, DNA–PK, ATRIP, ligase IV, Artemis, along with other proteins that have not been specifically discussed in this book.

14.8 Insights from molecular analysis of sporadic tumors

Analysis of cancer-cell genomes has undergone nothing short of a revolution over the last decade or so due to advancing genomic techniques (see Chapter 15 for discussion of some of the technologies). Prior to about 2008, analysis of mutations in cancer cells was essentially one gene at a time, for example, using polymerase chain reaction (PCR)-based amplification followed by manual DNA sequencing. Copy-number alterations could be tested on many genes in parallel using hybridization against various kinds of gene arrays, but the information was still quite limited. The advent of massively parallel sequencing (see Chapter 15) then completely transformed the field. When the approach was first developed, approximately one billion base pairs (1 GB) of DNA could be successfully sequenced in a single run. Within a few years (by 2012), the capacity increased dramatically to some 600 GB, roughly 100 times the length of a (diploid) human genome! Using this advanced technology, the first sequence of a single cancer genome was published in 2008, and since that time, many *tens of thousands of cancer genomes* have been sequenced and analyzed. Indeed, sequencing of tumors is already a clinical option that is valuable for certain cancers and will likely become part of the clinical standard of care for many or all cancers in the near future.

The most important outcome of these genomic studies to date is a great expansion in identification and understanding of genes that contribute causally to cancer. Note that such cancer genes come in different varieties: (1) genes in which knockout mutations inactivate function; (2) genes where overproduction or inappropriate expression contributes to cancer, and (3) mutations that create abnormal proteins such as the famous hybrid BCR-ABL kinase in certain forms of leukemia. It is also important to mention that many cancer genes are inactivated not by mutations but by epigenetic alterations that alter cytosine methylation and/or histone modification, reducing gene transcription and thus protein production.

Genomic analyses have now implicated several hundred genes as likely cancer genes, and only about 10% of these had been previously shown to cause an inherited predisposition to cancer like the cases discussed in the sections above. The number of cancer genes may be much higher than this, particularly if many genes have relatively modest effects. In any case, this large expansion in the number of cancer genes informs more sophisticated cancer-biology experimentation, tumor classification, and undoubtedly fuels advances in cancer therapy. In discussing this large number of cancer genes, it is worth pointing out the complexity of cancer, which in a very real sense comprises more than 100 different diseases. Based on the complete sequence of cancer genomes, it is becoming clear that essentially every cancer has a unique combination of mutations (see below) and distinct features even within a single tumor category.

Included in the expanded repertoire of cancer genes just mentioned are many that encode additional proteins in the repair pathways discussed in prior chapters. Furthermore, the importance of different repair pathways for different classes of cancer has been reinforced by this data. For example, mutations in homologous recombination functions are often found in sporadic breast and ovarian cancer, mutations in MMR and homologous recombination in sporadic colorectal cancer, and mutations in NER in prostate cancer. The implications of these findings for cancer therapy will be discussed in the next section.

DNA-sequence analyses of cancer cells must be done in parallel with normal tissue from the same patient in order to identify the somatic mutations that occurred in the development of the tumor (distinguishing them from mutations inherited from one or the other parent). Analyses of the numbers of mutations per tumor and the types of mutations have provided many surprises and useful conclusions. Amazingly, the number of mutations in a given human tumor genome was found to vary from just a handful to upward of a million! Furthermore, different tumor types have characteristic ranges of mutation frequencies. Many childhood tumors are at the low end of the spectrum, while mutagen-induced cancers such as lung cancer in smokers tend to have the very large numbers. Most types of cancers have ranges in between these extremes.

In addition to point mutations involving base substitutions and very small insertions or deletions, cancer cells also have much more dramatic genomic alterations. It has been estimated that the typical cancer genome has large-scale deletion or additions involving about 25% of the genome, along with sometimes quite numerous rearrangements such as chromosomal translocations.

Are the mutational processes in different cancers similar or distinct from each other? Put another way, do the types of mutations in different cancers differ from each other? Given the large number of sequenced mutations, scientists can now answer this question about the mutational "landscape" or "signature," and the results have been very informative. A recent study analyzed nearly five million mutations in more than 7000 cancers. Single base-substitution mutations were categorized into the six possible substitutions (based on the pyrimidine-containing strand), along with the identity of both the 5′ and 3′ base pair, giving a total of 96 combinations. Signatures were defined based on the preponderance of mutations that fit each of these 96 combinations, along with some additional analyses of small insertions and deletions. The scientists discovered 27 distinct mutational signatures in different cancer genomes (most cancers showed at least two such signatures). As an example, one signature involved a preponderance of C to T mutations in the sequence 5′-NCG-3′, which is very likely due to excessive deamination of the modified base 5-methylcytosine. While some signatures involved a single major type of mutation like the example just mentioned, another signature showed promiscuous mutations at every analyzed trinucleotide combination. Clearly, different mutational processes are at work in different cancers.

Some of the causes of these mutational signatures have been deduced, and others are not understood but currently under investigation. There is a signature attributed to tobacco smoke in lung cancer (which is highly correlated with smoking history), another signature associated with UV exposure in skin cancer, and yet another apparently caused by alkylating agents (found in relapsing cancers in patients who had previously been treated with alkylating agents for their primary tumor). Furthermore, some of the signatures are associated with repair or recombination defects of the tumor, or with known molecular processes. For example, two signatures appear to relate to the action of translesion DNA polymerases. As mentioned in Chapter 6, microsatellite instability is common in MMR-deficient tumors, and as expected, a signature including these small insertions and deletions was found in these kinds of tumors. Breast, ovarian, and pancreatic cancers with deficiencies in homologous recombination showed a signature that includes larger deletions with microhomologies at the deletion junction. As described in Section 11.10, these mutations can be generated by alternative NHEJ, a pathway that would be increased when homologous recombination fails to repair DSBs.

It is worth again considering that cancer formation generally requires the occurrence of multiple sequential mutations to inactivate certain functions/pathways and activate others. Many or perhaps all cancers can only acquire these multiple mutations when mutagenesis is increased due to mutagen exposure and/or genomic alterations that compromise repair, replication, or DDR pathways. In a broad sense, cancer is thus a genetic disease, and the formation of cancer in any individual is fundamentally an evolutionary process. The above mutational signatures reveal an important part of that evolutionary history. A high load of mutations can be generated, in some cases, by excessive mutagen exposure, and in other cases, by repair-pathway deficiencies that increase the frequency of a particular class of mutations. Presumably, most of the thousands or even millions of mutations so induced have little or no consequence in a particular tumor, but a critical few modulate the functions/pathways needed for cancer development.

The study of cancer genomics has also led to two complete surprises involving shocking examples of genome instability. The first surprise occurs in a small fraction of cancers (2%–3%) and involves massive rearrangements within just one or a couple of chromosomes. It is as if this chromosomal segment gets shattered into up to a thousand or more pieces and then the pieces get ligated back together in a haphazard fashion. The phenomenon is called chromothripsis, from the Greek "chromo" (color, representing stained chromosomes) and "thripsis" (shattering). One current model for these massive rearrangements is that a segregation error during cell division leads to entrapment of one or a couple chromosomes in micronuclei, which somehow leads to the shattering, followed by repeated nonhomologous end joining to reconstitute a contiguous (though rearranged) chromosome.

The second surprise was the finding, again in a small fraction of cancers, of a localized cluster of mutations. Up to hundreds or even thousands of base-substitution mutations are found on the same strand in a limited region of a chromosome. This phenomenon has been called kataegis, which is Greek for thunderstorm or shower (as in mutation shower). One current model for kataegis is that the process of synthesis-dependent strand annealing (Section 11.5) creates extensive regions of ssDNA, which is the preferred target for cellular cytidine deaminases. These are enzymes that deaminate cytosine into uracil within RNA and DNA, with cellular roles in the immune system and in battling infecting viral genomes (see Section 8.7). When one of these enzymes acts on ssDNA, mutations such as dC to dT substitutions are likely to occur. It is also worth noting that some of the mutation-signature patterns described above reflect, at least in part, the action of cytidine deaminases.

Both of these surprising findings of clustered mutation, along with others that have not been discussed, demonstrate that genome analysis of cancer cells can provide useful information about the evolutionary history of the individual cancer. Furthermore, they show that the multiple mutations that are needed for the development of cancer need not occur in isolation from one another. A single event of chromothripsis or kataegis can, by chance, alter two or more cancer genes simultaneously.

14.9 Anticancer therapy — traditional and gene based

Many traditional anticancer drugs damage cellular DNA. This includes agents that act directly on DNA, such as the platinum-based crosslinking agents (e.g., cisplatin, carboplatin, and oxaliplatin), the crosslinker mitomycin C, alkylating agents such as temozolomide and cyclophosphamide, and, of course, ionizing radiation. Others damage DNA indirectly by subverting DNA topoisomerases (e.g., doxorubicin, daunorubicin, etoposide, teniposide, CPT-11 [irinotecan], and topotecan; see Section 7.5). Finally, some anticancer drugs (nucleotide analogs and antimetabolites that modulate nucleotide pools) interfere with replication and repair processes, potentially causing indirect DNA damage from stalled replication forks. The general presumption is that all these agents kill cancer cells more effectively than normal cells because the cancer cells are undergoing more extensive growth and DNA replication, although other factors are certainly also at play. The damage caused by essentially all of the DNA-damaging anticancer drugs can be repaired, reversed, or tolerated by one or more of the pathways discussed earlier in this book. Thus, oncologists seek levels of chemotherapy that are sufficient to overwhelm the repair capabilities of the tumor cells but not so high as to cause excessive damage to normal tissues.

As you might infer, knowledge of the mutated cancer genes in repair/response pathways in particular tumors can be very helpful in choosing among traditional cancer chemotherapies. Perhaps the most dramatic example relates to testicular germ cell cancers, where more than 80% of patients can be cured with cisplatin-based chemotherapy. This remarkable response rate is due to a deficiency in a repair protein[4] that participates in both NER and DNA-crosslink repair; recent evidence suggests that the defect in interstrand-crosslink repair is responsible for the hypersensitivity to cisplatin. Conversely, cisplatin resistance occurs in some secondary tumors of various types due to overexpression of this same protein. A related example involves the subset of breast cancers with defects in genes involved in homologous recombination, including *BRCA1* and *BRCA2*. Again, these cancers were found to respond better than recombination-proficient cancers to DNA-crosslinking drugs like cisplatin, presumably reflecting the role of homologous recombination in DNA-crosslink repair (see Section 13.5).

Another example involves MMR-deficient cells, which are intrinsically more resistant to certain alkylating agents and antimetabolite 5-fluoropyrimidines. A likely molecular explanation for resistance to alkylating agents involves futile cycles of MMR on alkylated base pairs when the pathway is functional. Correspondingly, the subset of colorectal cancers that is deficient in MMR respond very poorly to treatments that include one of these agents (also see below). Yet another example is that certain tumors overexpress a particular DNA alkyltransferase, making them resistant to certain alkylating agents, while tumors with unusually low levels are hypersensitive to the alkylating agents (see Section 9.2). In all these cases and others,

---

[4] The protein complex XPF/ERCC1 that catalyzes one of the incision reactions in NER.

it is clear that identifying the precise defect(s) in the tumor can be a great help in selecting traditional chemotherapy agents that are more likely to control that tumor and avoiding others that are unlikely to be helpful. This is particularly true because generally only a subset of tumors within any class of cancers have the repair defect in question.

As described in Section 14.8, genomic approaches have identified a large number of new cancer genes, and scientists/oncologists are ardently pursuing some of these as new targets for chemotherapy. The field is vast and interesting, and largely beyond the scope of this chapter. However, a few issues are worth discussing.

One of the most successful areas for developing new chemotherapy involves the identification of protein-kinase inhibitors, including those involved in DDRs. The first big success story in this area involved the development of imatinib (Gleevec), an inhibitor of the ABL and KIT kinases. This drug is very effective against chronic myelogenous leukemia and gastrointestinal stromal tumors with activating mutations in these kinases. Preclinical studies have identified a variety of inhibitors of kinases involved in DDRs, including DNA–PK, ATR, and ATM (see Section 13.3), and one of the inhibitors of DNA–PK is currently in clinical trials.

Scientists are also pursuing the development of specific inhibitors of multiple repair and recombination proteins. As implied earlier, many cancers have alterations in one or more repair pathways, which provide opportunities to exploit the differences between cancer and normal cells in therapy. These findings immediately suggest that inhibitors of repair pathways could be highly synergistic with traditional (or newer) chemotherapeutic drugs that cause DNA damage that is repaired by that pathway. Scientists are actively searching for specific inhibitors of various repair proteins in different pathways to achieve this vision, and many promising preclinical studies have been published. One of the major challenges of this approach is that the initial inhibitors are generally not very specific and/or not very potent, and so cycles of drug development are necessary. In addition, inhibition of repair pathways could have an adverse unintended consequence: mutation rates could be increased, potentially promoting further metastasis or the development of resistance to chemotherapy. In spite of these hurdles, several small companies are now solely focused on the development of DNA-repair inhibitors for anticancer therapy.

In addition to the strategy of synergistic chemotherapy just discussed, there is also hope that some repair inhibitors might have anticancer efficacy by themselves. The basic idea is that inhibition of a repair pathway (call it A) might be sufficient to kill or inhibit cancer cells that lack some complementary repair pathway (call it B). That is to say, the cancer cells lacking pathway B rely on pathway A for survival, while the normal cells in that individual do not require pathway A because pathway B can suffice. This approach is often called a "synthetic-lethal" strategy, because the combination of two defects causes lethality (in this case, only in the cancer cells).

A well-known example of the synthetic-lethal approach involves inhibition of a family of enzymes called *poly* (*ADP ribose*) *polymer-ases*, or PARPs. PARPs exert their effects by covalently modifying target proteins with chains of ADP ribose, which modulates their activity. The major PARP, called PARP1, plays an important role in BER by binding to single-strand breaks and recruiting BER factors. Inhibition of PARP1 can thereby inhibit BER, leading to an accumulation of single-strand breaks. Multiple studies show that cells deficient in homologous recombination are hypersensitive to PARP inhibitors, presumably because the excess single-strand breaks are converted into DSBs during DNA replication, which cannot be repaired in these cells. Correspondingly, recombination-deficient tumors appear to show strong responses to PARP inhibitors. However, it is important to point out some of the complexities in the field of PARP inhibitors: (1) some of the PARP inhibitors are not very specific and inhibit other targets; (2) PARP enzymes play roles in many cellular functions, not just DNA repair; (3) PARP inhibitors vary greatly in how they inhibit the enzyme, and the ones that stabilize a PARP–DNA complex seem to be the best candidates for the synthetic-lethality approach discussed earlier; and (4) a number of repair and DDR defects beyond homologous recombination also appear to promote synthetic lethality with PARP inhibitors, perhaps expanding their utility.

We will close with one final strategy that cleverly exploits the molecular details of DNA-repair pathways. As discussed in Chapter 6, MMR reverses the vast majority of replication errors, and MMR-deficient cells therefore have very high mutation rates. In addition, MMR deficiency is involved in cancer development, particularly colorectal cancers (see Section 6.5). Because of their high mutation rate, MMR-deficient tumor cells have an unusually large number of amino acid substitutions and other aberrations throughout their cellular proteome (collection of all cellular proteins). Scientists discovered that this class of tumors is particularly responsive to immunotherapy, specifically an immune-checkpoint inhibitor called penbrolizumab (Keytruda®). Without going into molecular details about how the inhibitor works, the analogy is that the drug revs up the immune system by "releasing the brakes." Apparently, the very high level of aberrations in certain cellular proteins allows this revved-up immune response to more effectively attack the cancer. Two clinical results are particularly striking. First, while MMR-deficient colorectal cancers showed an excellent response, MMR-proficient colorectal cancer showed no response. Second, the positive response extended to MMR-deficient cancers of every type that was tested. In 2017, the Food and Drug Administration (FDA) approved this immunotherapy for all cancers that share the MMR deficiency, which is the first FDA approval based on a specific genetic profile across tumor types.

14.10 Summary of key points

- Cancer is nearly always dependent on the accumulation of multiple mutations, and many (perhaps all) cancers are able to accumulate multiple mutations because an early mutation in the cell lineage increases the frequency of all subsequent mutation events.
- Certain mutations in genes that encode replication proteins have been implicated in developmental defects.

- Many human diseases relate to inherited mitochondrial defects, including defects in the mitochondrial DNA replication and repair machineries.
- Expansions in nucleotide repeats cause more than three dozen human diseases (including HD), and these expansions are induced by disturbances in DNA replication, repair, and recombination.
- XP and CS, involving defects in NER or translesion DNA polymerase η, cause a high frequency of sunlight-induced cancers as well as neurological disturbances.
- Heterozygous mutations in genes involved in DNA MMR cause Lynch syndrome, involving a dramatic predisposition to colorectal cancer.
- Epigenetic silencing of MMR genes is involved in about 15% of sporadic colon and endometrial cancers, and MMR-defective cancers respond to chemotherapy in unique ways that allow clinicians to tailor anticancer regimens.
- A number of complex syndromes involving development and/or cancer are caused by inherited mutations in various DNA helicases, most notably the RecQ family helicases.
- Inherited mutations in DSB repair and DDR pathways cause a number of human disease syndromes with complex characteristics (including profound neurological defects and immune system disorders).
- Genomic analyses of cancer cells have provided remarkable insights, including a large increase in the number of genes implicated in cancer, the realization that cancer cells sometimes carry more than a million newly induced mutations, the identification of dozens of mutational "signatures" that imply distinct mutational processes in cancer, and the discovery of two totally unexpected and dramatic processes that lead to massive localized clusters of mutations.
- Understanding the specific defects of particular cancers has led to novel and effective anticancer therapy tailored to the individual tumor type.

Further Reading

Alexandrov, L. B., Nik-Zainal, S., Wedge, D. C., Aparicio, S. A., Behjati, S., Biankin, A. V., & Børresen-Dale, A.-L. (2013). Signatures of mutational processes in human cancer. *Nature, 500*(7463), 415–421.

Castel, A. L., Cleary, J. D., & Pearson, C. E. (2010). Repeat instability as the basis for human diseases and as a potential target for therapy. *Nat Rev Mol Cell Biol, 11*(3), 165–170.

Croteau, D. L., Popuri, V., Opresko, P. L., & Bohr, V. A. (2014). Human RecQ helicases in DNA repair, recombination, and replication. *Annu Rev Biochem, 83*, 519–552.

DePamphilis, M. L., & Bell, S. D. (2011). *Genome Duplication*. New York, NY: Garland Science.

Dietlein, F., Thelen, L., & Reinhardt, H. C. (2014). Cancer-specific defects in DNA repair pathways as targets for personalized therapeutic approaches. *Trends Genet, 30*(8), 326–339.

Friedberg, E. C., Walker, G. C., Siede, W., Wood, R. D., Schultz, R. A., & Ellenberger, T. (2006). *DNA Repair and Mutagenesis* (2nd ed.). Washington, DC: ASM Press.

Garraway, L. A., & Lander, E. S. (2013). Lessons from the cancer genome. *Cell, 153*(1), 17–37.

Gavande, N. S., VanderVere-Carozza, P. S., Hinshaw, H. D., Jalal, S. I., Sears, C. R., Pawelczak, K. S., & Turchi, J. J. (2016). DNA repair targeted therapy: The past or future of cancer treatment? *Pharmacol Ther, 160*, 65–83.

Holt, I. J., & Reyes, A. (2012). Human mitochondrial DNA replication. *Cold Spring Harb Perspect Biol, 4*(12), a012971.

Kelley, M. R., Logsdon, D., & Fishel, M. L. (2014). Targeting DNA repair pathways for cancer treatment: What's new? *Future Oncol, 10*(7), 1215–1237.

Maizels, N. (2015). G4-associated human diseases. *EMBO Rep, 16*(8), 910–922.

Mak, T. W., Saunders, M. E., & Jett, B. D. (2013). *Primer to the Immune Response* (2nd ed.). Amsterdam, Netherlands: Academic Cell Press.

McMurray, C. T. (2010). Mechanisms of trinucleotide repeat instability during human development. *Nat Rev Genet, 11*(11), 786–799.

Mirkin, S. M. (2007). Expandable DNA repeats and human disease. *Nature, 447*(7147), 932–940.

O'Driscoll, M. (2012). Diseases associated with defective responses to DNA damage. *Cold Spring Harb Perspect Biol, 4*(12), a012773.

Pearson, C. E., Edamura, K. N., & Cleary, J. D. (2005). Repeat instability: Mechanisms of dynamic mutations. *Nat Rev Genet, 6*(10), 729–742.

Puigvert, J. C., Sanjiv, K., & Helleday, T. (2016). Targeting DNA repair, DNA metabolism and replication stress as anti-cancer strategies. *FEBS J, 283*(2), 232–245.

Snell, R. G., MacMillan, J. C., Cheadle, J. P., Fenton, I., Lazarou, L. P., Davies, P., Shaw, D. J. (1993). Relationship between trinucleotide repeat expansion and phenotypic variation in Huntington's disease. *Nat Genet, 4*(4), 393–397.

Stratton, M. R., Campbell, P. J., & Futreal, P. A. (2009). The cancer genome. *Nature, 458*(7239), 719–724.

Zhao, X.-N., & Usdin, K. (2015). The repeat expansion diseases: The dark side of DNA repair. *DNA Repair, 32*, 96–105.

How did they test that?
Triplet-repeat expansion is the basis for Huntington's disease

Previous studies had suggested that Huntington's disease might be caused by expansion of CAG repeats in the *HTT* gene. Snell *et al.* (1993) measured the CAG repeat length in a large number of patients and controls, including family groups with multiple affected individuals. Using polymerase chain reaction (PCR) analysis (see Chapter 15 for discussion of this method), they found that nearly all of the individuals affected by Huntington's (left-hand panel, A lanes) had one unusually long repeat allele and one shorter allele, while the control individuals (N lanes) showed two alleles in the shorter, "normal" range. In this gel, the PCR products were denatured into ssDNA, and the S lanes contain DNA sequencing standards for reference; length standards in nucleotides are indicated in the scale at the bottom of the gel. Quantitative analysis showed that the longer repeat allele in the affected individuals had from about 35–70 copies of the CAG repeat, while normal chromosomes showed a clustering around 20 repeat copies (right-hand panel).[5] A striking correlation was found between the age of onset of disease symptoms and the number of copies of the CAG repeat in the longer allele (bottom figure; filled dots represent the longer repeat allele while the "+" signs represent the shorter (normal-size) allele in that same patient). Individuals with early-onset disease had 50–70 copies of the CAG repeat, while patients with the more common later onset (40+ years of age) tended to have about 40 repeat copies in their expanded allele. Analyses of transmission of the expanded alleles within kindreds showed that affected children of an affected parent tended to have a higher copy number in the expanded allele and earlier age of onset than their affected parent

---

[5] One affected individual was found with two abnormally long alleles, and about 4% of affected individuals showed two alleles in the normal range. These latter exceptional cases were attributed to sample error or misdiagnosis; several Huntington's-like diseases, unrelated to repeat expansion, have since been discovered.

(data not shown here; see discussion of anticipation in Section 14.3). Figures reproduced from Snell *et al.* (1993), with permission from Springer Nature; permission conveyed by Copyright Clearance Center, Inc.

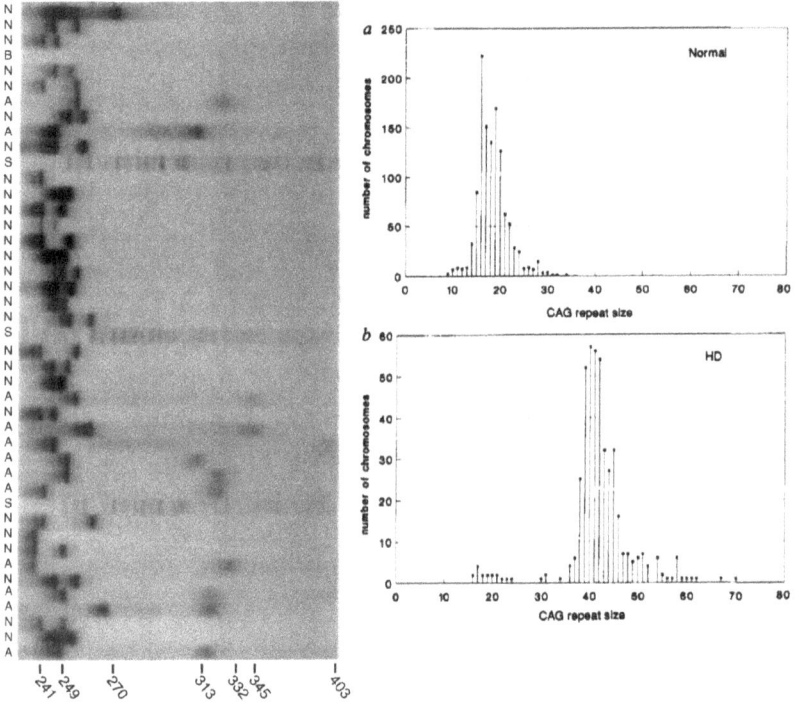

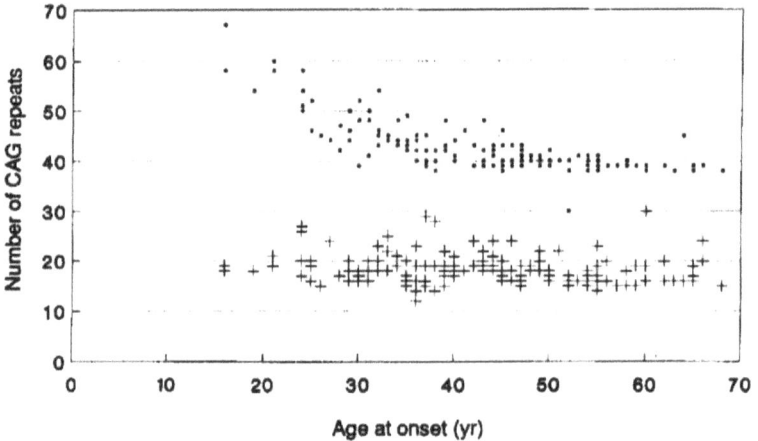

Chapter 15

Enzymes of DNA replication and repair fuel modern genomic technologies

Modern genomic technologies are utterly dependent on various enzymes and proteins from the processes of DNA replication and repair. Most importantly, a variety of DNA polymerases and DNA ligases are essential for the two cornerstones of this technology, the *polymerase chain reaction* (PCR) and DNA sequencing. In this chapter, we summarize aspects of modern genomic technologies, including key historical developments. We then close with a brief summary of the multitude of applications of the powerful DNA-sequencing approaches that have provided unfettered access to essentially all genomes, including those of single cells and those of organisms that have been extinct for millennia.

15.1 Commercialization of proteins and enzymes that act on DNA

In the 1970s and 1980s, enzymes and reagents related to molecular biology entered a rapid phase of commercialization. For many years, much of the attention was on restriction enzymes and DNA ligase, which together allowed extensive manipulation of DNA molecules purified from organisms. As described later, these enzymes allowed the construction of recombinant plasmids and viruses, providing the first opportunity to study defined segments of the genomes of higher organisms including humans.

As the years and decades went on, scientists found increasing uses for proteins and enzymes that act on DNA, and biotechnology supply companies responded by expanding the available reagents and also providing ever more sophisticated assays and "kits" to accomplish molecular manipulations of DNA (and RNA). Many of the proteins/enzymes that are now commercially available originate in the processes of DNA replication, repair, and recombination. As enumerated in Table 15.1, this includes DNA polymerases; DNA

Table 15.1. Replication and repair proteins used in biotechnology.

Protein/enzyme	Important activity
DNA polymerases*	Incorporate nucleotide residues into DNA utilizing preexisting primer and template. Many have 3′ to 5′ proofreading exonuclease, but exonuclease-deficient versions are also available. Some promote strand-displacement synthesis.
Thermostable DNA polymerases*	Same as above DNA polymerases but derived from a thermophilic organism so that the enzyme is active at high temperatures (e.g., for PCR).
Reverse transcriptases*	RNA-dependent DNA polymerases that incorporate deoxyribonucleotide residues using an RNA template.
DNA ligases*	Join together two DNA fragments, one with 5′-phosphate and the other with 3′-hydroxyl.
Terminal transferase	Template-independent DNA polymerase that adds deoxynucleotide residues to 3′-hydroxyl termini; often used to label or extend 3′ ends.
Polynucleotide kinase	Transfers phosphate from ATP to the 5′-hydroxyl termini of DNA and RNA fragments; can also exchange the terminal phosphate of a 5′-phosphate end; often used to label the ends of fragments.
Nucleoside kinases*	Transfer phosphate groups from one nucleoside triphosphate to the 5′-hydroxyl or 5′-phosphate ends of nucleosides/nucleotides; various transfer reactions depending on the enzyme.

Table 15.1. (*Continued*)

Protein/enzyme	Important activity
DNA nucleases*	Various exonucleases attack DNA from either the 5'- or 3'-end, with particular specificity for single- versus double-stranded DNA. Various DNA endonucleases cleave internal phosphodiester bonds, some with specificity for unusual bases and/or aspects of DNA structure.
RNase H	Degrades the RNA strand of an RNA–DNA duplex.
RecA protein	Catalyzes exchange of ssDNA fragment into homologous duplex DNA.
DNA topoisomerases*	Alter DNA topology by removing or introducing DNA supercoils and catalyzing other strand-passage reactions.
SSB/RPA proteins	Bind ssDNA regions; can promote or inhibit activity of a variety of DNA-acting enzymes.
Damage-recognition enzymes*	Variety of enzymes that recognize specific damaged or unusual bases and cleave either the phosphodiester backbone(s) near the damage or the N-glycosidic bond of the damaged residue.
AP endonucleases*	Cleave phosphodiester backbone immediately 5' from an abasic (AP) site.

Note: *Multiple enzymes from different organisms and/or with different properties are utilized depending on the situation.
AP, abasic; ATP, adenosine triphosphate; PCR, polymerase chain reaction.

ligases; nucleoside/nucleotide kinases; nucleases from replication, repair, and recombination pathways; DNA-binding proteins such as SSB and RPA; strand-exchange proteins like RecA; DNA topoisomerases; and many repair proteins that act on specific DNA lesions. Subsets of these proteins are used throughout the numerous modern genetic technologies, including those described in this chapter.

Many volumes have been published describing the wide variety of specific molecular assays that utilize these tools, and a description of these assays is beyond the scope of this chapter. As you read about the latest experimental findings in the primary literature, you will undoubtedly find many approaches that rely on the abovementioned proteins and enzymes. Also, for those readers who engage in laboratory experiments and use commercially available molecular biology kits, it is well worth looking closely into the components and reaction steps that occur in the kit mixtures. More likely than not, you will find one or more of our favorite enzymes and proteins from the processes of DNA replication, recombination, and repair.

15.2 The PCR revolution

It is difficult to overstate the importance of the PCR in the development of molecular biology and modern genomic technologies. PCR provides a simple, fast, and inexpensive means of amplifying a desired nucleic acid sequence a million-fold, starting with a crude DNA (or RNA) sample from essentially any organism or from material with a complex mix and/or fragments of nucleic acids. The PCR thus permitted the development of numerous useful assays and facilitated the detailed analysis of genes and transcripts, both in unitary fashion and in microtiter- and array-based systems that permitted hundreds or thousands of segments to be analyzed at once. In addition, as will be described in the next section, the PCR revolutionized the construction of recombinant DNA molecules, and in the section after that, led to powerful second-generation sequencing approaches.

The PCR was developed in 1983 and was recognized with the Nobel Prize in Chemistry a decade later (see Nobel Prize box below). As implied in the "chain reaction" part of its name, PCR directs the exponential amplification of the desired nucleic acid segment, as diagrammed in Figure 15.1. Two critical components of

1993 Nobel Prize in Chemistry

This prize was awarded to **Kary B. Mullis** and **Michael Smith** for developing chemical methods that greatly facilitated genetic engineering: the polymerase chain reaction (Mullis) and site-directed mutagenesis (Smith).

https://www.nobelprize.org/prizes/chemistry/1980/summary/

PCR are a pair of oligonucleotide primers that dictate the region to be amplified, along with a special DNA polymerase that carries out the repeated DNA synthesis events. Before going into the details, note the exponential nature of the reaction — the products of each cycle become the template for the next cycle, essentially doubling the amount of the desired nucleic acid in each step (within limits, e.g., the exhaustion of primers and deoxynucleotides, stability of the DNA polymerase). Theoretically, the number of product molecules should approach 2^n, where n is the number of cycles of the reaction (Figure 15.1). Thus, only 20 cycles should generate roughly a million (2^{20}) copies of product if the conditions are robust (in practice, efficiency is never 100% and somewhat more than 20 would generally be needed to reach the million-copy mark).

As described in the early chapters of this book, DNA polymerases require preexisting primers to initiate DNA synthesis. The PCR takes advantage of this requirement as a means to select the target region to be amplified. Two oligonucleotide primers, which hybridize ("anneal") to opposite strands at sites flanking the target region, must be provided in order for the amplification to be successful. In the first cycles, when the synthesis template is the original genomic DNA fragment, the polymerase can extend past the site where the second primer hybridizes (Figure 15.1, step 3). However, as synthesis products become new templates for additional reactions, the polymerase reaches a terminus at the 5' end of a previously used primer, and uniform-length products accumulate (Figure 15.1, later cycles).

The PCR, as we know it, would be impossible without the ability to generate oligonucleotide primers with a desired sequence. Fortunately, in the 1980s, when PCR was invented, chemists were also making great strides in the synthesis of oligonucleotides. By the end of the decade, the first automated system for oligonucleotide synthesis was developed. The capabilities of these systems expanded greatly in the 1990s, such that commercially available oligonucleotides became readily available. Parenthetically, the chemical synthesis methods permit the incorporation of numerous modified nucleotides, including natural modified bases such as methylated adenine or cytosine, bases that mimic important forms of DNA dam-

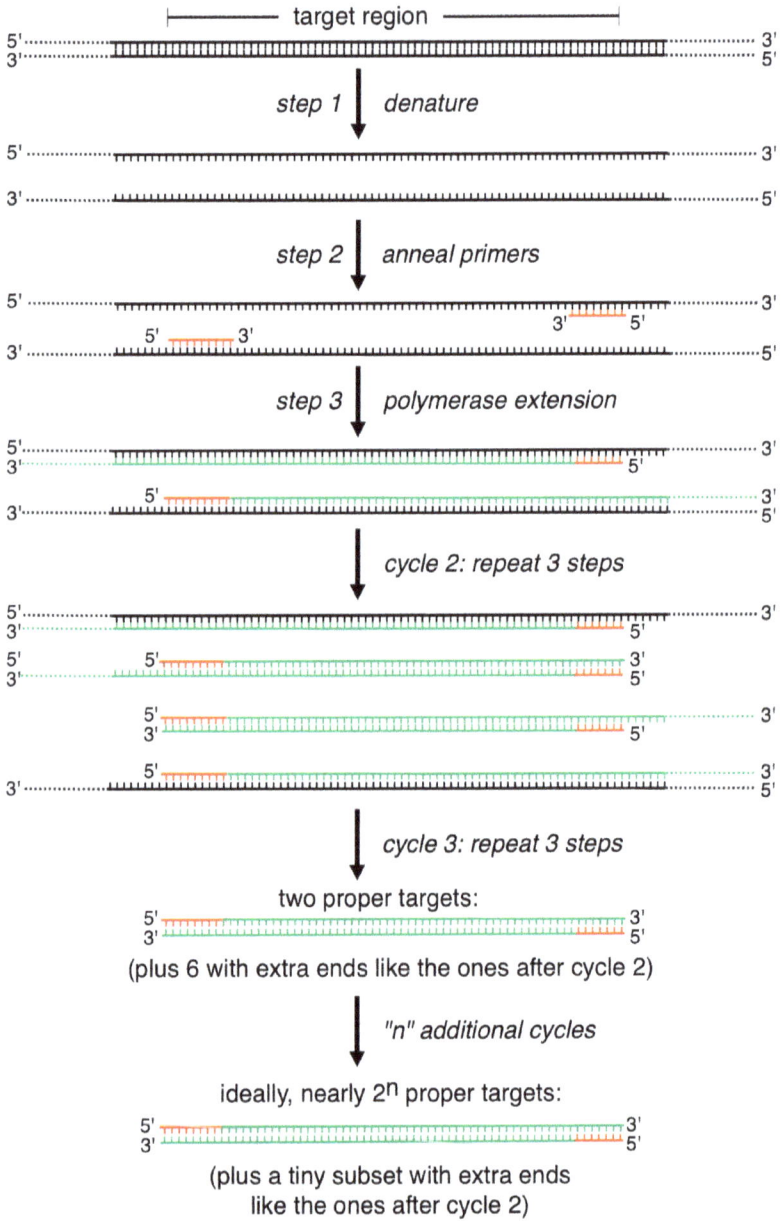

Figure 15.1. The polymerase chain reaction (PCR). In this diagram, PCR primers are indicated in red and newly synthesized DNA in green.

age, and even completely unnatural bases devised by chemists. Oligonucleotides containing forms of DNA damage have been heavily used in biochemical studies of DNA repair and DNA damage responses.

In order to perform a PCR reaction for a particular target segment, you first need to know the nucleic acid sequence where the primers will hybridize, so that the correct complementary oligonucleotides can be synthesized (or much more likely, ordered from your favorite company). You actually don't need to know the intervening sequence that will be amplified, although in a large majority of cases, you would also have this information from prior DNA-sequencing analyses (also see below). We will return to some of the considerations and limitations in primer design after we discuss the nature of the DNA polymerase and the amplification reaction.

A conceptual difficulty in developing the PCR was the fact that the product of a DNA polymerase reaction is a stable duplex DNA, with one template and one product strand. An oligonucleotide primer for the next cycle of synthesis would be unable to hybridize to this duplex molecule, preventing any subsequent round of DNA synthesis. Therefore, any successful PCR reaction would need to unwind the two strands, a step called DNA denaturation. The simplest way to denature a duplex DNA is by raising the temperature past the so-called melting temperature of DNA, which is dependent on both the length and the nucleotide composition of the DNA. The standard PCR reactions use this temperature-dependent DNA denaturation to separate the product strands, raising the reaction temperature to >90°C. This high temperature, however, causes a problem. The DNA polymerases discussed in the early chapters of this book — such as polymerases from phage T7, *Escherichia coli*, and eukaryotic organisms — are destroyed by such high temperatures. These enzymes therefore cannot survive a complete reaction cycle with its denaturation step, even if the polymerization steps were performed at much lower temperatures. The earliest iterations of PCR therefore required the addition of a new aliquot of DNA polymerase after each denaturation cycle, which was cumbersome.

Within a very short time after the invention of PCR, the procedure was greatly streamlined by incorporating DNA polymerase from an unusual "extremophile" organism that grows at very high temperatures. *Thermus aquaticus,* a thermophilic bacterial species that was isolated from a hot spring (roughly 90°C) in Yellowstone National Park, was found to encode a temperature-resistant DNA polymerase that easily survives the denaturation temperature in PCR reactions. Thus, the polymerase could survive many cycles of denaturation and only needed to be added at the start of the reaction.

Another advantage of thermostable polymerases is that they can synthesize DNA at relatively high temperatures (e.g., 70°C–80°C), which as it turns out is also advantageous for the PCR. Indeed, the entire sequence of PCR steps is generally carried out at temperatures in excess of 50°C. The optimal temperature for the step where primers hybridize to their targets depends on the length and sequence composition (G/C vs. A/T content) of the primers themselves. To understand the setting of the optimal temperature for this step, it is useful to again consider the melting temperature, but this time of the hybrid formed between the primer and the template DNA. Because this hybrid is much shorter than the duplex template discussed earlier, the melting temperature is significantly lower, generally in the range of 50°C–65°C or so. The primer can only anneal to its target template if the temperature is less than the melting temperature of this hybrid region. At the same time, the primer can also anneal to other nonspecific sites in the DNA that are similar but not identical to the desired target, for example, a sequence that is complementary to 13 or 14 nucleotides out of a 15-nucleotide primer. Because of the mismatch(es) in these nonspecific hybridization sites, they have a lower melting temperature. Thus, the optimal annealing temperature is just a couple degrees lower than the melting temperature of the desired "perfect" hybridization site, which is a temperature that is beyond the melting temperature of the nonspecific targets. If you set the annealing temperature too low, you often see undesirable products caused by such nonspecific hybridizations.

In practice, simple computer programs are available to predict the ideal annealing temperatures of any given oligonucleotide, based on its melting temperature when in duplex form. Recall that the PCR requires dual primers, with both primers acting in every reaction cycle. This necessitates that the two primers have relatively similar ideal annealing temperatures so they can function in a single annealing step. In practice, these and other considerations about primer design impose constraints on the precise location of the ends of a PCR product. Again, computer programs come to the rescue in identifying candidate primer pairs for amplifying any particular desired segment of DNA from a region of known sequence.

The thermostable DNA polymerase binds to the primer-template pair during the annealing step of PCR, but its polymerase activity is optimal at the higher temperature range of 70°C–80°C, which is generally used for the extension step. In summary, the three distinct steps in each PCR cycle are denaturation, annealing, and extension (Figure 15.1). A 25-cycle PCR will go through these three steps 25 times, generally with minor tweaking at the very beginning and end of the entire sequence. Some of the earliest PCR machines consisted of three temperature-controlled water baths, with a robotic arm that moved the reactions between these three at set time points. The current technology uses fixed tubes within a thermal-cycler device that rapidly changes temperatures by an electronic means called the "Peltier effect."

As alluded to above, the PCR is now used for countless applications in experimental science, clinical science, forensics, infectious disease, and other fields. In the realm of experimental science, two commonly used variants of the PCR are worth mentioning. Quantitative PCR (qPCR) is a PCR-based method for accurately measuring the starting quantities of particular target sequences, with versions for measuring both DNA and RNA. The method generally uses fluorescent dyes and recording devices that measure the PCR products that accumulate throughout the reaction cycles. You will sometimes hear this method called "real-time" PCR because the product accumulation is measured in real time, but this term should not be abbreviated because of the next commonly used variant. *Reverse transcriptase* PCR (RT-PCR) is a PCR-based method in which an initial RNA sample is converted into its DNA complement by use of the viral enzyme reverse transcriptase, a DNA polymerase that uses RNA templates. The converted DNA complements are then amplified by PCR, in some cases using qPCR to allow quantitation of RNA samples.

Isothermal PCR is a set of methods that use a single temperature to accomplish amplification in a much more rapid time frame than standard PCR. In these methods, a strand-displacing DNA polymerase, sometimes with the assistance of a DNA helicase, separates product strands from their templates. Otherwise, the basic principle is the same as in normal PCR, with the products from each round serving as templates for the next round. Isothermal PCR methods are used in a number of settings where the rapid turnaround and uniform temperature are particularly advantageous, for example, certain clinical tests and pathogen detection.

15.3 Constructing recombinant DNA molecules and synthetic biology

Cloning of DNA fragments into plasmid and viral vectors was a major advance that ignited the development of the field of molecular biology. In the early years of recombinant DNA (1970s and 1980s), scientists generally cleaved the target DNA with restriction enzymes and joined the desired fragment to a linearized plasmid or viral vector using DNA ligase (Figure 15.2A). Recombinant plasmids made in this manner were instrumental for countless studies involving DNA-sequence analysis, protein overproduction, and genome mapping, just to name a few.

The invention of PCR changed molecular biology in very dramatic ways. PCR allowed scientists to amplify, purify, and study DNA and RNA molecules without the need to clone them into recombinant plasmids. It was also used frequently as the first step in creating recombinant plasmids and other DNA constructs, because it allowed scientists to clone a desired fragment even if it was only a tiny fraction of the starting DNA, such as would be the case with a segment of a human chromosome within a crude preparation of human

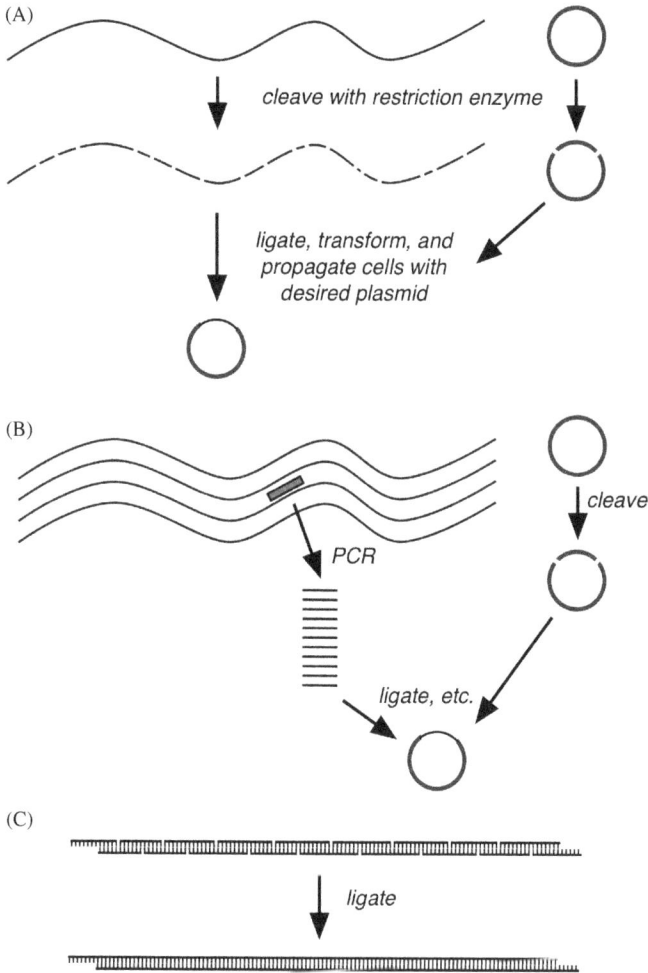

Figure 15.2. Common methods of constructing recombinant DNA molecules. See text for discussion of three widely used methods.

DNA (Figure 15.2B). Not only did PCR simplify and accelerate experimental approaches, it permitted the analysis of many nucleic acid segments that could not be cloned because they caused toxic effects on survival of the host organism.

Both restriction enzymes and PCR continue to be used extensively in the construction of recombinant DNA molecules. In addition, the synthesis of defined oligonucleotides has become increasingly more efficient, accurate, and economical, and so collections of partially complementary oligonucleotides are often ligated together to make long duplex molecules (Figure 15.2C). The modern field of synthetic biology utilizes all of these tools to engineer novel combinations of biological functions, either added to existing organisms or even for the creation of entirely new organisms. Synthetic biology is now making key contributions not only in experimental science, but also in commercial fields such as production of useful biochemicals, biosensors, industrial enzymes, biofuels, materials science, and even computational tasks and information storage. A culmination in synthetic biology was the construction and deployment of an entire "minimal" bacterial genome, with only those genes necessary for survival. The minimal chromosome was constructed sequentially by ligating designed oligonucleotides together. The ligated product, just over a million base pairs long, was then inserted into the cellular body of a bacterium that had been stripped of its own chromosome, and the resulting organism reproduced and could be grown in the laboratory!

15.4 First-generation DNA-sequencing technology

The determination of genomic DNA sequences is essential for many aspects of modern biology, and it is fascinating to appreciate how the technology has developed over the last half century. As will be described later, the time required to decipher a DNA sequence went from a few bases per year around 1970 to nearly a terabase (10^{12}) per day in 2018 (in a single machine), which translates to an improvement of about 14 orders of magnitude! Overall, this is a significantly quicker progression than the advances in microprocessor speeds/capacities according to the famous "Moore's law" in microelectronics (the revised version, that is, which posits a doubling every two years). It has also been estimated that the cost of sequencing a human genome dropped from nearly a billion dollars in the early 2000s to just a few thousand dollars in 2018.

Tracing direct-genomic DNA sequencing to its roots takes us back a half a century, to around 1970.[1] Using DNA polymerase and radioactively labeled nucleotides, Ray Wu and his colleagues were able to decipher the short sequence that forms the (single-stranded) tips of the bacteriophage λ genome. The methodology was laborious, and completion of this short sequence required several years of research (see "How did they test that" at the end of this chapter). Furthermore, this tiny bit of sequencing was only possible because the tips of the λ chromosome were in a single-stranded form that could be used as a template for DNA polymerase, with the adjoining 3′ end in the duplex segment serving as a built-in primer.

The first generally applicable DNA-sequencing technologies, which are referred to as first generation, were invented in the 1970s. Two different popular methods emerged, named after their inventors (see Nobel Prize box below). The Maxam/Gilbert method, which is also called chemical sequencing, required the production of a purified DNA fragment that was radiolabeled at only one end of a particular strand. This single-stranded fragment was subjected (in four separate reactions) to different chemical conditions that resulted in the cleavage of particular base residues (Figure 15.3A). One condition

---

1980 Nobel Prize in Chemistry

This prize (one half) was awarded to **Paul Berg** for his research on the biochemistry of nucleic acids and advances in recombinant DNA. The other half of the prize was awarded jointly to **Walter Gilbert** and **Frederick Sanger** for their development of nucleic acid sequencing technologies. *See "How did they test that?" at the end of this Chapter for attribution for the image of the sequencing gel.*

https://www.nobelprize.org/prizes/chemistry/1980/summary/

---

[1] At that time, RNA sequencing had actually progressed further than DNA sequencing, with the sequence of a specific tRNA being published in 1965 (Holley, R. W. *et al.*, 1965, Structure of a ribonucleic acid. *Science, 147,* 1462–1465).

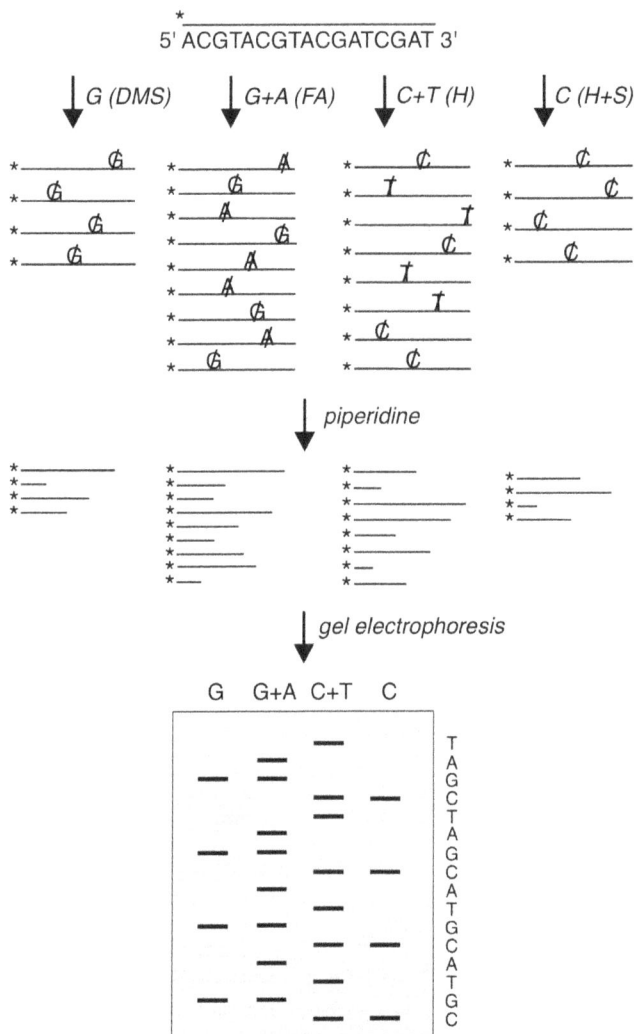

Figure 15.3. The Maxam/Gilbert method of DNA sequencing. A DNA fragment with a 5′-end label is first subjected to one of four chemical methods that damage the indicated bases: DMS, dimethylsulfoxide; FA, formic acid; H, hydrazine; H+S, hydrazine plus sodium chloride. The chemical conditions are such that only a small fraction of the indicated base(s) in any DNA fragment is damaged. Fragments with single damaged bases are diagrammed with "struck out" letters to indicate the damaged residue. The next step in the procedure is a treatment with hot piperidine, which results in cleavage of the backbone at the damaged base. Gel electrophoresis separates the fragments into discrete (radioactively labeled) bands, and comparison of the bands allows the assignment of sequence as shown to the right of the gel.

caused cleavage at G residues, the second at G and A, the third at C and T, and the fourth at C. The conditions were adjusted so that the extent of cleavage was quite low, roughly on the order of only one cleavage per fragment. The four reactions were then run in parallel on a high-resolution denaturing polyacrylamide gel capable of separating fragments that differ in size by only one base residue. After autoradiography of the gel, the bands in the four adjacent lanes were compared to read out the DNA sequence over a stretch of up to several hundred base pairs. While the Maxam/Gilbert method was used extensively for several years, it became obsolete due in part to its reliance on toxic chemicals, requirement for finicky reaction conditions, propensity for errors, and the relative simplicity and reliability of the competing method developed by Sanger.

The Sanger sequencing method uses DNA polymerase along with special chain-terminating nucleotides to generate DNA fragments of varying length corresponding to the positions of each of the four nucleotides (Figure 15.4). Chain-terminating deoxyribonucleotides are identical to normal deoxyribonucleotides except that they lack a suitable 3'-OH group (called dideoxyribonucleotides or ddNTPs). Because DNA polymerase adds incoming residues to the 3'-OH group, extension is blocked when a ddNTP is incorporated. In the original Sanger method, the desired DNA fragment was purified and added to a suitable primer that could trigger synthesis by DNA polymerase across the fragment of interest. Four parallel reactions were performed. All four had the four normal deoxyribonucleotides, but each also had a small amount of one of the four chain-terminating ddNTPs. The ratio of ddNTP to normal dNTP was such that DNA synthesis was terminated at any one position only about 1% of the time; this resulted in terminated fragments at each of the many locations of that particular nucleotide residue over a stretch of hundreds of nucleotides (Figure 15.4). The products were separated by denaturing polyacrylamide gel electrophoresis (again, with the four reactions side by side; Figure 15.4). The sequencing product bands in the gel were detected by means of radioactively labeled residues, which had either been incorporated during synthesis or were present on the starting primer (as indicated by the asterisk in Figure 15.4).

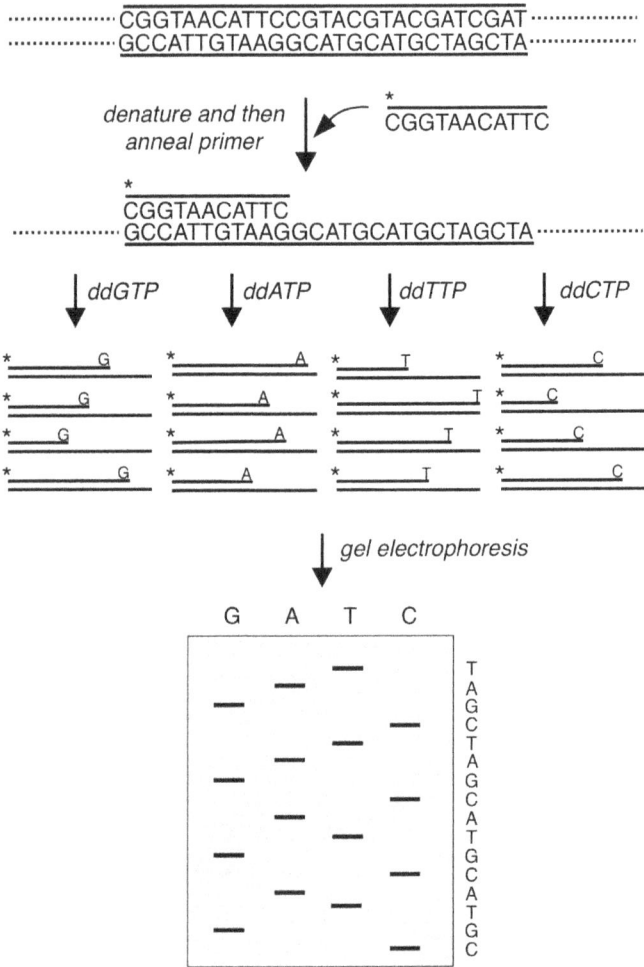

Figure 15.4. The Sanger method of DNA sequencing. This method uses DNA synthesis in the presence of chain-terminating nucleotides to determine DNA sequence. A radiolabeled primer allows fragment detection and also directs the sequencing reaction to the precise region of interest. Alternatively, radioactively labeled residues are incorporated during the synthesis reaction (not shown). The chain-terminating nucleotides are dideoxynucleotides, which lack the 3'-OH group for polymerase extension. The chain-terminating nucleotides are present in relatively low effective concentrations compared to normal deoxynucleotide triphosphates, so that termination occurs at only a small fraction of the relevant positions. The products are separated in a denaturing polyacrylamide gel, and the positions of the (radioactively-labeled) bands directly reveal the sequence (right of gel).

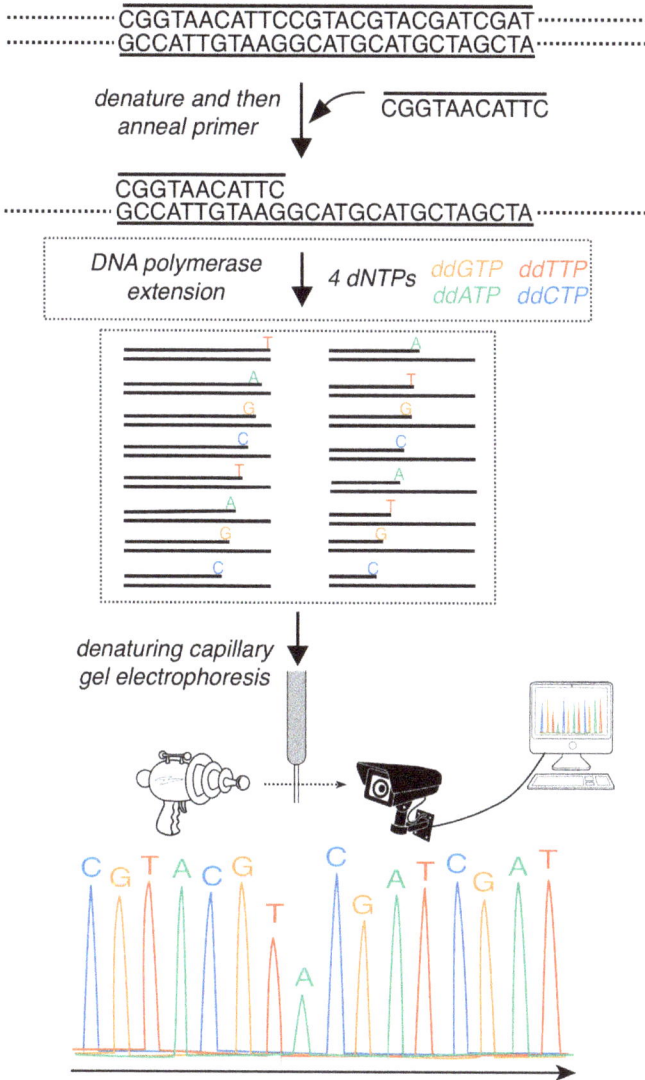

Figure 15.5. Dye-terminator sequencing. In this method, the dideoxynucleotides also carry distinct fluorescent dyes that allow detection of the terminated fragments. Sensitive detection of the dyes using laser technology occurs as the fragments exit from a denaturing capillary gel, which separates the fragments by length.

This original Sanger method was used to sequence the entire bacteriophage λ genome, which turned out to contain 48,502 base pairs (see "How did they test that" at the end of this chapter). The Sanger group, which had invented the method, published this landmark study in 1982.

The Sanger sequencing method was streamlined and improved in a number of ways over the years. One of the most important improvements, called dye-terminator sequencing, utilizes fluorescently labeled ddNTP chain terminators, with each of the four chain-terminating nucleotides having a different color dye (i.e., they each emit at different wavelengths). A second related improvement involved separation of the fragments through capillary electrophoresis instead of large and unwieldy polyacrylamide gels. Because each of the four chain-terminating nucleotides has a different color, all four can be distinguished from each other when they emerge at a fixed point from the capillary gel (Figure 15.5). This greatly simplifies the sequencing procedure and has led to further automation and advances, indeed setting the stage for second-generation sequencing approaches (see below). The more-streamlined version of the Sanger sequencing method was used to complete the first published genome sequence of an individual human (scientist J. Craig Venter) in 2007.

15.5 Next-generation DNA-sequencing technologies

While the Sanger method is still used to this date, newer sequencing methodologies have emerged and become dominant. These so-called second-generation or next-generation methods are characterized by the determination of the sequence of massive numbers of fragments in parallel. All of these methods circumvent the need for cloning of the DNA fragments in bacteria, which is generally necessary in the first-generation methods. In addition, the sequence outputs from the multiple parallel reactions are detected directly, with no need for separation methods such as gels or capillary electrophoresis.

We will first consider the general strategy used in most commercial second-generation sequencing methods before describing one of the methods in more detail (which will explain how these individual steps actually work). DNA from the desired source is first cleaved into small fragments, and special DNA oligonucleotide adapters are ligated onto each end (Figure 15.6A; the purposes of these ligated oligonucleotides will be discussed below). Next, individual fragments are fixed to particular addressable locations, for example, discrete locations on the surface of a glass slide (Figure 15.6B, top). Once they are localized, each isolated fragment is amplified via PCR into a small cluster of identical DNA molecules, which are maintained in the immediate vicinity of the original fragment by tricks that vary between platforms (Figure 15.6B, bottom). This PCR reaction relies on sequences within the ligated oligonucleotides to serve as the binding sites for oppositely oriented PCR primers. After this series of steps, the glass slide (or other matrix) now has a large collection of tiny clusters of PCR products, each of which originated from a different DNA fragment. The final step in the sequencing protocol is the actual DNA-sequencing reaction, where the identities of the bases within the PCR products are sequentially determined (Figure 15.6C). The nature of this determination varies between the commercially available platforms, but several utilize fluorescently labeled nucleotides and DNA polymerase incorporation, much like in the Sanger method. As in the revised Sanger method described earlier, different fluorescent dyes are used to distinguish the four different nucleotides (Figure 15.6C).

Second-generation sequencing methods generally read from 30 to about 500 bases in length in a single run. While this may not sound very impressive, many millions of fragments can be read in parallel at the same time (only four clusters are shown in Figure 15.6 for simplicity). Depending on the size of the genome being analyzed, this means that each segment of the genome will be represented in multiple different fragments, which is often referred to as "fold coverage." For accurate genome determination, something like 30-fold coverage is needed.

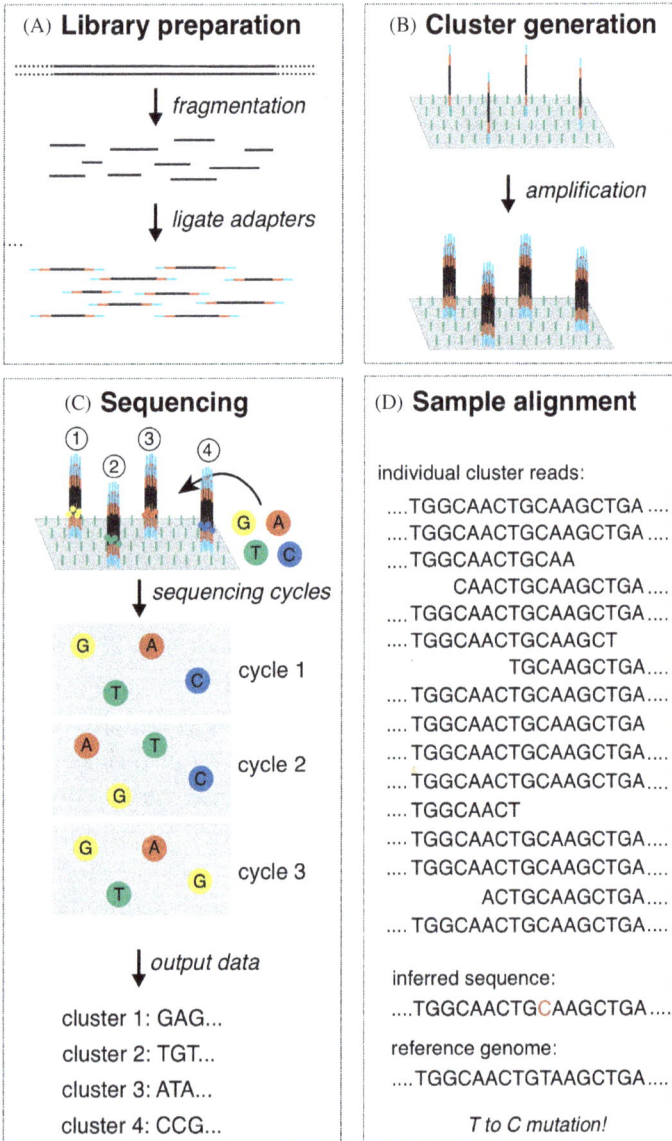

(A) Library preparation

↓ *fragmentation*

↓ *ligate adapters*

(B) Cluster generation

↓ *amplification*

(C) Sequencing

① ② ③ ④

G A
T C

↓ *sequencing cycles*

G A
T   C   cycle 1

A T
G   C   cycle 2

G A
T   G   cycle 3

↓ *output data*

cluster 1: GAG...
cluster 2: TGT...
cluster 3: ATA...
cluster 4: CCG...

(D) Sample alignment

individual cluster reads:

....TGGCAACTGCAAGCTGA....
....TGGCAACTGCAAGCTGA....
....TGGCAACTGCAA
        CAACTGCAAGCTGA....
....TGGCAACTGCAAGCTGA....
....TGGCAACTGCAAGCT
            TGCAAGCTGA....
....TGGCAACTGCAAGCTGA....
....TGGCAACTGCAAGCTGA
....TGGCAACTGCAAGCTGA....
....TGGCAACTGCAAGCTGA....
....TGGCAACT
....TGGCAACTGCAAGCTGA....
....TGGCAACTGCAAGCTGA....
            ACTGCAAGCTGA....
....TGGCAACTGCAAGCTGA....

inferred sequence:
....TGGCAACTGCAAGCTGA....

reference genome:
....TGGCAACTGTAAGCTGA....

T to C mutation!

Figure 15.6. Outline of next-generation sequencing. The steps in second-generation sequencing are broadly outlined. See text for more detailed discussion. Parts of this figure were modified from a brochure entitled "An Introduction to Next-Generation Sequencing Technology" from Illumina, Inc. (www.illumina.com/technology/next-generation-sequencing.html).

A major challenge that needed to be solved for second-generation sequencing was the development of computational methods to assemble these millions of different sequence reads into a complete genome, using this high level of coverage and the overlapping segments between the various sequence reads. The bioinformatics involved in determining and analyzing genomic sequences is a field unto itself, well beyond the scope of this discussion.

To illustrate a very small window of this sequence alignment, Figure 15.6D shows a modest number of overlapping individual sequence reads of a hypothetical genome site. Most of the individual reads extend outward in both directions from this 17-base region, while a small number of reads have one or the other end within the region. Note the overlapping nature of the reads, and the fact that the ends of the individual reads vary. The alignment of these individual reads determines an inferred sequence of the genome from which the DNA samples arose (Figure 15.6D). In cases where the genome sequence of that species (or a parental strain) has already been determined, mutations can be recognized as deviations between the reference sequence and the newly determined sequence (Figure 15.6D, bottom, T to C mutation).

Before leaving this general discussion, it is worth mentioning a second important feature of the oligonucleotide primers that were added to the original fragment in the paragraph above. A unique "barcode" or "index" is generally encoded in the oligonucleotides. If 12 different DNA samples are to be sequenced, each is tagged with a different barcode. Then, when the millions of individual sequence reads are obtained, the reads are computationally sorted into 12 different bins based on their barcode (which is revealed in the DNA-sequencing readouts). The DNA-sequence assembly is performed only after this sorting step. The barcodes thereby allow more efficient and economical use of the sequencing platforms in cases where the capacity of a single sequencing run in the platform is higher than necessary to achieve the desired genome fold-coverage, which is often the case.

Let us now dive into a more detailed description of the steps in the dominant platform for second-generation sequencing at the time of publication, which will illustrate how some of the amazing tricks in the generic description above actually work. The current dominant platform is that of Illumina/Solexa, with instrument names such as MiSeq, MiniSeq, NextSeq, HiSeq, and HiSeq X (often followed by a number). The various instruments vary in capacity as well as detailed features, but they all use the methodology described here. The sequencing process begins with random fragmentation of the DNA and ligation of appropriate oligonucleotides to both the 5′ and 3′ ends of each fragment (Figure 15.7A). To follow the subsequent steps, let's call the oligonucleotide ligated to the 3′ side "oligo A" and the one ligated to the 5′ side "oligo B." This pool of DNA fragments is usually PCR-amplified in bulk to increase the overall DNA amounts (Figure 15.7A; note that this amplification is prior to the step where DNA fragments are attached to a matrix — a second DNA amplification will be described later). The amplified fragments now have the sequences of the ligated oligonucleotides at their two ends. A gel purification step is used to purify a collection of fragments that have an appropriate size for the subsequent steps.

Individual fragments are then attached to the surface within a flow cell by hybridization between oligonucleotides covalently bound to the flow cell (call them "oligo C") and part of the sequence introduced by the ligated oligo A on the fragments (Figure 15.7B, top image). The Illumina platform uses a special kind of PCR amplification, called bridge amplification or cluster amplification, which allows the products to stay localized in a cluster near the site where the first fragment bound. In the first round of PCR, the DNA polymerase uses the 3′ end of the flow-cell-bound oligo C to synthesize a complement of the fragment that initially hybridized to that oligonucleotide (Figure 15.7B, step 1). As the first PCR cycle proceeds, denaturation of the resulting duplex releases the initially bound DNA fragment, but the complement that was just synthesized is still bound to the flow cell by means of the covalently attached oligo C (which constitutes the 5′ end of that fragment). The opposite end of this bound complement strand has the sequence determined by oligo B (actually the complement of that sequence), which had been ligated to the 5′ end of the original fragment. The flow cell has a second kind of oligonucleotide bound throughout, let's call it "oligo D," and oligo

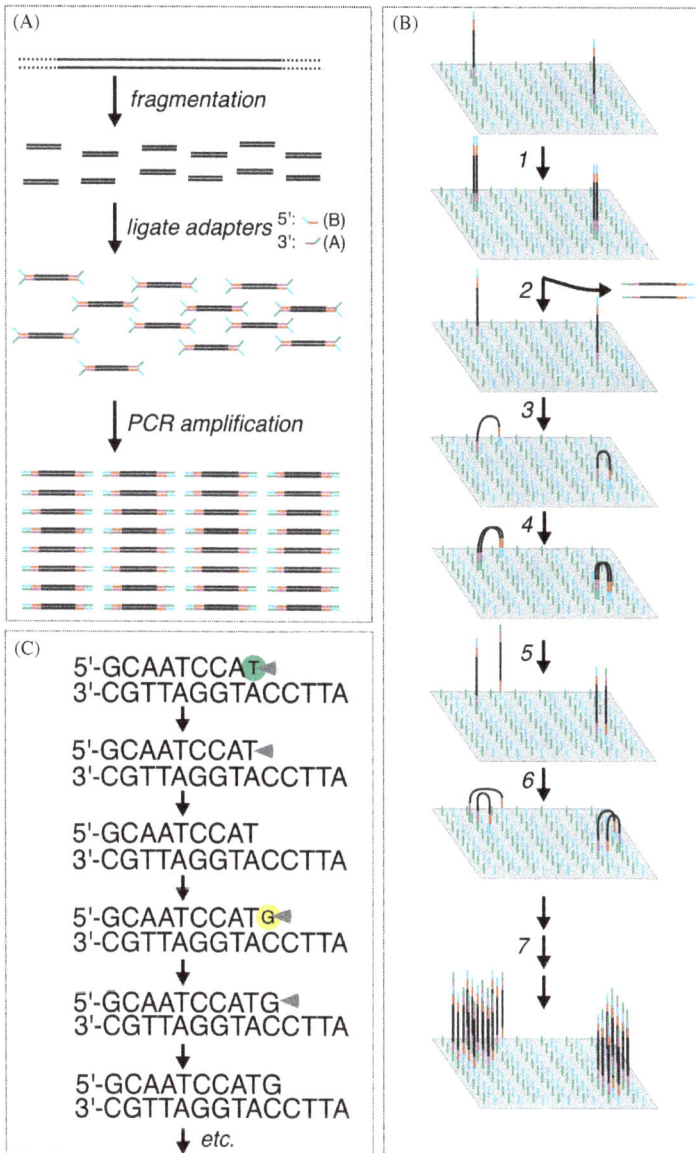

Figure 15.7. Key steps in the Illumina sequencing strategy. Library preparation involves DNA fragmentation, adapter ligation, and polymerase chain reaction (PCR) amplification (panel A). Bridge amplification allows the generation of localized clusters of PCR products, with each cluster derived from a single fragment (panel B). The actual sequencing reaction involves the use of reversible nucleotide terminators in which both the fluorescent dye (colored circles) and the terminator blockage (gray wedge) are reversible (panel C). See text for further discussion.

D is complementary to a stretch of the far end of the first synthesis product. Thus, the extended strand forms a bridge between the two different flow-cell-bound oligonucleotides (Figure 15.7B, step 3). Now, the process can essentially repeat itself, with DNA polymerase synthesizing back in the other direction (Figure 15.7B, step 4). Subsequent rounds of denaturation lead to an ever-expanding cluster of DNA bridges in the immediate vicinity of the original DNA-binding event (Figure 15.7B, steps 6 and 7). At a larger scale, the flow cell ends up with millions of these localized clusters, each originating from a single fragment (Figure 15.7B shows only two such clusters).

The Illumina platform uses a "sequencing by synthesis" method for determining the DNA sequence of these clusters, using four different fluorophores to distinguish the four different bases (Figure 15.7C). A proprietary aspect of the Illumina platform is the nature of the sequencing chemistry. The four fluorescently labeled bases are special in two ways, both involving a form of reversibility. First, the fluorophores (indicated by colored circles in the figure) can be released from the incorporated nucleotide residues after the CCD camera visualizes them. This allows the instrument to detect which of the four residues are incorporated, but then to wash away this signal in preparation for determining the next nucleotide. Second, the nucleotides are "reversible terminators," with a reversible group attached to the 3'-OH that is normally used for extension by DNA polymerase (indicated by a gray wedge in the figure). This group blocks elongation by DNA polymerase during the current cycle of sequencing and ensures that only a single nucleotide is incorporated in that cycle. However, after the nucleotide is visualized, the terminating group is released, just like the fluorophore, in preparation for the next cycle of the sequencing reaction. This reversible-terminator strategy is critical because incorporation by two or more nucleotides in a single cycle would muddle accurate DNA sequencing. The overall flow of each cycle is therefore incorporation of a single nucleotide, followed by CCD imaging of the incorporated residue as one of four colors, and then release of the fluorophore and terminating group. This cycle is repeated over and over again for each nucleotide residue that is determined.

At a larger scale across the flow cell, millions of clusters flash once in each cycle of sequencing, with one of the four colors that identify which of the four bases has just been incorporated in each particular cluster (as in Figure 15.6C). The instrument detects and records the sequence of flash colors from every single cluster, which is fed into the alignment end of the platform to either assemble a new genome or align the reads to a known genome (as in Figure 15.6D). One example of an application illustrated in the figure is that the source DNA might reveal a single-base mutation compared to the reference genome (T to C).

Individual instruments and applications within the Illumina/ Solexa platform have some important variations that are commonly used. One is the ability to read DNA sequence from both ends of each DNA fragment, called paired-end sequencing. This method improves the accuracy of the sequencing and has some other technical advantages. A second variation is the ability, in some situations, to conduct the sequencing without the initial library amplification, which can avoid certain PCR artifacts. Most importantly, some genomic regions amplify poorly in the PCR and therefore become underrepresented when a library is PCR amplified prior to sequencing.

A large majority of the DNA sequences determined in labs across the world today arise from second-generation sequencing methods like those just described. However, third-generation approaches are emerging and are already having some impact. The third-generation platform that is currently leading the field in the commercial realm is the PacBio RS, which was introduced in 2010. A key advantage of this platform over second-generation sequencing is its ability to generate sequence reads of several thousand bases, which is particularly useful for completing the sequences of novel genomes where there is no starting reference genome.

A unifying feature of third-generation approaches is that they allow determination of the DNA sequence of single molecules, rather than amplified clusters of molecules, avoiding PCR reactions. With regard to the kind of fluorescently based sequencing described earlier, third-generation approaches require incredibly sensitive fluorescence imaging so that incorporation of just a single nucleotide molecule is reliably detectable. Alternative methods are also under development, including methods that rely on detection of electrical current, methods that utilize a signal involving the DNA polymerase that is extending the DNA fragment, and methods that determine DNA sequence as the molecule transits through a pore in a motor protein.

15.6 Expansive applications of high-throughput DNA sequencing

The ability to sequence incredibly complex DNA (and RNA) samples efficiently and economically has opened up avenues of research that were previously unthinkable. In this closing section, we will summarize a few of these approaches and touch on future directions that now seem within reach.

Numerous variants of the high-throughput sequencing platform have allowed scientists to probe cellular biology in increasingly sophisticated ways. There are now methods that allow a semiquantitative analysis of: (1) all RNA molecules in a cell; (2) all RNA molecules that are undergoing transcription at any one time; (3) all RNA molecules that are undergoing translation at any one time; (4) various forms of noncoding small RNAs in the cell; or (5) mRNA subsets related to specific diseases such as cancer or cardiotoxicity or specific cellular pathways. Other high-throughput sequencing techniques are focused on chromosomal biology, allowing scientists to identify sites of DNA methylation, sites of protein–DNA interactions, and regions of DNA–DNA contacts that reflect higher-order chromosome structure. Finally, single-cell DNA sequencing is rapidly advancing and promises to allow scientists to track cell lineage in development and/or cancer progression. The strategy is that the progressive accumulation of somatic mutations, which occur during mitotic growth and division, act as identifying markers to allow the tracing of cell genealogy through a progression of cell cycles.

Turning to clinical applications, the insight gained from high-throughput sequencing of cancer-cell DNA was already discussed in the context of human health (see Section 14.8). Recall that one of the uses of this information involves more accurate targeting of chemotherapy that is suited for a particular tumor (see Section 14.9). Many other applications relating to clinical medicine and human health have also emerged. Genetic screening for inherited-risk genes such as the *BRCA* genes has been possible for some time with first-generation sequencing approaches, but newer high-throughput approaches are now being used to search for subtler and multigenic combinations of genes that can be used to assess risk and tailor therapy in diseases including those of the cardiovascular system. Based largely on the explosion of human sequencing data, the number of "personalized drugs" in clinical use has exploded over the last decade (Figure 15.8).

Only a small fraction of the human genome encodes for proteins, with the remainder representing the regions between genes and the introns that are spliced out of mRNA before translation. The protein-coding regions are sometimes called exons, to distinguish them from introns, and the collection of all protein-coding sequences is called "the exome." About 85% of the known disease-causing variants in humans lie within the 2% of the genome constituting the

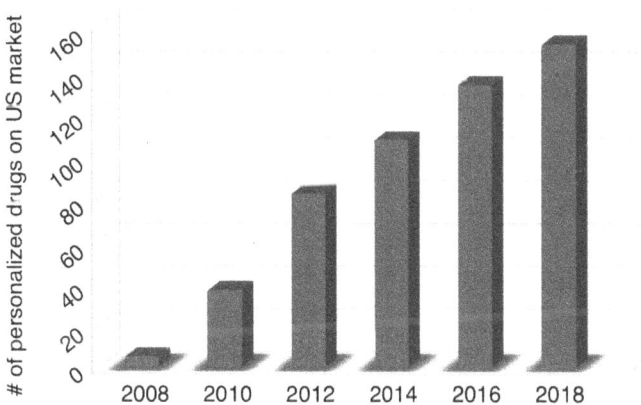

Figure 15.8. The growth in usage of personalized drugs in the United States. The number of drugs in clinical usage that are provided with information about how the patients' genetic makeup can affect their response. Data from Regalado, 2018.

exome. Thus, many of the applications in the clinical setting focus on this 2% in an approach called "whole-exome sequencing."

Another application relates to the screening of cancer by DNA-sequence analysis of cells and fragmented DNA in the blood, searching for mutant DNA sequences common in cancer. Early studies have begun to show promise, and it seems quite possible that accurate screening methods of this type will be available within the next decade or two and have a large impact on early detection.

As mentioned in Chapter 1, the human body has more microbial cells than human cells. The collection of microbes in locations such as the gut, skin, and mouth consist of a huge variety of species and variants. High-throughput DNA sequencing is now used to characterize and catalog this collection of microbes, which is called the "microbiome." Recent research has shown that the nature of the microbiome influences health outcomes such as obesity, heart disease, inflammatory bowel disease, cancer, and others.

Related to the topic of human health, the diversity of the human genome that has emerged from sequencing the genomes of thousands of human individuals is remarkable. Each individual genome was found to contain on the order of four-million single-nucleotide variants from what is called the consensus sequence (the reference human genome sequence consisting of the most common base at each position). In addition, individual genomes had a few-hundred-thousand short insertions or deletions. Of course, a large majority of these variations were found in noncoding regions of the genome, but even so, individuals were found to have hundreds of mutations that should cause loss-of-function alterations in genes. Morbidity and mortality from these mutations is greatly mitigated by the fact that the human genome is diploid, as most of these genes are in the heterozygous state with one functional and one mutant copy.

High-throughput sequencing is also impacting the study of the evolution of the human species. One of the most dramatic developments was the sequencing of Neanderthal genomes. Sequences of modern human individuals show clear remnants of Neanderthal and other more-ancient human genomes, showing intermixing of early hominid lineages in the emergence of modern man.

Totally novel applications and insights from high-throughput sequencing are constantly emerging in the literature, and here are just a few in closing. Recently, scientists were able to infer the pathway for biosynthesis of a brand-new antibiotic by means of sequence analysis of a community of microbes. They have never isolated the microorganism, and don't even know which one it comes from, but nonetheless were able to pull out the information to make the antibiotic. Another example is the discovery of the probable cause of colony-collapse disorder, in which a large fraction of honeybee hive colonies in the United States were lost. High-throughput sequencing of affected and unaffected hives strongly implicated a particular bee virus in the colony collapse. It seems clear that countless novel advances using high-throughput sequencing approaches will be limited only by the imagination of researchers and clinicians, particularly as the power and accuracy of sequencing methodologies continue to improve.

15.7 Summary of key points

- Numerous proteins and enzymes involved in DNA replication and repair are commercially available as key reagents in modern biotechnology.
- PCR revolutionized biological research and genetic technologies by allowing the exponential amplification of desired DNA segments at will.
- Recombinant DNA molecules have been central in the progress of molecular biology, and more recently synthetic biology; the process generally involves manipulation of DNA using restriction enzymes, DNA ligases, PCR, and annealing of synthetic DNA strands.
- The technology for sequencing DNA has progressed faster than the advances in microprocessor speeds/capacities, going from a speed of a few bases per year (circa 1970) to nearly 10^{12} bases per day (circa 2018).
- First-generation sequencing (Maxam–Gilbert and Sanger methods) required individual reactions and analyses for each DNA fragment, and was limited to a few hundred base pairs.

- Next-generation sequencing methods determine the sequence of massive numbers of fragments in parallel and require computational bioinformatics to assemble millions of sequence reads into complete (or nearly complete) genomic sequences.
- Modern high-throughput DNA sequencing has revolutionized many fields of biological and evolutionary research and clinical practice; novel and important applications continue to emerge.

Further Reading

Ansorge, W. J. (2009). Next-generation DNA sequencing techniques. *N Biotechnol, 25*(4), 195–203.

Bartlett, J. M., & Stirling, D. (2003). A short history of the polymerase chain reaction. *PCR Protocols* (pp. 3–6). Springer.

Gibson, D. G., Glass, J. I., Lartigue, C., Noskov, V. N., Chuang, R.-Y., Algire, M. A., Moodie, M. M. (2010). Creation of a bacterial cell controlled by a chemically synthesized genome. *Science, 329*(5987), 52–56.

Maxam, A. M., & Gilbert, W. (1977). A new method for sequencing DNA. *Proc Natl Acad Sci USA, 74*(2), 560–564.

McPherson, M., & Moller, S. (2006). *PCR. The Basics*. Garland Science, Taylor and Francis.

Regalado, A. (2018). *Look How Far Precision Medicine Has Come*. Cambridge, MA: MIT Technology Review.

Reuter, J. A., Spacek, D. V., & Snyder, M. P. (2015). High-throughput sequencing technologies. *Mol Cell, 58*(4), 586–597.

Sanger, F., & Coulson, A. R. (1978). The use of thin acrylamide gels for DNA sequencing. *FEBS Lett, 87*(1), 107–110.

Sanger, F., Coulson, A. R., Hong, G. F., Hill, D. F., & Petersen, G. B. (1982). Nucleotide sequence of bacteriophage lambda DNA. *J Mol Biol, 162*(4), 729–773.

Van Dijk, E. L., Auger, H., Jaszczyszyn, Y., & Thermes, C. (2014). Ten years of next-generation sequencing technology. *Trends Genet, 30*(9), 418–426.

Wu, R., & Taylor, E. (1971). Nucleotide sequence analysis of DNA: II. Complete nucleotide sequence of the cohesive ends of bacteriophage λ DNA. *J Mol Biol, 57*(3), 491–511.

Zhang, J., Chiodini, R., Badr, A., & Zhang, G. (2011). The impact of next-generation sequencing on genomics. *J Genet Genomics, 38*(3), 95–109.

How did they test that?
DNA sequence of bacteriophage λ

Bacteriophage λ played a pivotal role in the development of DNA-sequencing technologies. λ DNA has two short single-stranded cohesive ends, which can base pair to form a circle (top panel; ABC and A'B'C' represent the complementary end sequences). Wu and Taylor (1971) deduced their sequence by meticulously analyzing incorporation of radioactively labeled nucleotides by DNA polymerase; the nucleotides were radioactive either in the base (e.g., G° ppp) or the α-phosphate (e.g., Gp*pp). They measured the number of incorporated radioactive nucleotides (per λ DNA) of each of the four nucleotide types, either singly or in combinations. The table shows a small portion of their data, which identifies the first few residues on each end.[2] With only Ap*pp, only one molecule was incorporated (left end), so only the first residue is A (the second must be either G, C, or T). Nearest neighbor analysis involved digesting the product with a nuclease that leaves 3'-phosphate groups, which showed that G was the (5') neighbor of A. On the right end, three consecutive G's were incorporated when radioactive deoxyguanosine triphosphate (dGTP) was added. Two additional G's were incorporated when deoxyadenosine triphosphate (dATP) was added, implying that two G's follow the A on the left end. See whether you can follow the logic that they used for each of the inferred sequences (the diagrams below the table should help). Many more combinations and permutations were needed before they could deduce the entire 12-base sequence on each end (final inferred end sequences displayed at bottom left). Notice that the inferred sequences of the two ends are indeed complementary to each other.

The Sanger sequencing method allowed determination of longer stretches in a simpler procedure (see Section 15.4). Sanger *et al.* (1982) determined the complete sequence of phage λ DNA by cloning random λ DNA fragments into the DNA of M13 virus, which

---

[2] Previous data had identified which residue was incorporated first at each end (A on left and G on right).

can produce ssDNA. Every fragment could be sequenced using a single primer that binds just upstream of the cloning site that was used. By combining sequence information from hundreds of (over-lapping) clones, the entire λ sequence of 48,502 base pairs was deduced. The sample gel shown is a comparable analysis of a segment of a different bacteriophage; reproduced from Sanger and Coulson (1978), with permission from John Wiley and Sons; permission conveyed by Copyright Clearance Center, Inc.

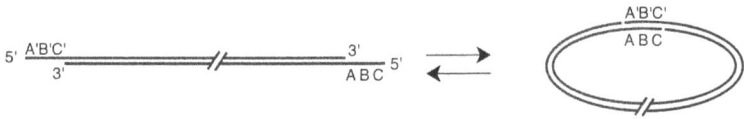

dNTP added	dNTP incorporated	Products after digestion	Inferred sequence left	Inferred sequence right
Ap*pp	1.0 p*A	1 Gp*	Gp*A (13'-12')	
G°ppp	3.1 pG°	2.1 G°p, 1.0 G°		G°pG°pG (1-3)
Gp*pp	3.2 p*G	3.0 Gp*		Gp*Gp*Gp*G (0-3)
Gp*pp, A°ppp	4.8 p*G, 1 pA°	3.8 Gp*, 0.9A°p*	A°p*Gp*G (12'-10')	Gp*Gp*Gp*G (0-3)

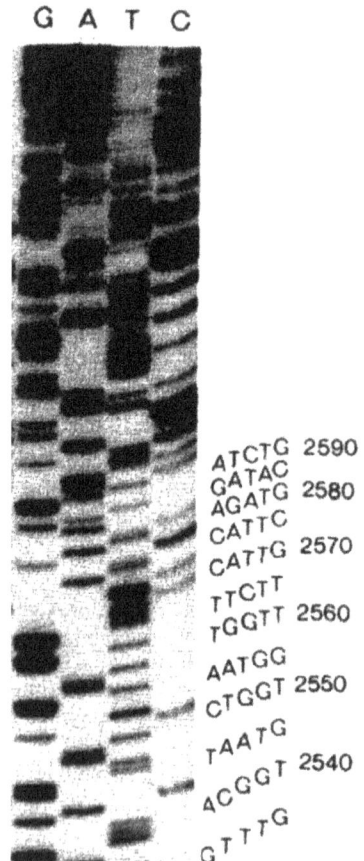

nuclease digestion example:

↓ ↓ ↓ ↓
......Gp*Gp*Gp*G ⟶ 3 Gp* (radioactive)
G (non-radioactive)

nucleotides incorporated:

5'
Left: 3' CCCGCCGCTGGA⋯⋯
1' 3' 5' 7' 9' 11'(G)
13'

0
(G)1   3   5   7   9 11
Right: ⋯⋯GGGCGGCGACCT 3'
5'

inferred end sequences:

Left: 5' GGGCGGCGACCT⋯⋯⋯

Right: ⋯⋯⋯CCCGCCGCTGGA 5'

Appendix

Table 1. Replication fork proteins in various species[1]

	Phage T7	*E. coli*[2]	*S. cerevisiae*[3]	*H. sapiens*[4]
Replicative DNA polymerase(s)[5]	gp5 + thioredoxin	DNA Pol III (α, ε, θ)	DNA Pol δ, DNA Pol ε	DNA Pol δ, DNA Pol ε
Replisome architectural proteins	NA[6]	χ, Ψ[7]	Ctf4	AND1[8]
Clamp	NA[6]	β	PCNA	PCNA
Clamp loader	NA[6]	γ complex (τ, γ, δ, δ')	RFC (Rfc1/2/3/ 4/5)	RFC (RFC1/2/ 3/4/5)

(Continued)

---

[1] Some of the entries in this table are from www.UniProt.org.

[2] The *E. coli* proteins with Greek names are also known by their gene product names as follows: α, DnaE; ε, DnaQ; θ, HolE; χ, HolC; Ψ, HolD; β, DnaN; τ, DnaX (long form); γ, DnaX (short form); δ, HolA; δ', HolB.

[3] Protein names in yeast usually consist of three letters (first letter only capitalized) followed by a single number; protein complex names often consist of three or four capital letters without a number.

[4] Protein names in humans often consist of three upper-case letters followed by a single number; protein complex names often consist of three or four capital letters without a number.

[5] See Appendix Table 2 for subunit composition of eukaryotic DNA polymerases.

[6] NA, not applicable – no such protein in the bacteriophage T7 system.

[7] The χ and Ψ proteins associate with the clamp loader complex, and so they are sometimes considered as part of that complex; they are not required for clamp loading.

[8] AND1 is also called WDHD1 or WD repeat and HMG-box DNA-binding protein 1.

Table 1. (*Continued*)

	Phage T7	*E. coli*[2]	*S. cerevisiae*[3]	*H. sapiens*[4]
Primase	gp4	DnaG	DNA Pol α/ primase[5]	DNA Pol α/ primase[5]
Helicase	gp4	DnaB	CMG complex (MCM, Cdc45, GINS)[9]	CMG complex (MCM, CDC45, GINS)[10]
Initiator protein	NA[6]	DnaA	Orc1/2/3/ 4/5/6	ORC1/2/3/ 4/5/6
Helicase loading at origin	NA[6]	DnaC	Cdc6 Cdt1	CDC6 CDT1
ssDNA binding protein	gp2.5	SSB	RPA (Rpa70/ Rpa32/ Rpa14)	RPA (RPA70/ RPA32/ RPA14)
Okazaki fragment trimming	gp6	DNA Pol I	Fen1 Dna2	FEN1 DNA2
Okazaki fragment fill in	gp5+ thioredoxin	DNA Pol I	DNA Pol δ	DNA Pol δ
Okazaki fragment sealing	gp1.3	DNA ligase	DNA ligase I	DNA ligase I

[9] The six subunits of the MCM complex in yeast are Mcm2, Mcm3, Mcm4, Mcm5, Mcm6 and Mcm7; the four subunits of the GINS complex are Psf1, Psf2 Psf3 and Sld5.
[10] The six subunits of the MCM complex in humans are MCM2, MCM3, MCM4, MCM5, MCM6 and MCM7; the four subunits of the GINS complex are PSF1, PSF2 PSF3 and SLD5.

Table 2. Subunit composition of eukaryotic replicative DNA polymerases[11]

DNA polymerase	Subunit descriptive name (*S. cerevisiae*)	Alternative name(s) (*S. cerevisiae*)	Subunit descriptive name (human)	Alternative name(s) (human)
α/primase	α catalytic subunit	Pol1, p180, Cdc17	α catalytic subunit	POLA1, p180
	α subunit B	Pol2, p86	α subunit B	POLA2, p70
	primase small subunit[12]	Pri1, p48	primase small subunit[12]	PRIM1, p49
	primase large subunit[12]	Pri2, p58	primase large subunit[12]	PRIM2, p58
δ	δ catalytic subunit	Pol3, Cdc2	δ catalytic subunit	POLD1, p125
	δ small subunit	Pol31, Hus2	δ subunit 2	POLD2, p50
	δ subunit 3	Pol32	δ subunit 3	POLD3, p68
			δ subunit 4	POLD4, p12
ε	ε catalytic subunit A	Pol2	ε catalytic subunit A	POLE1
	ε subunit B	Dpb2	ε subunit 2	POLE2
	ε subunit C	Dpb3	ε subunit 3	POLE3, p17
	ε subunit D	Dpb4	ε subunit 4	POLE4, p12

---

[11] Some of the entries in this table are from www.UniProt.org.

[12] Both subunits form the primase catalytic site.

Table 3. Mismatch repair proteins

	E. coli	Eukaryotes
Strand discrimination	Dam methylase	PCNA RFC (RFC1/2/3/4/5)
Mismatch recognition	MutS	MutSα (MSH2/MSH6) MutSβ (MSH2/MSH3)
Matchmaker	MutL	MutLα (MLH1/PMS2)[13]
Helicase	UvrD (Helicase II)	
Strand nicking	MutH	MutLα (MLH1/PMS2)[13]
Exonuclease(s)	ExoI, ExoV, RecJ, ExoX	ExoI
DNA polymerase	DNA Pol III	DNA Pol δ
Other involved proteins	SSB clamp (β) clamp loader (γ complex) DNA ligase	RPA (RPA70/RPA32/RPA14) DNA ligase

Table 4. Base excision repair proteins

	E. coli	Eukaryotes
N-glycosylase examples	Ung (uracil) Fpg (MutM) (damaged purines including 8-oxoG) MutY (A from 8-oxoG:A) (at least 8 enzymes total)[14]	UNG1 (uracil) UNG2 (uracil) OGG1 (8-oxoG) MUTYH (A from 8-oxoG:A) (at least 11 enzymes total)[14]
AP endonuclease	Exo III (XthA) Endo IV (Nfo)	APEX1 (Apn2 in *Saccharomyces cerevisiae*)
dRP lyase	Fpg	DNA Pol β
Resynthesis	DNA Pol I	DNA Pol β
Nick sealing	DNA ligase	DNA ligase I or III

---

[13] In *Saccharomyces cerevisiae*, MutLα is composed of MLH1 and PMS1 (the latter is an ortholog of human PMS2).

[14] See Table 6-1 in Friedberg *et al*, 2006, in the Further Reading list of Chapter 10 for a more complete list of *E. coli* DNA glycosylases. See Tables 6-2 and 6-3 in Friedberg *et al*, 2006, for lists of DNA glycosylases from *S. cerevisiae* and from humans; Krokan and Bjoras, 2013 (Further Reading list of Chapter 10) also list and discuss the human DNA glycosylases.

Table 5. Nucleotide excision repair proteins

	E. coli	*S. cerevisiae*[15]	*H. sapiens*[15]
Damage recognition	UvrA UvrB	Rad14 Rad4/Rad23 RPA (Rpa70/ Rpa32/Rpa14)	XPA XPC/HHR23B RPA (RPA70/ RPA32/RPA14) UV-DDB (DDB1/ DDB2(XPE))[16]
Damage-specific endonuclease	UvrC (5′, 3′)	Rad2 (3′) Rad1/Rad10 (5′)	XPG (3′) XPF-ERCC1 (5′)
DNA unwinding	UvrD (Helicase II)	TFIIH (Rad25, Rad3, 8 additional subunits)	TFIIH (XPB, XPD, 8 additional subunits)
Resynthesis	DNA Pol I	DNA Pol ε, DNA Pol δ PCNA RFC (Rfc1/2/3/4/5) RPA (Rpa70/ Rpa32/Rpa14)	DNA POL ε, DNA POL δ PCNA RFC (RFC1/2/3/4/5) RPA (RPA70/ RPA32/RPA14)
Nick sealing	DNA ligase	DNA ligase I	DNA ligase I
Additional protein(s) for TC-NER	TRCF (Mfd)	TRCF (Rad26)	TRCF (CSB) CSA[17] UVSSA[17] XAB2[17] HMGN1[17]

---

[15] The *Saccharomyces cerevisiae* and *Homo sapiens* proteins that are evolutionarily and functionally related are lined up in this table.

[16] UV-DDB is a specific factor for recognition of pyrimidine dimers. *S. cerevisiae* lacks this factor, but does have photolyase which is lacking in humans.

[17] The additional four factors in TC-NER found in mammals have not been found in *S. cerevisiae*.

Table 6. Homologous recombination proteins

	E. coli	*S. cerevisiae*	*H. sapiens*
Preparing 3' ends	RecBCD	MRX (Mre11/ Rad50/Xrs2)	MRN (MRE11/RAD50/NBS1)
RMP proteins	RecBCD RecO	Rad52 Rad54	BRCA2 RAD54
Strand-invasion proteins	RecA	Rad51 Dmc1 (meiosis)	RAD51 DMC1 (meiosis)
Rad51 paralogs		Rad55 Rad57	RAD51B RAD51C RAD51DF XRCC2 XRCC3
ssDNA coverage	SSB	RPA (Rpa70/ Rpa32/Rpa14)	RPA (RPA70/RPA32/ RPA14)
Annealing	RecO	Rad52	RAD52
Branch-migration helicases[18]	RuvAB RecG RecQ	Sgs1 Rad5	BLM WRN RECQ4 FANCM SMARCAL1
Holliday junction resolution	RuvABC	Mus81/Mms4 Yen1 Slx1/Slx4	MUS81/EME1 Gen1 SLX1/SLX4
Holliday junction dissolution		Sgs1/Top3/ Rmi1	BLM/TOP3α/RMI1

---

[18] The branch-migration helicases play varied and complex roles, some stimulatory and some inhibitory, in different recombination and recombination-related replication pathways (and some also in fork regression pathways). These detailed roles are currently under intense investigation in many laboratories.

Index

www.ingramcontent.com/pod-product-compliance
Lightning Source LLC
Chambersburg PA
CBHW050534190326
41458CB00007B/1784